PAXINOS & WATSON

The Rat Brain

IN STEREOTAXIC COORDINATES

PAXINOS & WATSON
The Rat Brain
IN STEREOTAXIC COORDINATES

George Paxinos

Prince of Wales Medical Research Institute
The University of New South Wales
Sydney, Australia

g.paxinos@unsw.edu.au
www.powmri.edu.au/staff/paxinos.htm

Charles Watson

Division of Health Sciences
Curtin University of Technology
Perth, Australia

c.watson@curtin.edu.au

ELSEVIER
ACADEMIC
PRESS

Amsterdam   Boston   Heidelberg   London   New York   Oxford
Paris   San Diego   San Francisco   Singapore   Sydney   Tokyo

Elsevier Academic Press
30 Corporate Drive, Suite 400, Burlington, MA 01803, USA
525 B Street, Suite 1900, San Diego, California 92101-4495, USA
84 Theobald's Road, London WC1X 8RR, UK

This book is printed on acid-free paper. ∞

Cover Design by Yvette Paxinos.
Book Design by Lewis Tsalis.
CD-ROM Design by Paul Halasz.

Library of Congress Cataloging-in-Publication Data
Application submitted

British Library Cataloguing in Publication Data
A catalogue record for this book is available from the British Library

ISBN: 0-12-088472-0
ISBN: 0-12-088572-7 (CD-ROM)

For all information on all Elsevier Academic Press publications
visit our Web site at www.books.elsevier.com

PRINTED IN THE UNITED STATES OF AMERICA
04   05   06   07   08   09   9   8   7   6   5   4   3   2   1

We dedicate this book to Kosta Theodore Paxinos and Anwen Angharad Williams.

Contents

Preface

In the first four editions of this atlas, we relied on a coronal section set that had some significant limitations; the section frequency proved over time to be too wide, sections did not always appear at regular intervals, and a few damaged sections had been replaced with sections from another brain. The fifth edition is based on a new coronal set which includes 161 sections from a single brain at regular 120μm intervals. This edition of the atlas is not simply an incremental improvement on the previous edition, but a completely new and far more comprehensive map of the rat brain.

Although the fifth edition features a different coronal section set, readers can be assured that the stereotaxic coordinates in the new atlas match those in previous editions. We have increased the scope of the atlas by incorporating new anatomical concepts where appropriate, and have once again delineated and named some areas not previously recognised.

Over the past decade our efforts at mapping the brain have been greatly enhanced by the availability of sections stained with a wide range of different chemical markers. A further contribution to the accuracy of our maps has been the knowledge we have we have gained from comparative neuroanatomical studies. One of us (GP) has published atlases of the human (Paxinos and Huang, 1995; Mai *et al.* 2004), monkey (Paxinos *et al.* 2000), and mouse brain (Paxinos and Franklin, 2004), and both of us are part of a team that is in the final stages of preparing an atlas of the chicken brain (Puelles *et al.* in press). Each of these atlas projects has provided us with new insights that have enhanced our ability to interpret the anatomy of the rat brain.

Acknowledgements

We thank Hongqin Wang for outstanding technical assistance from the inception to the completion of the project. We are indebted to Lewis Tsalis for speed, accuracy and brilliance in construction of diagrams and designing this book; to Paul Halasz for his imagination in construction of the CD-ROM; to Julia Tsalis for accurate labelling of diagrams; to Hongmei Liu for technical assistance, and to Yvette Paxinos for the cover design.

We acknowledge with gratitude the intellectual contribution to delineations made by Yuri Koutcherov (hypothalamus), Konrad Talbot (hippocampus), Nicola Palomero-Gallagher and Karl Zilles (cortex), Brent Vogt (cingulate cortex), Jan Vogt (cerebellum, precerebellar nuclei and vestibular nuclei), George Alheid (basal forebrain), Pascal Carrive (periaqueductal gray and hypothalamus), Glenda Halliday (substantia nigra and VTA), Ellen Covey and Manolo Malmierca (auditory system), Joel Elmquist (hypothalamus), Ann Goodchild and David Hopkins (rostroventrolateral and caudoventrolateral medulla), Jose DeOlmos (amygdala), Henk Groenewegen (thalamus), Joseph Travers (orofacial motor nuclei), John Mitrofanis (zona incerta), Pierre-Yves Risold (septum), Miklos Palkovits (paralemniscal nuclei), Harvey Karten (pretectal area), Chip Gerfen (basal ganglia), Terry Furlong (hypothalamus), Jean Buettner-Ennever (oculomotor nuclei), Marina Bentivoglio (parafascicular nucleus).

George Paxinos acknowledges the support he has received from the Australian National Health and Medical Research Council (he holds an NHMRC Principal Research Fellowship), as well as assistance from the Clive and Vera Ramaciotti Foundation, the Rebecca Cooper Foundation, and the Brennan Foundation.

We have appreciated the intelligent and enthusiastic support from our Elsevier editor Johannes Menzel. His patience and consideration have made a real difference to the successful completion of this project. We also thank Maureen Twaig and other Elsevier staff for their willingness to help in solving production problems.

Features of the Fifth Edition

- 161 coronal diagrams based on a single brain

- Diagrams spaced at constant 120 μm intervals giving scientists the most comprehensive and convenient atlas of the rat brain

- The most accurate stereotaxic reference system available

- Outlines of figures and brain structures in blue, but labels and leader lines in black for increased clarity of delineations

- All delineations re-examined in the light of recent findings

- Delineations of brain structures have been made with reference to sections stained for Nissl substance, AChE, parvalbumin, calbindin, calretinin, SMI-32, tyrosine hydroxylase, and NADPH diaphorase (Paxinos *et al.* 1999a,b)

- Extensive use was made of reference works, including the third edition of *The Rat Nervous System* (Paxinos, 2004) and other recent neuroanatomical literature

- Spinal cord drawings from the atlas of Molander and Grant(1995)

- Diagrams available on CD-ROM for printing.

Introduction

There are many reasons why the rat is the most commonly selected subject for research in mammalian neuroscience. First, rats are the right size: neither too small for accurate stereotaxic localization of discrete brain areas, nor too large for cost-effective laboratory management. Second, rats are generally hardy animals and are resistant to infections. Third, a number of inbred strains are available commercially, so that animals of consistent size can be used for stereotaxic procedures.

When the first edition of *The Rat Brain in Stereotaxic Coordinates* was published in 1982, it was the first atlas to be based on the flat skull position. It offered a choice of bregma, lambda, or the midpoint of the interaural line as the reference point. Although the coordinates were developed from study of adult male Wistar rats with weights ranging from 270 to 310 g, the atlas can be successfully used with male or female rats, with weights ranging from 250 to 350 g (Paxinos *et al.*, 1985).

With each new edition of the atlas, we have attempted to improve the accuracy of our delineations and have incorporated new findings on brain anatomy. However, our work has been hampered by the fact that our original series of coronal sections suffered from a number of limitations. First of all, our primary sections series showed sections at 0.5 mm intervals, which is insufficient to adequately represent all major structures in the brain for modern research purposes. Although we later attempted to better illustrate some areas by using some intervening sections, we could never fully compensate for the wide section interval in the primary series. In addition, we lost some sections in some areas of the brain and were forced to interpolate sections from another brain to compensate for the missing sections.

We were aware that the only real solution to these problems was to replace the coronal section series with a new section set based on shorter intervals and with all sections taken from the one brain. The new coronal section series is presented in the present (fifth) edition of the atlas. It shows diagrams of sections taken at regular intervals of 0.12 mm. Having constant intervals between the sections shown in the atlas diagrams eliminates one of the annoying features of many brain atlases – the fact that when the reader turns a page they do not know how far they have advanced along the prime axis. All sections are from the one brain.

The sections in our new coronal series were stained with cresyl violet or with methods to demonstrate AChE or NADPH diaphorase because we found that these three methods were compatible with using fresh (unfixed) tissue, a requirement for deriving an accurate stereotaxic grid. However, we consistently used other markers to confirm our delineations (Paxinos *et al.*, 1999a, Paxinos *et al.*, 1999b).

We have once again been greatly assisted by the suggestions of many colleagues in the delineation of structures. We welcome further advice that might improve the accuracy of our diagrams in the future. Please email us on *g.paxinos@unsw.edu.au* or *c.watson@curtin.edu.au*.

The present book will be followed by a comprehensive publication, which will include accompanying photographs and revised diagrams of sagittal and horizontal sections.

Methods

A fresh brain from a male 290 g Wistar rat was frozen, and coronal sections were cut at 40 mm thickness. The sections were cut at right angles to the horizontal plane joining bregma and lambda.

Stereotaxic surgery

We placed an anesthetized rats in a Kopf small-animal stereotaxic instrument, and the incisor bar was adjusted until the heights of lambda and bregma were equal. This flat-skull position was achieved when the incisor bar was lowered

3.3 ± 0.4 mm below horizontal zero (Table 1). Because the point of intersection of the lambdoid and sagittal sutures is variable, we have chosen to define lambda as the midpoint of the curve of best fit along the lambdoid suture (see skull diagram). This redefined reference point is considerably more reliable than the true lambda (the point of intersection of the sagittal and lambdoid sutures), and it is located 0.3 ± 0.3 mm anterior to the interaural line. We defined bregma as the point of intersection of the sagittal suture with the curve of best fit along the coronal suture. When the two sides of the coronal suture meet the sagittal suture at different points, bregma usually falls midway between the two junctions. The anteroposterior position of bregma was 9.1 ± 0.3 mm anterior to the coronal plane passing through the interaural line, but for the brain represented in this atlas bregma is deemed to lie at 9.0 mm. The top of the skull at bregma and lambda was 10.0 ± 0.2 mm dorsal to the interaural zero plane. To confirm the stereotaxic orientation of sections in the brain used for this atlas, reference needle tracks were made perpendicular to the horizontal and coronal planes. One horizontal needle insertions perpendicular to the coronal plane were made from the posterior of the brain at 4.0 mm above the interaural line and was 2.0 mm lateral to the midline. The reference track from the horizontal needle appears as a small hole in coronal sections.

Following surgery, the rat was decapitated and the whole head frozen on dry ice. The frozen skull was then prised off the frozen brain, and the brain was carefully mounted on the stage of microtome so that the sections would be cut in the coronal stereotaxic plane.

Every third section was used for preparation of the atlas diagrams, so that the interval between atlas diagrams is 0.12 mm. Exceptions to this rule are found in the region rostral to the rostrum of the corpus collosum (Interaural AP 11.28) and in the region of the medulla caudal to the inferior olive (Interaural AP -5.76 mm). In these two regions, sections were selected for presentation in the atlas at 0.24 mm intervals. Finally, the olfactory bulb is depicted at only three representative levels.

Histological methods

The 'atlas' sections were stained with either cresyl violet or for the demonstration of AChE on an alternate basis, so that cresyl violet sections are 0.24 mm apart and AChE sections are also 0.24 mm apart. The two sections that intervene between each 'atlas' section were stained with cresyl violet or for the presence or AChE or NADPH diphorase, according to the following sequence:

1. Cresyl violet 'atlas' section
2. AChE intervening section
3. NADPH intervening section
4. AChE 'atlas' section
5. Cresyl violet intervening section
6. NADPH diaphorase intervening sections

This sequence was repeated throughout the series of coronal section. This arrangement ensures that every 'atlas' section is accompanied by two adjacent sections, each of a different stain. For example, the 'atlas' AChE section described as number four above is preceded by an NADPH diaphorase section and followed by a cresyl violet section. This arrangement gave us maximum information for each 'atlas' section from the three stains. Staining was carried out on the same day as section cutting.

All sections, whether 'atlas' or intervening, were photographed on 4"x5" black and white negatives and printed on 36"x24" photographic paper. Each 'atlas' section was then covered with a sheet of 'Mylar' tracing film and outlines of structures were drawn in pencil. The final pencil drawings were scanned and then digitized using Adobe Illustrator.

Quality of Sections

In some cases, the sections were slightly stretched or compressed in the process of cutting and mounting on slides. We have compensated for this by constructing diagrams which represent, as best we can judge from the study of adjacent sections, the original shape of the brain section. In the worst cases, the 'atlas' section was so badly damaged that we have taken our drawing from an adjacent section.

Cresyl Violet Staining

Slides were immersed for 5 min in each of the following: xylene, xylene, 100% alcohol, 100% alcohol, 95% alcohol, and 70% alcohol. They were dipped in distilled water and stained in 0.5% cresyl violet for 15-30 min. They were differentiated in water for 3-5 min and then dehydrated through 70% alcohol, 95% alcohol, 100% alcohol, and 100% alcohol. They were then put in xylene and coverslipped.

To make 500 mL of 0.5% cresyl violet of about pH 3.9, mix 2.5 g of cresylecht violet (Chroma Gesellschaft, Postfach 11 10, D-73257, Kongen, Germany, Fax number: 49-7024-82660), 300 mL of water, 30 mL of 1.0 M sodium acetate (13.6 g of granular sodium acetate in 92 mL of water), and 170 mL of 1.0 M acetic acid (29 mL of glacial acetic acid added to 471 mL of water). Mix this solution for at least 7 days on a magnetic stirrer, then filter.

AChE Histochemistry

The method for the demonstration of AChE followed the procedures of Koelle and Friedenwald (1949) and Lewis (1961). Slides were incubated for 15 h in 100 mL of stock solution (see below) to which had been added 116 mg of S-acetylthiocholine iodide and 3.0 mg ethopropazine (May & Baker). The slides were rinsed with tap water and developed for 10 min in 1% sodium sulphide (1.0 g in 100 mL of water) at pH 7.5. They were then rinsed with water and immersed in 4% paraformaldehyde in phosphate buffer for 8 h, and then allowed to dry. Subsequently, they were dehydrated for 5 min in 100% alcohol, then immersed in xylene and coverslipped with Permount. The stock solution was a 50 mM sodium acetate buffer at pH 5.0 which was made 4.0 mM with respect to copper sulphate and 16 mM with respect to glycine. This was done by adding 6.8 g of sodium acetate, 1.0 g of copper sulphate crystals, and 1.2 g of glycine to 1.0 L of water and lowering the pH to 5.0 with HCl. We found that fresh, unfixed tissue from the frozen brains showed a substantially stronger reaction for both stains than tissue fixed with formalin, paraformaldehyde, glutaraldehyde, or alcohol.

NADPH diaphorase

The sections were washed in phosphate buffer for 10 minutes and incubated in 10 ml of a phosphate buffer solution containing 0.0125% nitroblue tetrazolium, 0.05% NADPH, 0.5% Triton X-100, and 1 mM magnesium chloride. The pH of the solution was adjusted to 7.6. The sections were incubated at 4°C for 48 hours. The incubation was stopped with a wash in phosphate buffer.

Photography and drawings

Photography

The photographs of stained brain sections were taken with a Nikon Multiplot macrophotographic apparatus using 4"x5" Kodax Plus X film. High contrast paper was used to print the photographs of Nissl sections, whereas lower contrast paper was used to print the photographs of AChE and NADPH sections.

Drawings

Drawings, which later formed the basis of the figures, were made by tracing the photographs of sections. We drew only the right side of each section and derived the outline of structures on the left side by mirror image construction using Adobe Illustrator.

Fiber tracts in the drawings are outlined by solid lines, and nuclei and cell groups are outlined by broken lines. In general, each abbreviation is placed in the center of the structure to which it relates; where this is not possible, the abbreviation is placed alongside the structure and a leader line is used. The abbreviations for fiber tracts and fissures are almost always positioned on the left side of the figure, and the abbreviations for nuclei and other cell groups are generally positioned on the right side. The outlines of the ventricles and aqueduct are filled in with solid color.

Stereotaxic Reference System

The stereotaxic reference system is based on the flat skull position, in which bregma and lambda lie in the same coronal plane. Two coronal and two horizontal zero-reference planes are referred to in these drawings. One reference coronal plane cuts through bregma and the other cuts through the interaural line. Similarly, one horizontal plane is at the level of bregma on the top of the skull and the other is at the level of the interaural line. Lambda is usually located 0.3 mm anterior to the interaural line, and it can be used as an alternative reference point in conjunction with the dorsoventral coordinate of bregma. The position of the stereotaxic reference points and planes are indicated on the skull diagram. The stereotaxic reference grid shows 0.2 mm intervals.

Drawings of coronal brain sections

In each of the coronal drawings, the large number at the bottom left shows the anteroposterior distance of the section from the vertical coronal plane passing through the interaural line. The large number at bottom right shows the anteroposterior distance of the plate from a vertical coronal plane passing through bregma. Note that these two coronal planes are 10 mm apart, so the two numbers on any one plate add up to 10 mm. The small numbers on the left margin show the dorsoventral distance from the horizontal plane passing through the interaural line. The numbers on the right margin show the dorsoventral distance from the horizontal plane passing through bregma and lambda on the surface of the skull. The numbers on the top and bottom margins show the distance of structures from the midline sagittal plane.

290 g Male Wistar

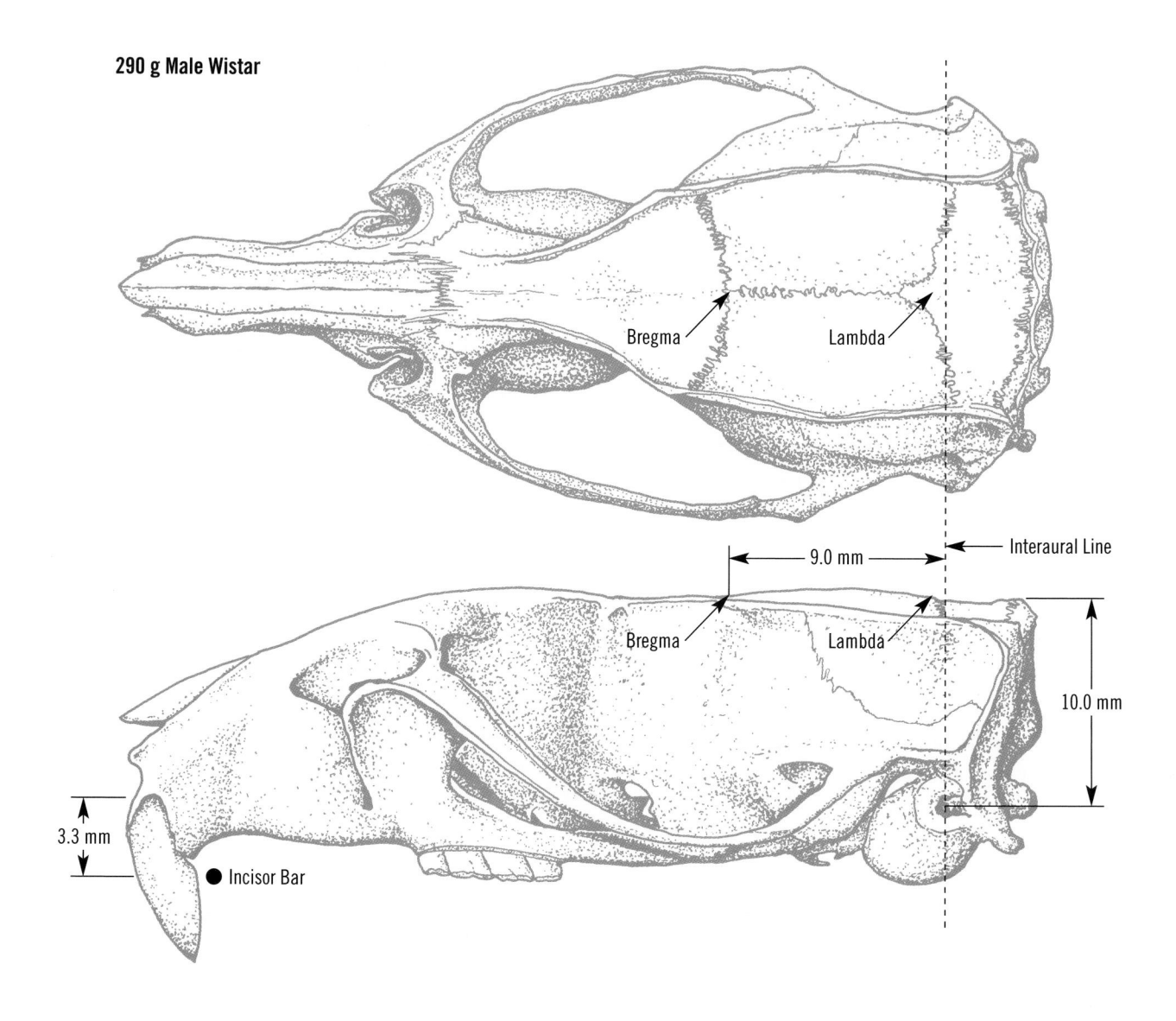

Skull Diagram Dorsal and lateral views of the skull of a 290 g Wistar rat. The positions of bregma, lambda and the plane of the interaural line are shown above the lateral view. The distance between the horizontal plane passing through the interaural line is shown on the right of the lateral view. The distance between the incisor bar and the horizontal plane passing through the interaural line is shown on the left of the lateral view. Lambda (midpoint of the curve of best fit along the lambdoid suture) is 0.3 mm anterior to the coronal plane passing through the interaural line.

Accuracy of the stereotaxic coordinates

In almost all cases, the potential error in defining the position of any point in the brain is less than 0.5 mm. Although we used medium-sized (average 290 g) male Wistar rats in the construction of this atlas, we recognize that researchers often use animals of different sex, strain, and weight. Because of this, we have estimated the error that may occur if this atlas is used with female Wistar rats, male hooded (Long Evans) rats, male Sprague Dawley rats of 300-g weight, juvenile (180 g) Wistar rats, and mature (436 g) Wistar rats. The results of these estimations are shown in Table 1 (reproduced from Paxinos *et al.*, 1985).

It is evident from these studies that no substantial stereotaxic error will occur when rats of different sex and strain are chosen, provided that the rats are of similar weight to those on which the atlas is based (290 g). For example, for rats of different sex and strain but of similar weight, the anteroposterior distance between the interaural line and bregma is between 9.0 and 9.4 mm. Similarly, the dorsoventral distance between the interaural line and the surface of the skull at bregma and lambda is very stable (9.8-10.1 mm). By contrast, craniometric data for juvenile (180 g) and mature (436 g) Wistar rats differ substantially from those of other groups. The anteroposterior distance between the interaural line and bregma is 7.7 mm in the juvenile

Table 1 Craniometric and stereotaxic data (means + S.D.) for rats of different sex, strain and weight

Subject	Mean weight (g)*	AP I – B (mm)	AP I – L (mm)	DV I – B (mm)	AP I – Acb (mm)**	AP B – ac (mm)**	AP I – 7n (mm)**	DV I – incisor bar (mm)
'Atlas' Wistar	290	9.1 ± 0.3	0.3 ± 0.3	10.0 ± 0.2	11.7	0.0	-1.3	-3.3 ± 0.4
Coronal plates	300	9.2	0.2	10.1				
Sagittal plates	270	8.9	0.0	10.0				
Horizontal plates	290	9.1	0.2	10.1				
Female Wistar	282	9.3 ± 0.2	0.5 ± 0.3	10.0 ± 0.1	11.6	0.1	-1.2	-3.2 ± 0.5
Hooded	290	9.4 ± 0.4	0.3 ± 0.6	9.8 ± 0.2	11.9	0.0	-1.2	-3.9 ± 0.6
Sprague	299	9.0 ± 0.2	0.7 ± 0.2	10.1 ± 0.1	11.7	0.1	-1.2	-3.9 ± 0.5
Juvenile Wistar	180	7.7 ± 0.4	-0.4 ± 0.3	9.9 ± 0.2	10.2	-0.1	-1.6	-2.0 ± 0.4
Mature Wistar	436	9.7 ± 0.3	0.6 ± 0.3	10.7 ± 0.4	12.4	-0.1	-0.8	-2.7 ± 0.3

* S.D.s ≤ 20g.
** S.D.s ≤ 0.4 mm.

ac, anterior commissure; Acb, accumbens nucleus; AP, anterior-posterior; B, bregma; DV, dorsal-ventral; 7n, facial nerve; I, interaural line; L, lambda.
Reprinted with permission from *J. Neuroscience Methods*. **13** (1985) 139-143.

and 9.7 mm in the mature rats (9.0 mm in 290-g male rats). Lambda is 0.4 mm posterior to the interaural line in the juvenile rats and 0.6 mm anterior to this line in the mature rats (0.3 mm anterior in 290-g rats). Unexpectedly, the dorsoventral distance between the interaural line and bregma for juvenile rats (9.9 mm) was almost the same as that of 290-g rats (10.0 mm). In the mature rats, the interaural line to bregma vertical distance was 10.7 mm. In female rats, as well as in hooded, juvenile (180 g), mature (436 g) and 290-g Wistar rats, bregma was found to be above the most forward crossing fibers of the anterior commissure. This is the point at which the posterior limbs of the anterior commissure appear. These data confirm the observation of Whishaw *et al.* (1977) that bregma is more stable than the interaural line for positioning of electrodes in brain structures close to, or anterior to, bregma. However, data from insertion of needles aimed at the level where the facial nerve leaves the facial genu show that the interaural reference point is more stable than bregma for localization of such posterior structures. Therefore, if juvenile or mature rats are used, greater accuracy can be achieved if bregma is used as the reference point for work with rostral structures and the interaural line for work with caudal structures.

A further improvement in accuracy can be obtained by taking into account the actual location of the accumbens nucleus and the genu of the facial nerve. In agreement with Slotnick and Brown (1980), we noticed that coordinates of structures were closer to target if the coordinates given by the interaural and bregma reference systems were averaged.

No atlas or stereotaxic instrument will compensate for using bregma and lambdoid points inappropriately. These reference skull marks for bregma is the midpoint of the curve of best fit along the coronal suture, and the reference skull mark for lambda is the midpoint of the curve of best fit along the lambdoid suture. These two reference marks are not necessarily the points of intersection of these sutures with the midline suture.

Nomenclature and the construction of abbreviations

The need for a stable neuroanatomical nomenclature to accurately and efficiently convey information between neuroscientists is obvious. However, many terms or abbreviations are still used to describe a single structure, and, in some cases, the same term or abbreviation is used for completely different structures. We urge all researchers to consider the merits of our system of nomenclature because it is systematic and derived after extensive consultations with neuroanatomy experts.

In considering the merit of a particular name over synonyms, we have chosen terms that have been ratified by modern usage, particularly usage by experts in that field. We have used anglicized versions of terms rather than older latinized versions wherever possible, and we have in all but a handful of cases avoided the use of eponyms.

Neuroscience communities concerned with different systems have developed identical abbreviations for completely different structures; for example SO may stand for both supraoptic nucleus and superior olive, SC for suprachiasmatic nucleus and superior colliculus, and IC for inferior colliculus and internal capsule. In dealing with the entire nervous system (as increasingly more researchers do) these parochial abbreviation schemes are serious obstacles to communication of data. An additional complication arises when homologous structures are nonetheless named or abbreviated differently in different species. We have made an effort to establish homologies and are using the same abbreviations for homologous structures in atlases of the rat (Paxinos and Watson, 1986), mouse (Franklin and Paxinos, 1996), monkey (Paxinos *et al.*, 2000), human (Paxinos *et al.*, 1990; Mai *et al.*, 2004), and chicken (Puelles *et al.*, in press).

The importance of following a logical system of abbreviations is shown by the fact that a term such as the accumbens nucleus can be (and probably has been) abbreviated about 20 different ways. We used the abbreviation Acb in this as well as all our other atlases. We have adopted the same rule for other all other structures, and so we have maintained the same abbreviation across mammalian and avian species, homologies permitting.

Our abbreviations have been constructed on the basis of the following principles:

1. The abbreviations represent the order of words as spoken in English (e.g., DLG = dorsal lateral geniculate nucleus) rather than the order in which they appear in older latinised terms.

2. The general principle used in the abbreviations of the names of elements in the periodic table was followed: the capital letter representing the first letter of a word in a nucleus is followed by a lower case letter most characteristic of that word (not necessarily the second letter; e.g., Mg = magnesium; Rt = reticular thalamic nucleus).

3. All nuclei and other cell groupings (such as cortical areas) begin with a capital letter, except some cranial nerve nuclei which begin with a number followed by capital N (for 'nucleus'). All fibre bundles begin with lower case letters except some cranial nerves which begin with a number followed by lower case n (for 'nerve'). Thus, there is no necessity in for the letter "N" to been used to point out that a structure is a nucleus, except in the case of some cranial nerves. Similarly, there is no need for the letter "t" to be used to denote a fiber tracts.

4. Compound names of nuclei have a capital letter for each part (e.g., LPGi = lateral paragigantocellular nucleus.

5. If a word occurs in the names of a number of structures, it is almost always given the same abbreviation (e.g., Rt = reticular thalamic nucleus; RtTg = reticulotegmental nucleus of the pons). Exceptions to this rule are made for well-established abbreviations such as VTA.

6. Abbreviations of brain regions are omitted where the identity of the region in question is clear from its position (CMn = centromedian thalamic nucleus; not CMnTh).

7. Arabic numerals are used instead of Roman numerals in identifying cranial nerves and nuclei (as in the Berman, 1968, atlas), layers of the cortex, and layers of the spinal cord. While the spoken meaning is the same, the detection threshold is lower, ambiguity is reduced, and they are easier to position in small spaces available on diagrams.

To assist in the recognition of brain structures, the labels for cell groups are placed on the right hand side of the diagram, and the labels for fibre bundles and nerves are placed on the left hand side of the diagram. However, where structures are crowded, we have placed some cell group labels on the left hand side for reasons of clarity.

The basis of delineation of structures

For the fifth edition, we completely reviewed our delineations of all areas of the brain. Our primary guide was an extensive collection of histochemically stained sections (monoclonal antibodies and enzyme-based stains - Paxinos *et al.*, 1999a,b). We have also made extensive use of other publications from our laboratory (Paxinos, 1995; Paxinos and Huang, 1995; Paxinos *et al.*, 1994), the atlas prepared by Swanson (2004), as well as many authoritative studies published in major journals such as the *Journal of Comparative Neurology.*

We present below a brief account of the basis of delineation of structures. We have not repeated here the rationale for the delineation of structures presented in the second edition (Paxinos and Watson, 1996). Readers will be better able to judge the suitability of our delineations when our photographs are published as part of the comprehensive (three cardinal planes) edition of this atlas in 2005.

Olfactory System

Refer to Shipley *et al.* (2004) and de Olmos *et al.* (1978) for a general description of the olfactory system.

Intermediate endopiriform nucleus (IEn)

We have given this name to an area ventral to the dorsal endopiriform nucleus (DEn). Both DEn and IEn are deep to the piriform cortex. We previously included this area in DEn (RBSC4); Swanson sometimes calls it DEn and sometimes includes it in the deep layer of the piriform cortex. The cells in this area are relatively sparse and smaller than those in DEn. Both DEn and IEn are NADPH positive (Fig 88 Paxinos *et al.*, 1999a), but IEn has parvalbumin positive elements (Fig 89 Paxinos *et al.*, 1999a).

Basal Ganglia and Basal Forebrain

Refer to Heimer *et al.* (1995) and Gerfen (2004) for a general description of the basal ganglia and to de Olmos *et al.* (2004) for a discussion of the substantia innominata and extended amygdala. Immunoreactivity for parvalbumin and the neurofilament protein SMI-32 identifies the ventral pallidum (Paxinos *et al.*,1999a). We retained the term substantia innominata and identified dorsal, ventral (as in Grove, 1988), and basal components with the assistance of George Alheid. The basal component is marked by some positivity in tyrosine hydroxylase but is negative for SMI-32 (although surrounding areas are positive).

The concept of ventral pallidum, first proposed by Heimer and his associates, has been the guiding principle for structure/function relations of the basal forebrain (Barragan and Ferreyra-Moyano, 1995; Heimer *et al.*, 1997).

The researchers at the University of Virginia and Universidad Nacional de Cordoba have carved out of the substantia inominata another big territory, the sublenticular extended amygdala (Alheid *et al.*, 1995). Paxinos and Franklin (2001) have used the new scheme for their mouse brain atlas. We retained the name substantia inominata (for the part remaining after the territory of the ventral pallidum has been defined) in keeping with earlier editions of this atlas. The dorsal substantia inominata roughly corresponds to the sublenticular extended amygdala, central part, while the ventral substantia inominata corresponds to the sublenticular extended amygdala, medial part. The area previously called fundus striati resembles the striatum proper in some respects and the accumbens shell in others. Given that the use of the term fundus striati creates problems with primate homologues, we followed the advice of George Alheid and called it the lateral accumbens shell. The remaining accumbens is delineated in accordance with Zaborszky *et al.* (1985) and Heimer *et al.* (1991). We followed de Olmos *et al.* (1995) in the identification of the interstitial nucleus of the posterior limb of the anterior commissure (IPAC).

Substantia nigra

We used the primate terminology for the dorsal and ventral tiers of the substantia nigra and also for the part of the VTA which is called the parabrachial pigmented nucleus. The names of the primate SN subdivisions were developed earlier than those subsequently used in rodent studies, and the identified primate subdivisions are consistent with the degeneration patterns seen in Parkinson's disease (Glenda Halliday, personal communication, 2004). The reticular part of the substantia nigra can be divided into a ventrolateral and a dorsomedial component on the basis of parvalbumin and calbindin distribution (Paxinos *et al.*, 1999b). The remainder of the substantia nigra and the ventral tegmental area were delineated according to the work of McRitchie *et al.* (1996).

Navicular nucleus (Nv)

We have renamed the area in the basal forebrain which we previously called the semilunar nucleus. The reason is that the name semilunar nucleus has priority in the avian literature and it refers to a completely different structure in birds. The existence of the semilunar nucleus was established on the basis of NADPH-diaphorase histochemistry (Paxinos *et al.*,1999a). We acknowledge assistance of R. Harlan and P-Y. Wang in the identification of this structure (Ahima and Harlan, 1990; Wang and Zhang, 1995).

Globus pallidus external and internal parts (GPE, GPI)

We have used the terms internal and external instead of medial and lateral in relation to the parts of globus pallidus to be consistent with the primate literature.

Lateral stripe of the striatum (LSS)

LSS is a dense band of cells in Nissl stained sections. The area is negative in calbindin sections and lighter stained than the LAcbSh and striatum in TH (the distinction is very clear in Fig. 87 of Paxinos *et al.*, 1999a).

Dorsal and ventral parts of the claustrum (DCl and VCl)

We have identified distinct dorsal and ventral parts of the claustrum in AChE stained sections. The dorsal part is positive for AChE and the ventral part is negative.

Ventral Tegmental Area

We have identified the rat homologue of the human parapeduncular nucleus (Paxinos & Huang, 1995) but have named it the parainterfascicular nucleus. We suggest that the new term be also used for the human given its more descriptive nature. With the identification of the parainterfascicular in the rat, the entire VTA (in most levels) is represented by specifically name component parts. This avoids the problem of previous editions of this atlas where the label VTA was placed only on what we now call parainterfascicular nucleus, giving the impression that it alone was the VTA. The VTA in our view consists of the paranigral, the parainterfascicular, the parabrachial pigmented nuclei and the VTAR.

Septum, Hypothalamus, and Neurosecretory Nuclei

Refer to Simerly (2004), Armstrong (2004), Risold (2004), and Oldfield and McKinley (2004) for a general description of these structures. Jutting ventrolaterally from the anterodorsal preoptic nucleus is a strip which is negative for parvalbumin which we have called the alar nucleus. The alar nucleus displays substance P positive cell bodies but little reactivity in its

neuropil (Larsen, 1992). In the preoptic area we followed Simerly (2004) and Simerly *et al.* (1984) except for the identification of the ventromedial and ventrolateral preoptic nuclei, for which we followed Elmquist *et al.* (1996) and Sherin *et al.* (1996). The compact part of the medial preoptic nucleus is negative for substance P (Harding *et al.*, 2004).

In the lateral hypothalamus we identified a ventrolateral hypothalamic nucleus on the basis of NADPH-diaphorase reactivity (Paxinos *et al.*, 1999a). This nucleus is caudal to the ventrolateral preoptic nucleus and dorsal to the supraoptic nucleus. The ventral part of dorsomedial nucleus is marked by densely stained cell bodies and terminals in NADPHdiaphorase preparations (Paxinos *et al.*, 1999a). The gemini nucleus is a conspicuous nest of a NADPH-diaphorase cell bodies (Paxinos *et al.*, 1999a). The parasubthalamic nucleus is present in the rat (Wang and Zhang, 1995), but it is not as impressive as the homologous structure seen in the mouse brain (Paxinos and Franklin, 2001). The arcuate nucleus was delineated according to the work of Magoul *et al.* (1994). See Paxinos and Watson (1986) for the identification of the striohypothalamic, magnocellular lateral hypothalamic, terete, and subincertal nuclei.

Lateral hypothalamus (PLH, TuLH, PeFLH, JPLH)

We have given names to different regions that are now recognized as comprising the lateral hypothalamus. The features characteristic of lateral hypothalamus (particularly the population of large cells) are not limited to the area lateral to the fornix. The orexin and hypocretin containing cells are not confined to the classically defined LH area, but are also found medial to the fornix. Swanson (2004) correctly extended the lateral hypothalamus medial to the fornix, but still identified the fornix as another boundary from where various areas emanate. The region both medial and lateral to the fornix when stimulated electrically induces attack by a cat on a rat (Paxinos, Bandler and Flynn, unpublished observations), indicating that the PeFLH behaves in a unitary fashion as it concerns this behavior. The components of the lateral hypothalamus in our scheme are as follows.

PLH – 'peduncular part of the lateral hypothalamus'
TuLH – 'tuberal part of the lateral hypothalamus'
PeFLH – perifornical part of the lateral hypothalamus'
JPLH – justaparaventricular part of the lateral hypothalamus'

Posterior hypothalamus, dorsal area (PHD)

This is an area previously identified as PHA in our atlas.

Arcuate nucleus (ArcMP)

ArcMP is ACHE positive and this distinguishes it from DM

Episupraoptic nucleus (ESO)

We named this nucleus on the basis of its location. Its rostral pole begins at the caudal pole of the ventrolateral preoptic nucleus; ESO can be found in Figs 39-45.

Paraterate nucleus (PTe)

This nucleus is located within the ventrolateral hypothalamic tract (Swanson, 2004) and rostral, dorsal and lateral to the terate hypothalamic nucleus.

Amygdala and Bed Nucleus of Stria Terminalis

Refer to de Olmos *et al.* (2004) for a general description of the amygdala and the bed nucleus of the stria terminalis. The anterodorsal part of the medial nucleus of the amygdala and the basomedial nucleus are defined by the presence of intense NADPH-diaphorase reactivity (Paxinos *et al.*, 1999a). The lateral part of the central nucleus of the amygdala is marked by the presence of tyrosine hydroxylase fibers and AChE negativity (Paxinos *et al.*, 1999a).

Reticulostrial nucleus (RtSt)

We have named this nucleus for its postion between the stria terminalis and the reticular nucleus. In calretinin stained sections it has a densely positive neuropil whereas the reticular nucleus has a pale neuropil. In parvalbumin sections, RtSt is positive in neuropil while the reticular nucleus is positive for cells and neuropil (streaky and spotty). RtSt is largest at anterior pole of the thalamus. In calbindin sections, the stria terminalis is positive whereas the RtSt is negative (Fig 160, Paxinos *et al.*, 1999a). Its medial part is negative and lateral is positive in parvalbumin (Fig 166, Paxinos *et al.*, 1999a).

Rostral amygdalopiriform area (RAPir)

The rostral amygdalopiriform area is a distinct region between PLCo and Pir has a dense layer 2 in lateral two thirds but much less dense in medial third. We had outlined but not labeled this structure in previous editions of our atlas. This is the area which Swanson (2004) calls the posterior amygdalo cortical area, a term which we did not adopt because it does not fit well with the names we had already given to surrounding areas.

Thalamus

Refer to Groenewegen and Witter (2004) for a general description of thalamic nuclei. See Paxinos and Watson (1986) for the identification of the ethmoid, retroethmoid, subgeniculate, and precommissural nuclei. We have reverted to the use of the term ventral posterior nucleus, parvicellular part (Paxinos and Watson, 1982) for the nucleus that we previously named the gustatory nucleus of the thalamus (Paxinos and Watson, 1986). We made this change on the advice of Clifford Saper that gustatory input is more medial in this nucleus, and autonomic-related inputs can be found at more lateral parts of this structure (Yasui *et al.*, 1989).

Retroreuniens area (RRe)

We gave the name RRe to a region dorsal to the PH, ventral to CM, and medial to VPPC and SPF. Caudally, RRe merges with the periventricular gray matter.

Paraxiphoid (PaXi)

The paraxiphoid nucleus (PaXi) lies between the xiphoid nucleus (Xi) of the thalamus medially and the zona incerta laterally. It appears to be part of a belt separating hypothalamus from thalamus. In Fig 264 of Paxinos *et al.*, 1999a, an NADPH positive belt can be seen to extend from the reticular nucleus to zona incerta and paraxiphoid.

Ventral limitans thalamic nucleus (Vli)

This nucleus is a thin sheet between subparafascicular, parvicellular part and the medial lemniscus. We so named it to complement the posterior limitans thalamic nucleus. Palkovits has observed CGRP positivity in this nucleus (personal communications, 2004).

Hippocampal Region

Refer to Witter and Amaral (2004) for a general description of the hippocampal region.

We now distinguish a dorsal and a ventral subiculum. We have labeled the transition area of dorsal and ventral subiculum as STr. We have drawn the borders of the presubiculum and parasubiculum so as to reach the white matter as explained in Haug (1976) and Mulders *et al.* (1997).

The entorhinal parcellation scheme of Insausti *et al.* (1997) is appealing because each of the cytoarchitecturally distinct divisions has a different pattern of connections as they detail in their paper. The parcellation recognizes that the medial and lateral sectors of the entorhinal area are separated by two intermediate sectors obvious in our Nissl preparations. Insausti *et al.* specifically identify six entorhinal fields: (1) an amygdalopiriform cortex which they termed amygdalo-entorhinal transition field; (2) a medial entorhinal field (MEnt) equivalent to the ventromedial entorhinal area of Krettek and Price (1977); (3) a caudomedial entorhinal field (CEnt), which is the classic medial entorhinal area; (4) a ventral intermediate entorhinal field (VIEnt) equivalent to the caudal ventrolateral entorhinal field of Krettek and Price (1977); (5) a dorsal intermediate entorhinal field (DIEnt); and (6) a dorsolateral entorhinal field (DLEnt). The last two fields together are equivalent to the dorsolateral entorhinal field of Krettek and Price (1977).

The postsubicular area was identified on the basis of the work of Van Groen *et al.* (1992).

Cerebral cortex

There have been two comprehensive cortical parcellation schemes in recent decades. The first notable one was presented by Zilles (1985) and was constructed on the original stained sections of the earlier editions of our atlas. In the second edition of our atlas we used the cortical parcellations of Zilles (1985). The second comprehensive cortical delineation scheme was presented by Swanson (1992). The Zilles (1985) delineations differ significantly from the Swanson (1992, 2004) scheme. The atlas of chemical markers (Paxinos *et al.*, 1999a,b) enabled us to make a decision on the strengths of the two schemes. On this basis we have retained many of the features of the sensory, motor, and insular areas proposed by Zilles (1985). However, we have curtailed the rostral spread of Zilles's occipital areas and delineated the sensory representation of the trunk region and temporal association area in line with Swanson (1992, 2004).

We have retained the perirhinal cortex at caudal levels (along with Zilles, 1985) because there is a characteristic NADPH-diaphorase reactivity associated with this area. Palomero-Gallagher and Zilles (2004) have recently completed a substantial revision of the Zilles and Wree (1995) plan especially in the non-sensory parietal regions and we have followed their lead. Strong parvalbumin immunoreactivity is present in layer 4 of the primary somatosensory cortex. SMI-32 immunoreactivity formed distinctive patches in layer 4 of the barrel field and forelimb and hindlimb region.

The primary auditory area was identified on the basis of reduced calbindin immunoreactivity in the deep layers. All the auditory areas were marked by the presence of SMI- 32 positive cells in the superficial layers (Paxinos *et al.*, 1999a,b).

AChE marked the location of the prelimbic and agranular insular cortices. NADPH-diaphorase assisted in defining the agranular insular, perirhinal, and retrosplenial granular cortices. Additionally, NADPH-diaphorase immunoreactivity indicated the ventral part of the medial entorhinal cortex.

Calretinin immunoreactivity assisted in delineation of the lateral entorhinal cortex where the outer part of layer one is densely stained.

The dorsolateral orbital cortex was delineated in accordance with the work of Ray and Price (1992). We use the term frontal association cortex for the frontal cortex that others allocated to the secondary motor cortex (Swanson, 1992, 2004; Zilles, 1985). This designation is in agreement with microstimulation data (Neafsey et al., 1986).

Retrosplenial cortex (RSD and RSG)

The retrosplenial dysgranular cortex (RSD) was previously named by us the retrosplenial agranular cortex, but we have changed it on the advice of Brent Vogt. The retrosplenial granular cortex (RSG) is divided into three areas a, b, and c. Some authors refer to the RSG as area 24, according to the original scheme of Brodmann.

Reticular Formation

Refer to Jones (1995) for a general description of the reticular formation. The intermediate reticular zone was first identified in the rat (Paxinos and Watson, 1986), but is seen to advantage in the human brain (Paxinos and Huang, 1995). The intermediate reticular zone at levels of the caudal pole of the facial nerve nucleus is marked by NADPH-diaphorase positive cells. The lateral paragigantocellular nucleus is conspicuous in NADPH-diaphorase preparations (Paxinos et al., 1999b). We have identified the parapyramidal nucleus as the cell group dorsolateral to the pyramidal tract, which is outlined but not named in the second edition of this atlas (Paxinos and Watson, 1986). The identification of the epifascicular nucleus is based on the description of this nucleus in the human brain (Paxinos and Huang, 1995).

Conterminal nucleus (Ct)

We have identified the conterminal nucleus in the medulla close to the inferior olive. This group was originally identified by Olzewski and Baxter (1954) and is clearly shown in the human brain stem atlas (Paxinos and Huang, 1995). The nucleus is seen as two separate AChE positive cell groups, one lateral to the inferior olive (caudal pole of IOA) and a second group medial to the IOA.

Periaqueductal and Periventricular Gray

Refer to Keay and Bandler (2004) for a general description of the periaqueductal gray. The boundaries of periaqueductal gray cell columns were drawn according to Carrive (1993), Carrive and Paxinos (1994), and Paxinos and Huang (1995). We identify the rodent homologue of the human pleoglial periaqueductal gray in Figs 77-80. Two nuclei lateral to the central gray pars alpha were identified on the basis of SMI-32 immunoreactivity – central gray pars beta and central gray pars gamma (Paxinos et al., 1999b).

Lithoid nucleus (Li)

We have named this nucleus for the Greek word for a stone. This word was particularly applied to an elongated stone used in the ancient Olympics. Li is a prominent group of large cells in the dorsal part of the rostral PVG. While generally ovoid in cross section, the medial and lateral sides are in Figures 71-73 parallel to each other. Li lies medial to MCPC caudally, and medial to RPF rostrally. It is ventral to the PrC and dorsal to the fasciclus retroflexus. More caudally, it is dorsal to Dk. It can be readily identified in a horizontal section (Fig 105, Paxinos and Watson, 1998).

Tegmental Nuclei

For the identification of the anterior tegmental, microcellular tegmental, subpeducular tegmental, rabdoid, and epirubrospinal nuclei, see Paxinos and Watson (1986).

Epipeduncular nucleus (EpP)

A small but distinctive group of large cells below the peripeduncular nucleus and above the cerebral peduncle (Fig 78-80) has no home in the surrounding nuclei. We have named this group the epipedunclular nucleus.

Reticular tegmental nucleus, L part (RtTgL)
In Figs 112-114 the rat homologue of the subnuclues L of the LtRt nucleus Olzewski and Baxter (1954) can been identified (see also Paxinos and Huang, 1995).

Raphe Nuclei

We identified the raphe nuclei on the basis of 5-hydroxytryptamine sections prepared by G. Halliday and I. Tork (see also Harding *et al.*, 2004). We identified the raphe interpositus nucleus on the basis of the work of Buettner-Ennever *et al.* (1988).

Locus Coeruleus and Brainstem Catecholamine Cell Groups

Refer to Aston-Jones (2004) for the delineation of the locus coeruleus. We delineated the catecholamine cell groups by following Hökfelt *et al.* (1984) with assistance from our own tyrosine hydroxylase preparations (Paxinos *et al.*, 1999b).

Brainstem Nuclei Associated with Taste, Respiratory, Cardiovascular and Other Autonomic Functions

Refer to Saper (1995) and Norgren (1995) for a general description of these nuclei.

Nucleus of the solitary tract
The posterior part of the nucleus of the solitary tract was delineated in accordance with the work of Whitehead (1990), Herbert *et al.* (1990), McRitchie (1992), and Altschuler *et al.* (1989). The rostral part of the nucleus of the solitary tract was difficult to delineate, but we recognize a rostrolateral subnucleus on the basis of NADPH-diaphorase positivity.

Medullary respiratory groups and the Botzinger complex
These areas were delineated in accordance with Ellenberger *et al.* (1990), Kanjhan *et al.* (1995), and Cox and Halliday (1993).

Parabrachial nucleus
The parabrachial nucleus is delineated in accordance with Fulwiler and Saper (1984), Herbert *et al.* (1990), Whitehead (1990), and Herbert and Saper (1990). The external part of the lateral parabrachial nucleus and medial parabrachial nucleus are marked by NADPH-diaphorase positive cells and fibers (Paxinos *et al.*, 1999b).

Oromotor Nuclei

Refer to Travers (2004) for a description of the oromotor nuclei.

Precerebellar Nuclei and Red Nucleus

Refer to Ruigrok (2004) for a general description of these structures. Within what has been previously called the pararubral area there is a circumscribed cell group which we called the pararubral nucleus. We have named the large cells above the lateral lemniscus the epilemniscal nucleus (Paxinos *et al.*, 1999b).

Cerebellum

The identification of lobules, fissures, and deep cerebellar nuclei is based on the work of Voogd (2004) and Swanson (2004).

Somatosensory System

Refer to Tracey (2004b) for a general description of the somatosensory system. The general basis of delineation of these structures is described in Paxinos and Watson (1996). However, we followed Marfurt and Rajchert (1991) for the borders of the spinal trigeminal nucleus.

Trigeminal transition zone (5Tr)
We identify the NADPH-diaphorase positive area medial to the principal sensory trigeminal nucleus as the 5Tr given it juxtaposition between the trigeminal and the parabrachial nuclei.

The trigeminosolitary zone (5Sol)

The trigeminosolitary zone commences caudal to the trigeminal transition zone, and extends as far caudal as the level of the area postrema. The rostral part of this zone was previously identified by Paxinos and Huang (1995) in the human and named the subsolitary nucleus. We note that in the medulla, much like the in thalamus (VPM, VPPC – Lundy and Norgren, 2004), there is a progression of functional areas from the trigeminal concerned with somatosensory function to the solitary concerned with gustatory function from receptors of the same peripheral structures.

Matrix (Mx)

Paxinos and Huang (1995) identified in the human the pericuneate and peritrigeminal matrix. We observed a similar structure in the rat and identified it as the residual region after the solitary, trigeminosolitary, cuneate and parvicellular reticular nuclei are accounted for.

Cuneate nucleus, rotund part (CuR)

We note that as in primates, the rat has a rotund part in the cuneate nucleus which almost certainly represents the forepaw area.

Visual System

Refer to Sefton *et al.* (2004) for a general description of the visual system. The ventral tegmental visual relay zone was identified on the basis of the work of Giolli *et al.* (1985). The intergeniculate leaf was delineated on the basis of the work of Morin and Blanchard (1995).

Auditory System

Refer to Malmierca and Merchan (2004) for a general description of the auditory system. We used Faye-Lund and Osen (1985) as well as Malmierca and Merchan (2004) for the identification of areas of the inferior colliculus. The medial geniculate was delineated according to the work of LeDoux *et al.* (1985). The nucleus of the central acoustic tract has been identified in the cat, and Ellen Covey has delineated this structure in our atlas. For additional details on the basis of delineation of the components of the auditory system refer to Paxinos and Watson (1986).

Nucleus of the commissure of the inferior colliculus (Com)

The commissure of the inferior colliculus is populated by many cells. On the advice of Ellen Covey we have named this the nucleus of the commissure of the inferior colliculus.

Dorsal cochlear nucleus

We have adopted new abbreviations for the layers of the dorsal cochlear nucleus: DCDp is the dorsal cochlear deep core; DCFu is the dorsal cochlear fusiform layer; DCMo is the dorsal cochlear molecular layer.

Periolivary horn (POH)

The periolivary region has a dorsolateral protrusion positive in parvalbumin (Paxinos *et al.*, 1999b), which we termed the periolivary horn (POH).

References

Ahima, R. S., and Harlan, R. E. (1990). Charting of Type II glucocorticoid receptor-like immunoreactivity in the rat central nervous system. Neuroscience 39, 579-604.

Alheid, G. F., de Olmos, J. S., and Beltramino, C. A. (1995). Amygadala and extended amygdala. In G. Paxinos (Ed.), The Rat Nervous System, 2nd ed. (pp. 495-578). Academic Press, San Diego.

Altschuler, S. M., Bao, X., Bieger, D., Hopkins, D. A., and Miselis, R. R. (1989). Viscerotopic representation of the upper alimentary tract in the rat: Sensory ganglia and nuclei of the solitary and spinal trigeminal tracts. J. Comp. Neurol. 283, 248-268.

Armstrong, W.E. (2004). Hypothalamic supraoptic and paraventricular nuclei. In G. Paxinos (Ed.), The Rat Nervous System, 3rd ed., Elsevier Academic Press, San Diego, pp 369-388.

Aston-Jones, G. (2004). Locus coreuleus, A5 and A7 noradrenergic cell groups. In G. Paxinos (Ed.), The Rat Nervous System, 3rd ed., Elsevier Academic Press, San Diego, pp 259-294.

Barragan, E. I., and Ferreyra-Moyano, H. (1995). Ventrostiopallidal functional interconnections with cortical and quasi-cortical regions. Brain Res. Bull. 37, 329-336.

Berman, A. L. (1968). The Brainstem of the Cat: A Cytoarchitectonic Atlas with Stereotaxic Coordinates. University of Wisconsin Press, Madison.

Buettner-Ennever, J. A., Cohen, G., Pause, M., and Fries, W. (1988). Raphe nucleus of the pons containing omnipause neurons of the oculomotor system in the monkey, and its homologue in man. J. Comp. Neurol. 267, 307-321.

Carrive, P. (1993). The periaqueductal gray and defensive behavior functional representation and neuronal organization. Behav. Brain Res. 58, 27-47.

Carrive, P., and Paxinos, G. (1994). The supraoculomotor cap: A region revealed by NADPH diaphorase histochemistry. NeuroReport 5, 2257-2260.

Cox, M., and Halliday, G. M. (1993). Parvalbumin as an anatomical marker for discrete subregions of the ambiguous complex in the rat. Neurosci. Lett. 160, 101-105.

De Olmos, JS, Beltramino, CA and Alheid, G. (2004). Amygdala and extended amygdala of the rat: a cytoarchitectonical, fibroarchitectonical, and chemoarchitectonical survey. In G. Paxinos (Ed.), The Rat Nervous System, 3rd ed., Elsevier Academic Press, San Diego, pp 510-601.

De Olmos, J., Hardy, H., and Heimer, L. (1978). The afferent connections of the main and the accessory olfactory bulb formation in the rat: An experimental HRP study. J. Comp. Neurol. 181, 213-244.

Ellenberger, H.H., Feldman, J.L., and Zhan, W.-Z. (1990). Subnuclear organization of the lateral tegmental field in the rat. II: Catecholamine neurons and ventral respiratory group. J. Comp. Neurol. 294, 212-222.

Elmquist, J. K., Scammell, T. E., Jacobson, C. D., and Saper, C. B. (1996). Distribution of Foslike immunoreactivity in the rat brain following intravenous lipopolysaccharide administration. J. Comp. Neurol. 371, 1-19.

Faye-Lund, H., and Osen, K. K. (1985). Anatomy of the inferior colliculus in rat. Anat. Embryol. 171, 1-20.

Fulwiler, C.E., and Saper, C. B. (1984). Subnuclear organization of the efferent connections of the parabrachial nucleus in the rat. Brain Res. Rev. 7, 229-259.

Gerfen, C.R. (2004). Basal ganglia. In G. Paxinos (Ed.), The Rat Nervous System, 3rd ed., Elsevier Academic Press, San Diego, pp 458-509.

Giolli, R.A., Blanks, R.H.I., Torigoe,Y., and Williams, D. D. (1985). Projections of medialterminal accessory optic nucleus, ventral tegmental nuclei, and substantia nigra of rabbitand rat as studied by retrograde axonal transport of horseradish peroxidase. J.Comp. Neurol. 232(1), 99-116.

Grant G and Koerber, RH. (2004). Spinal cord cytoarchitecture. In G. Paxinos (Ed.), The Rat Nervous System, 3rd ed., Elsevier Academic Press, San Diego, pp 121-128.

Groenewegen, HJ and Witter, MP. (2004). Thalamus. In G. Paxinos (Ed.), The Rat Nervous System, 3rd ed., Elsevier Academic Press, San Diego, pp 408-457.

Grove, E. A. (1988). Efferent connections of the substantia innominata in the rat. J. Comp. Neurol. 277, 347-364.

Harding, A, Paxinos, G, Halliday, G. (2004). The serotonin and tachykinin systems. In G. Paxinos (Ed.), The Rat Nervous System, 3rd ed., Elsevier Academic Press, San Diego, pp 1203-1256.

Haug, F.-M. S. (1976). Sulphide silver pattern and cytoarchitectonics of parahippocampal areas in the rat. Advances in Anatomy, Embryology, and Cell Biology 52: 1-73.

Heimer, L., Zahm, D. S., Churchill, L., Kalivas, P., and Wohltmann C. (1991). Specificity in the projection patterns of accumbal core and shell in the rat. Neuroscience 41, 89-125.

Heimer, L., Zahm, D.S., and Alheid, G.F. (1995). Basal ganglia. In G. Paxinos (Ed.), The Rat Nervous System, 2nd ed. (pp. 579-628). Academic Press, San Diego.

Heimer, L., Harland, R.E., Alheid, G.F., Garcia, M.M., and de Olmos, J. (1997). Substantia inominata: a notion which impedes clinical-anatomical correlations in neuropsychiatric disorders. Neuroscience 76, 957-1006.

Herbert, H., Moga, M., and Saper, C. (1990). Connections of the parabrachial nucleus with the nucleus of the solitary tract and the medullary reticular formation in the rat. J. Comp. Neurol. 293, 540-580.

Hökfelt, T., Martensson, R., Bjorklund, A., Kleinau, S., and Goldstein, M. (1984). Distributional maps of tyrosine-hydroxylase-immunoreactive neurons in the rat brain. In A. Bjorklund and T. Hökfelt (Eds.), Handbook of Chemical Neuroanatomy, Vol. 2, Part 1. Elsevier, Amsterdam.

Holstege, G. (2004). Central nervous system control of micturition. In G. Paxinos (Ed.), The Rat Nervous System, 3rd ed., Elsevier Academic Press, San Diego, pp 321-331.

Insausti, R., Herrero, M.T., and Witter, M.P. (1997) Entorhinal cortex of the rat: cytoarchitectonic subdivisions and the origin and distribution of cortical efferents. Hippocampus 7: 146: 183.

Jones, B. E. (1995). Reticular formation: Cytoarchitecture, transmitters, and projections. In G. Paxinos (Ed.), The Rat Nervous System, 2nd ed., Academic Press, San Diego, pp. 155-171.

Kanjhan, R., Lipski, J., Kruszewska, B., and Rong, W. (1995). A comparative study of presympathetic and Bötzinger neurons in the rostral ventrolateral medulla (RVLM) of the rat. Brain Res. 699, 19-32.

Krettek, J.E. and Price, J.L. (1977) Projections from the amygdaloid complex and adjacent olfactory structures to the entorhinal cortex and to the subiculum in the rat and cat. Journal of Comparative Neurology 172: 723-752.

Koelle, G.G., and Friedenwald, J.S. (1949). A histochemical method for localizing cholinesterase activity. Proc. Soc. Exp. Biol. Med. 70, 617-622.

Larsen, P.J. (1992). Distribution of substance P-immunoreactive elements in the preoptic area and the hypothalamus of the rat. J. Comp. Neurol. 316, 287-313.

LeDoux, J.E., Ruggiero, D.A., and Reis, D.J. (1985). Projections to the subcortical forebrain from anatomically defined regions of the medial geniculate body in the rat. J. Comp. Neurol. 242, 182-213.

Lewis, P.R. (1961). The effect of varying the conditions in the Koelle method. Bibltheca Anat. Vol. 2, Karger, Basel, 11-20.

Lundy, R.F Jr and Norgren, R. (2004). Gustatory system. In G. Paxinos (Ed.), The Rat Nervous System, 3rd ed., Elsevier Academic Press, San Diego, pp 890-921.

Magoul, R., Ciofi, P. and Tramu, G. (1994). Visualization of an efferent projection route of the hypothalamic rat arcuate nucleus through the stria terminalis after labeling with carbocyanine dye (DiI) or proopiomelanocortin-immunohistochemistry. Neurosci. Lett. 172,134-138.

Mai, J. K., Assheuer, J., and Paxinos, G. (2004). Atlas of the Human Brain. Elsevier Elsevier Academic Press, San Diego.

Malmierca, M.S. and Merchan, M.A. (2004). Auditory system. In G. Paxinos (Ed.), The Rat Nervous System, 3rd ed., Elsevier Academic Press, San Diego, pp 996-1081.

Marfurt, C., and Rajchert, D. M. (1991). Trigeminal primary afferent projections to "Non-Trigeminal" areas of the rat central nervous system. J. Comp. Neurol. 303, 489-511.

McRitchie, D. A. (1992). Cytoarchitecture and chemical neuroanatomy of the nucleus of the solitary tract: Comparative and experimental studies in the human and the rat. Unpublished Ph.D. thesis, Univ. of New South Wales.

McRitchie, D.A., Hardman, C.D. and Halliday, G. M. (1996). Cytoarchitectural distribution of calcium binding proteins in midbrain dopaminergic regions of rats and humans. J. Comp. Neurol. 364, 121-150.

Molander, C., and Grant, G. (1995). Spinal cord cytoarchitecture. In G. Paxinos (Ed.), TheRat Nervous System, 2nd ed., Academic Press, San Diego, pp. 39-44.

Morin, L. P., and Blanchard, J. (1995). Organization of the hamster intergeniculate leaflet: NPY and ENK projections to the suprachiasmatic nucleus, intergeniculate leaflet and posterior limitans nucleus. Visual Neurosci. 12, 57-67.

Mulders, W.H.A.M., West, M.J., and Slomianka, I. (1997) Neuron numbers in the presubiculum, parasubiculum, and the entorhinal area of the rat. Journal of Comparative Neurology 385: 83-94.

Neafsey, E. J., Bold, E. L., Haas, G., Hurley-Gius, K. M., Quirk, G., Sievert, C. F., and Terreberry, R. R. (1986). The organization of the rat motor cortex: A microstimulation mapping study. Brain Res. Rev. 11, 77-96.

Oldfield, BJ and McKinley, MJ. (2004). Circumventricular organs. In G. Paxinos (Ed.), The Rat Nervous System, 3rd ed., Elsevier Academic Press, San Diego, pp 389-407.

Olszelwski, J., and Baxter, D. (1954). Cytoarchitecture of the Human brain Stem. Karger, Basel.

Palomero-Gallagher, N and Zilles, K. (2004). Isocortex. In G. Paxinos (Ed.), The Rat Nervous System, 3rd ed., Elsevier Academic Press, San Diego, pp 728-760.

Paxinos, G. (Ed.) (2004). The Rat Nervous System, 3rd ed., Elsevier Academic Press, San Diego.

Paxinos, G., and Huang, X.-F. (1995). Atlas of the Human Brainstem. Academic Press, San Diego.

Paxinos, G., and Watson, C. (1982) The Rat Brain in Stereotaxic Coordinates. Academic Press, Sydney.

Paxinos, G., and Watson, C. (1986). The Rat Brain in Stereotaxic Coordinates, 2nd ed. Academic Press, San Diego.

Paxinos, G. and Watson C. (1997). The Rat Brain in Stereotaxic Coordinates, Compact 3rd ed. CD-Rom. Academic Press, San Diego.

Paxinos, G., Kus, L, Ashwell, K, and Watson, C. (1999a). Chemoarchitecture of the Rat Forebrain. Academic Press, San Diego.

Paxinos, G., Carrive, P., Wang, H., and Wang P-Y. (1999b). Chemoarchitecture of the Rat Brainstem. Academic Press, San Diego.

Paxinos, G., Ashwell, K.W., and Tork, I. (1994). Atlas of the Developing Rat Nervous System, 2nd ed, Academic Press, San Diego.

Paxinos, G, Huang, X-F, and Toga A.W. The Rhesus Monkey Brain in Stereotaxic Coordinates, Academic Press, San Diego.

Paxinos, G., Tork, I., Halliday, G., and Mehler, W. R. (1990). Human homologs to brainstem nuclei identified in other animals as revealed by acetylcholinesterase. In G. Paxinos (Ed.), The Human Nervous System, Academic Press, San Diego, pp. 149-202.

Paxinos, G., Watson, C., Pennisi, M., and Topple, A. (1985). Bregma, lambda and the interaural midpoint in stereotaxic surgery with rats of different sex, strain and weight. J. Neurosci. Meth. 13, 139-143.

Paxinos, G. and Franklin, K (2003). The Mouse Brain in Stereotaxic Coordinates. 3rd ed, Elsevier Academic Press, San Diego.

Puelles, L., Martinez-de-la-Torre, M, Martinez, S., Watson, C., and Paxinos, G. (in press). The Chick Brain in Stereotaxic Coordinates, Elsevier Academic Press, San Diego.

Ray, J. P. and Price, J. L. (1992). The organization of the thalamocortical connections of the mediodorsal thalamic nucleus in the rat, related to the ventral forebrain-prefrontal cortex topography. J. Comp. Neurol. 323, 167-197.

Ribeiro-Da-Silva, A. (2004). Substantia gelatinosa of the spinal cord. In G. Paxinos (Ed.), The Rat Nervous System, 3rd ed., Elsevier Academic Press, San Diego, pp 129-148.

Risold, PY. (2004). The septal region. In G. Paxinos (Ed.), The Rat Nervous System, 3rd ed., Elsevier Academic Press, San Diego, pp 602-636.

Ruigrok, TJH. (2004). Precerebellar nuclei and red nucleus. In G. Paxinos (Ed.), The Rat Nervous System, 3rd ed., Elsevier Academic Press, San Diego, pp 167-204.

Saper, CB. (2004). Central autonomic system. In G. Paxinos (Ed.), The Rat Nervous System, 3rd ed., Elsevier Academic Press, San Diego, pp 761-794.

Sefton, AJ, Dreher, B and Harvey, A. (2004). Visual system. In G. Paxinos (Ed.), The Rat Nervous System, 3rd ed., Elsevier Academic Press, San Diego, pp 1082-1164.

Scremin, OU. (2004). Cerebral vascular system. In G. Paxinos (Ed.), The Rat Nervous System, 3rd ed., Elsevier Academic Press, San Diego, pp 1165-1202.

Sherin, J. E., Shiromani, P. J., McCarley, R. W., and Saper, C. B. (1996). Activation of ventrolateral preoptic neurons during sleep. Science 271, 216-219.

Shipley, MT, Ennis, M and Puche, A. (2004). Olfactory system. In G. Paxinos (Ed.), The Rat Nervous System, 3rd ed., Elsevier Academic Press, San Diego, pp 922-963.

Simerly, RB. (2004). Anatomical substrates of hypothalamic integration. In G. Paxinos (Ed.), The Rat Nervous System, 3rd ed., Elsevier Academic Press, San Diego, pp 336-368.

Simerly, R. B., Swanson, L. W., and Gorski, R. A. (1984). Demonstration of a sexual dimorphism in the distribution of serotonin-immunoreactive fibers in the medial preoptic nucleus of the rat. J. Comp. Neurol. 225, 151-139.

Slotnick, B. M., and Brown, D. L. (1980). Variability in the stereotaxic position of cerebral points in the albino rat. Brain Res. Bull. 5, 135-139.

Swanson, L. (2004). Brain maps: structure of the rat brain, 3rd ed. Elsevier Academic Press, Amsterdam.

Tracey, D. (2004). Ascending and descending pathways in the spinal cord. In G. Paxinos (Ed.), The Rat Nervous System, 3rd ed., Elsevier Academic Press, San Diego, pp 149-164.

Tracey, D (2004). Somatosensory system. In G. Paxinos (Ed.), The Rat Nervous System, 3rd ed., Elsevier Academic Press, San Diego, pp 795-814.

Travers, JB. (2004). Oromotor nuclei. In G. Paxinos (Ed.), The Rat Nervous System, 3rd ed., Elsevier Academic Press, San Diego, pp 295-320.

Van Groen, T., and Wyss, J. M. (1992). Connections of the retrosplenial dysgranular cortex in the rat. J. Comp. Neurol. 315, 200-216.

Vidal, P and Sans, A. (2004). Vestibular system. In G. Paxinos (Ed.), The Rat Nervous System, 3rd ed., Elsevier Academic Press, San Diego, pp 964-995.

Vogt, BA, Vogt, L and Farber, NB. (2004). Cingulate cortex and disease models. In G. Paxinos (Ed.), The Rat Nervous System, 3rd ed., Elsevier Academic Press, San Diego, pp 704-760.

Voogd, J. (2004). Cerebellum. In G. Paxinos (Ed.), The Rat Nervous System, 3rd ed., Elsevier Academic Press, San Diego, pp 205-243.

Waite, PME. (2004). Trigeminal sensory system. In G. Paxinos (Ed.), The Rat Nervous System, 3rd ed., Elsevier Academic Press, San Diego, pp 815-852.

Wang, P. Y., and Zhang, F. C. (1995). Outlines and Atlas of Learning Rat Brain Slides. Westnorth University Press, China.

Whishaw, I. Q., Cioe, J. D. D., Previsich, N., and Kolb, B. (1977). The variability of the interaural line vs the stability of bregma in rat stereotaxic surgery. Physiol. Behav. 19, 719-722.

Whitehead, M. C. (1990). Sibdivisions and neruon types of the nucleus of the solitary tract in the hamster. J. Comp. Neurol. 310, 554-574.

Willis, WD, Westlund, KN and Carlton, SM. (2004). Pain system. In G. Paxinos (Ed.), The Rat Nervous System, 3rd ed., Elsevier Academic Press, San Diego, pp 853-889.

Witter, MP and Amaral, DG. (2004). Hippocampal formation. In G. Paxinos (Ed.), The Rat Nervous System, 3rd ed., Elsevier Academic Press, San Diego, pp 637-703.

Yasui, Y., Saper, C., and Cechetto, D. (1989). Calcitonin gene-related peptide immunoreactivity in the visceral sensory cortex, thalamus, and related pathways in the rat. J. Comp. Neurol. 290, 487-501.

Zaborszky, L., Alheid, G. F., Beinfeld, M. C., Eidens, L. E., Heimer, L., and Palkovits, M. (1985). Cholecystokinin innervation of the ventral striatum: A morphological and radioimmunological study. Neurosci. 14, 427-453.

Zilles, K. (1985). The Cortex of the Rat: A Stereotaxic Atlas. Springer-Verlag, Berlin.

List of Structures

Names of the structures are listed in alphabetical order. Each name is followed by abbreviation of the structure.

1st cerebellar lobule (lingula)   1Cb
2b cerebellar lobule   2bCb
2nd and 3rd cerebellar lobules   2/3Cb
2nd cerebellar lobule   2Cb
3rd and 4th cerebellar lobules   3/4Cb
3rd ventricle   3V
4th and 5th cerebellar lobules   4/5Cb
4th cerebellar lobule   4Cb
4th ventricle   4V
5th cerebellar lobule   5Cb
6a cerebellar lobule   6aCb
6b cerebellar lobule   6bCb
6c cerebellar lobule   6cCb
6th cerebellar lobule   6Cb
7th cerebellar lobule   7Cb
8th cerebellar lobule   8Cb
9th cerebellar lobule, a   9aCb
9th cerebellar lobule, a and b   9a,bCb
9th cerebellar lobule, b   9bCb
9th cerebellar lobule, c   9cCb
10th cerebellar lobule (nodule)   10Cb

A

A1 noradrenaline cells   A1
A1 noradrenaline cells/C1 adrenaline cells   A1/C1
A11dopamine cells   A11
A12 dopamine cells   A12
A14 dopamine cells   A14
A2 noradrenaline cells   A2
A4 noradrenaline cells   A4
A5 noradrenaline cells   A5
A7 noradrenaline cells   A7
abducens nucleus   6N
accessory abducens nucleus   6Acs
accessory abducens nucleus   6Acs
accessory abducens/facial nucleus   Acs6/7
accessory nerve nucleus   11N
accessory neurosecretory nuclei   ANS
accessory olfactory bulb   AOB
accessory olfactory tract   aot

accessory optic tract   aopt
accumbens nucleus   Acb
accumbens nucleus, core   AcbC
accumbens nucleus, rostral pole   AcbR
accumbens nucleus, shell   AcbSh
acoustic radiation   ar
acoustic stria   as
agranular insular cortex   AI
agranular insular cortex, dorsal part   AID
agranular insular cortex, posterior part   AIP
alar nucleus   Al
alveus of the hippocampus   alv
ambiguus nucleus   Amb
ambiguus nucleus, compact part   AmbC
ambiguus nucleus, loose part   AmbL
amygdalohippocampal area   AHi
amygdalohippocampal area, posteromedial part   AHiPM
amygdaloid fissure   af
amygdaloid intramedullary gray   IMG
amygdalostriatal transition area   ASt
anbiguus nucleus, subcompact part   AmbSC
angular thalamic nucleus   AngT
ansa lenticularis   al
ansoparamedian fissure   apmf
anterior amygdaloid area   AA
anterior amygdaloid area, dorsal part   AAD
anterior amygdaloid area, ventral part   AAV
anterior cerebral artery   acer
anterior commissural nucleus   AC
anterior commissure   ac
anterior commissure, anterior part   aca
anterior commissure, intrabulbar part   aci
anterior commissure, posterior part   acp
anterior hypothalamic area   AH
anterior hypothalamic area, central part   AHC
anterior lobe of pituitary   APit
anterior olfactory nucleus, dorsal part   AOD
anterior olfactory nucleus, external part   AOE
anterior olfactory nucleus, ventroposterior part   AOVP
anterior perifornical nucleus   APF
anterior pretectal nucleus   APT
anterior pretectal nucleus, dorsal part   APTD
anterior pretectal nucleus, ventral part   APTV
anterior spinal artery   asp
anterior tegmental nucleus   ATg
anterodorsal preoptic nucleus   ADP
anterodorsal thalamic nucleus   AD

anterovent thalamic nucleus, dorsomedial part   AVDM
anteroventral preoptic nucleus   AVPO
anteroventral thalamic nucleus, ventrolateral part   AVVL
aqueduct   Aq
arcuate hypothalamic nucleus, dorsal part   ArcD
arcuate hypothalamic nucleus, lateral part   ArcL
arcuate hypothalamic nucleus, lateroposterior part   ArcLP
arcuate hypothalamic nucleus, medial part   ArcM
area postrema   AP
artery   a
ascending fibers of the facial nerve   asc7
azygous pericallosal artery   azp

B

B9 serotonin cells   B9
Barrington's nucleus   Bar
basal nucleus (Meynert)   B
basilar artery   bas
basolateral amygdaloid nucleus   BL
basolateral amygdaloid nucleus, anterior part   BLA
basolateral amygdaloid nucleus, posterior part   BLP
basolateral amygdaloid nucleus, ventral part   BLV
basomedial amygdaloid nucleus   BM
basomedial amygdaloid nucleus, posterior part   BMP
bed nucleus of stria terminalis, fusiform part   Fu
bed nucleus of stria terminalis, supracapsular division   STS
bed nucleus of stria terminalis, supracapsular division, lateralpart   STSM
bed nucleus of stria terminalis, supracapsular division, medial part   STSL
bed nucleus of the anterior commissure   BAC
bed nucleus of the stria terminalis   BST
bed nucleus of the stria terminalis, intermediate division   STI
bed nucleus of the stria terminalis, intraamygdaloid division   STIA
bed nucleus of the stria terminalis, lateral division   STL
bed nucleus of the stria terminalis, lateral division, dorsal part   STLD
bed nucleus of the stria terminalis, lateral division, juxtacapsular part   STLJ
bed nucleus of the stria terminalis, lateral division, ventral part   STLV

bed nucleus of the stria terminalis, medial division   STM
bed nucleus of the stria terminalis, medial division, anterior part   STMA
bed nucleus of the stria terminalis, medial division, anterolateral part   STMAL
bed nucleus of the stria terminalis, medial division, posterior part   STMP
bed nucleus of the stria terminalis, medial division, posterointermediate part   STMPI
bed nucleus of the stria terminalis, medial division, posterolateral part   STMPL
bed nucleus of the stria terminalis, medial division, posteromedial part   STMPM
bed nucleus of the stria terminalis, medial division, ventral part   STMV
blood vessel   bv
brachium of the inferior colliculus   bic
brachium of the superior colliculus   bsc
brachium pontis (stem of middle cerebellar peduncle)   bp

C

C1 adrenaline cells   C1
C1 adrenaline cells and A1 noradrenaline cells   C1/A1
C2 adrenaline cells   C2
C3 adrenaline cells   C3
caudal interstitial nucleus of the medial longitudinal fasciculus   CI
caudal linear nucleus of the raphe   CLi
caudal periolivary nucleus   CPO
caudomedial entothinal cortex   CEnt
caudoventral respiratory group   CVRG
caudoventrolateral reticular nucleus   CVL
cell bridges of the ventral striatum   CB
central amygdaloid nucleus, medial division, anteroventral part   CeMAV
central amygdaloid nucleus   Ce
central amygdaloid nucleus, capsular part   CeC
central amygdaloid nucleus, lateral division   CeL
central amygdaloid nucleus, medial division   CeM
central amygdaloid nucleus, medial posteroventral part   CeMPV
central canal   CC
central cervical nucleus of the spinal cord   CeCv
central gray   CG
central gray of the pons   CGPn

central gray, alpha part   CGA
central gray, gamma part   CGG
central gray, nucleus O   CGO
central medial thalamic nucleus   CM
central tegmental tract   ctg
cerebellar white matter   cbw
cerebellum   Cb
cerebral cortex   Cx
cerebral peduncle   cp
choroid plexus   chp
cingulate cortex, area 1   Cg1
cingulate cortex, area 2   Cg2
cingulum   cg
circular nucleus   Cir
cochlear root of the vestibulocochlear nerve   8cn
commissural nucleus of the inferior colliculus   Com
commissural stria terminalis   cst
commissure of the inferior colliculus   cic
commissure of the superior colliculus   csc
conterminal nucleus   Ct
copula of the pyramis   Cop
corpus callosum   cc
crus 2 of the ansiform lobule   Crus2
cuneate fasciculus   cu
cuneate nucleus   Cu
cuneate nucleus, rotundus part   CuR
cuneiform nucleus   CnF
cuneiform nucleus, dorsal part   CnFD
cuneiform nucleus, intermediate part   CnFI
cuneiform nucleus, ventral part   CnFV

D

decussation of the superior cerebellar peduncle   dscp
decussation of the trapezoid body   tzd
deep cerebral white matter   dcw
deep gray layer of the superior colliculus   DpG
deep mesencephalic nucleus   DpMe
deep white layer of the superior colliculus   DpWh
dentate gyrus   DG
dorsal 3rd ventricle   D3V
dorsal acoustic stria   das
dorsal cochlear nucleus   DC
dorsal cochlear nucleus, deep core   DCDp
dorsal cochlear nucleus, fusiform layer   DCFu
dorsal cochlear nucleus, molecular layer   DCMo
dorsal cortex of the inferior colliculus   DCIC
dorsal corticospinal tract   dcs
dorsal endopiriform nucleus   DEn

dorsal fornix   df
dorsal hippocampal commissure   dhc
dorsal hypothalamic area   DA
dorsal hypothalamic nucleus   Do
dorsal intermediate entorhinal cortex   DIEnt
dorsal lateral geniculate nucleus   DLG
dorsal lateral olfactory tract   dlo
dorsal longitudinal fasciculus   dlf
dorsal motor nucleus of vagus   10N
dorsal nucleus (Clarke)   D
dorsal nucleus of the lateral lemniscus   DLL
dorsal paragigantocellular nucleus   DPGi
dorsal part of claustrum   DCl
dorsal peduncular cortex   DP
dorsal peduncular pontine nucleus   DPPn
dorsal raphe nucleus   DR
dorsal raphe nucleus, dorsal part   DRD
dorsal raphe nucleus, lateral part   DRL
dorsal raphe nucleus, ventral part   DRV
dorsal spinocerebellar fibres and olivocerebellar fibres   dsc/oc
dorsal subiculum   DS
dorsal tegmental bundle   dtg
dorsal tegmental decussation   dtgd
dorsal tegmental nucleus, central part   DTgC
dorsal tegmental nucleus, pericentral part   DTgP
dorsal tenia tecta   DTT
dorsal tenia tecta layer 1   DTT1
dorsal terminal nucleus of the accessory optic tract   DT
dorsal transition zone   DTr
dorsal tuberomammillary nucleus   DTM
dorsolateral entorhinal cortex   DLEnt
dorsolateral orbital cortex   DLO
dorsolateral periaqueductal gray   DLPAG
dorsolateral pontine nucleus   DLPn
dorsomedial hypothalamic nucleus   DM
dorsomedial hypothalamic nucleus, dorsal part   DMD
dorsomedial periaqueductal gray   DMPAG
dorsomedial pontine nucleus   DMPn
dorsomedial spinal trigeminal nucleus   DMSp5
dorsomedial spinal trigeminal nucleus, dorsal part   DMSp5D
dorsomedial spinal trigeminal nucleus, ventral part   DMSp5V
dysgranular insular cortex   DI

E

ectorhinal cortex   Ect
ectotrigeminal nucleus   E5
Edinger-Westphal nucleus   EW
epipeduncular nucleus   EpP
ependyma and subependymal layer   E
epilemniscal nucleus   ELm
epimicrocellular nucleus   EMi
episupraoptic nucleus   ESO
ethmoid thalamic nucleus   Eth
external capsule   ec
external cortex of the inferior colliculus   ECIC
external cortex of the inferior colliculus. layer 1   ECIC1
external cuneate nucleus   ECu
external globus pallidus   EGP
external medullary lamina   eml
external plexiform layer of the accessory olfactory bulb   EPlA
external plexiform layer of the olfactory bulb   EPl

F

facial motor nucleus, accessory part   7Acs
facial nerve   7n
facial nucleus   7N
facial nucleus, dorsal intermediate subnucleus   7DI
facial nucleus, dorsolateral subnucleus   7DL
facial nucleus, dorsomedial subnucleus   7DM
facial nucleus, lateral subnucleus   7L
facial nucleus, ventromedial subnucleus   7VM
fasciculus retroflexus   fr
fasciola cinereum   FC
field CA1 of the hippocampus   CA1
field CA2 of the hippocampus   CA2
field CA3 of the hippocampus   CA3
fimbria of the hippocampus   fi
flocculus   Fl
forceps major of the corpus callosum   fmj
forceps minor of the corpus callosum   fmi
fornix   f
frontal assocn cortex   FrA

G

gelatinous layer of the caudal spinal trigeminal nucleus   Ge5
gemini hypothalamic nucleus   Gem
genu of the corpus callosum   gcc

gigantocellular reticular nucleus   Gi
gigantocellular reticular nucleus, alpha part   GiA
gigantocellular reticular nucleus, ventral part   GiV
glomerular layer of the accessory olfactory bulb   GlA
glomerular layer of the olfactory bulb   Gl
gracile fasciculus   gr
granular cell layer of the olfactory bulb   GrO
granular insular cortex   GI
granular layer of the dentate gyrus   GrDG
granule cell layer of cochlear nuclei   GrC
granule cell layer of the accessory olfactory bulb   GrA
granule cell layer of the cerebellum   GrCb
gustatory thalamic nucleus   Gus

H

habenular commissure   hbc
hilus of the dentate gyrus   Hil
hippocampal fissure   hif
hypoglossal nucleus   12N
hypoglossal nucleus, geniohyoid part   12GH

I

indusium griseum   IG
inferior cerebellar peduncle (restiform body)   icp
inferior cerebellar peduncle decussation   icpd
inferior colliculus   IC
inferior olive, beta subnucleus   IOBe
inferior olive, cap of Kooy of the medial nucleus   IOK
inferior olive, dorsal nucleus   IOD
inferior olive, dorsomedial cell column   IODMC
inferior olive, dorsomedial cell group   IODM
inferior olive, medial nucleus   IOM
inferior olive, principal nucleus   IOPr
inferior olive, subnucleus A of medial nucleus   IOA
inferior olive, subnucleus B of medial nucleus   IOB
inferior olive, ventrolateral protrusion   IOVL
inferior salivatory nucleus   IS
infralimbic cortex   IL
infundibular reccess   IRe
infundibular stem   InfS
interanterodorsal thalamic nucleus   IAD
interanteromedial thalamic nucleus   IAM
intercalated amygdaloid nucleus, main part   IM
intercalated nuclei of the amygdala   I
intercalated nucleus of the medulla   In

intercrural fissure   icf
interfascicular nucleus   IF
intermediate acoustic stria   ias
intermediate endopiriform nucleus   IEn
intermediate geniculate nucleus   IntG
intermediate interstitial nucleus of the medial longitudinal fasciculus   II
intermediate reticular nucleus   IRt
intermediate reticular nucleus, alpha part   IRtA
intermediate white layer of the superior colliculus   InWh
intermediodorsal thalamic nucleus   IMD
intermediomedial cell column of the spinal cord   IMM
intermedioventral thalamic commissure   imvc
intermedius nucleus of the medulla   InM
internal arcuate fibers   ia
internal capsule   ic
internal globus pallidus (intrapeduncular nucleus)   IGP
internal medullary lamina   iml
internal plexiform layer of the olfactory bulb   IPl
interoculomotor nucleus   I3
interpeduncular fossa   IPF
interpeduncular nucleus   IP
interpeduncular nucleus, apical subnucleus   IPA
interpeduncular nucleus, caudal subnucleus   IPC
interpeduncular nucleus, dorsolateral subnucleus   IPDL
interpeduncular nucleus, dorsomedial subnucleus   IPDM
interpeduncular nucleus, intermediate subnucleus   IPI
interpeduncular nucleus, lateral subnucleus   IPL
interpeduncular nucleus, rostral subnucleus   IPR
interpedunculotegmental tract   ipt
interposed cerebellar nucleus, anterior part   IntA
interposed cerebellar nucleus, dorsolateral hump   IntDL
interposed cerebellar nucleus, posterior part   IntP
interstitial basal nucleus of the medulla   IB
interstitial nucleus of Cajal   InC
interstitial nucleus of Cajal. shell region   InCSh
interstitial nucleus of the posterior limb of the anterior commissure   IPAC
interstitial nucleus of the posterior limb of the anterior commissure, central part   IPACC

interstitial nucleus of the posterior limb of the anterior commissure, lateral part   IPACL
interstitial nucleus of the posterior limb of the anterior commissure, medial part   IPACM
interstitial nucleus of the vestibulocochlear nerve   I8
intertrigeminal nucleus   I5
interventricular foramen   IVF
intradecussational nucleus of the decussation of the superior cerebellar peduncle   ID
islands of Calleja   ICj
islands of Calleja, major island   ICjM

J

juxtaolivary nucleus   JxO
juxtaparaventricular part of lateral hypothalamus   JPLH
juxtarestiform body   jx

K

Killiker-Fuse nucleus   KF

L

lacunosum moleculare layer of the hippocampus   LMol
lambdoid septal zone   Ld
lat amygdaloid nucleus   La
lateral accumbens shell   LAcbSh
lateral amygdaloid nucleus, dorsolateral part   LaDL
lateral amygdaloid nucleus, ventrolateral part   LaVL
lateral amygdaloid nucleus, ventromedial part   LaVM
lateral cerebellar nucleus, parvicellular part   LatPC
lateral cervical nucleus of the spinal cord   LatC
lateral entorhinal cortex   LEnt
lateral habenular nucleus, lateral part   LHbL
lateral habenular nucleus, medial part   LHbM
lateral hypothalamic area   LH
lateral lemniscus   ll
lateral mammillary nucleus   LM
lateral olfactory tract   lo
lateral orbital cortex   LO
lateral parabrachial nucleus, central part   LPBC
lateral parabrachial nucleus, crescent part   LPBCr
lateral parabrachial nucleus, dorsal part   LPBD
lateral parabrachial nucleus, medial part   LPBM
lateral parabrachial nucleus, superior part   LPBS

lateral parabrachial nucleus, ventral part   LPBV
lateral paragigantocellular nucleus   LPGi
lateral paragigantocellular nucleus. alpha part   LPGiA
lateral parietal association cortex   LPtA
lateral periaqueductal gray   LPAG
lateral posterior thalamic nucleus   LP
lateral posterior thalamic nucleus, laterocaudal part   LPLC
lateral posterior thalamic nucleus, laterorostral part   LPLR
lateral posterior thalamic nucleus, mediocaudal part   LPMC
lateral posterior thalamic nucleus, mediorostral part   LPMR
lateral recess of the 4th ventricle   LR4V
lateral reticular nucleus   LRt
lateral reticular nucleus, parvicellular part   LRtPC
lateral reticular nucleus, subtrigeminal part   LRtS5
lateral septal nucleus, dorsal part   LSD
lateral septal nucleus, intermediate part   LSI
lateral septal nucleus, ventral part   LSV
lateral spinal nucleus   LSp
lateral stripe of the striatum   LSS
lateral superior olive   LSO
lateral terminal nucleus of the accessory optic tract   LT
lateral vestibular nucleus   LVe
lateroanterior hypothalamic nucleus   LA
laterodorsal tegmental nucleus   LDTg
laterodorsal tegmental nucleus, ventral part   LDTgV
laterodorsal thalamic nucleus   LD
laterodorsal thalamic nucleus, dorsomedial part   LDDM
laterodorsal thalamic nucleus, ventrolateral part   LDVL
lateroventral periolivary nucleus   LVPO
layer 1 of cortex   1
layer 1a of cortex   1a
layer 1b of cortex   1b
layer 4 of cortex   4
layer 5 of cortex   5
layer 5a of cortex   5a
layer 5b of cortex   5b
layer 6 of cortex   6
layer 6a of cortex   6a
layer 6b of cortex   6b
layers 3 and 4 of cortex   3/4
lemina terminalis   LTer

linear nucleus of the medulla   Li
lithoid nucleus   Lth
locus coeruleus   LC
longitudinal fasciculus of the pons   lfp
lucidum layer of the hippocampus   Lu

M

magnocellular nucleus of the lateral hypothalamus   MCLH
magnocellular nucleus of the posterior commissure   MCPC
magnocellular preoptic nucleus   MCPO
mammillary peduncle   mp
mammillary recess of the 3rd ventricle   MRe
mammillotegmental tract   mtg
mammillothalamic tract   mt
marginal zone of the medial geniculate   MZMG
matrix region of the medulla   Mx
medial accessory oculomotor nucleus   MA3
medial amygdaloid nucleus   Me
medial amygdaloid nucleus, ant dorsal   MeAD
medial amygdaloid nucleus, anteroventral part   MeAV
medial amygdaloid nucleus, posterodorsal part   MePD
medial amygdaloid nucleus, posteroventral part   MePV
medial cerebellar nucleus, caudomedial part   MedCM
medial cerebellar nucleus, dorsolateral protuberance   MedDL
medial corticohypothalamic tract   mch
medial eminence, external layer   MEE
medial eminence, internal layer   MEI
medial entorhinal cortex   MEnt
medial entorhinal cortex, ventral part   MEntV
medial forebrain bundle   mfb
medial forebrain bundle, 'a' component   mfba
medial forebrain bundle, 'b' component   mfbb
medial geniculate nucleus   MG
medial geniculate nucleus, dorsal part   MGD
medial geniculate nucleus, medial part   MGM
medial geniculate nucleus, ventral part   MGV
medial habenular nucleus   MHb
medial lemniscus   ml
medial lemniscus decussation   mld
medial longitudinal fasciculus   mlf
medial mammillary nucleus, lateral part   ML

medial mammillary nucleus, medial part   MM
medial mammillary nucleus, median part   MnM
medial orbital cortex   MO
medial parabrachial nucleus   MPB
medial paralemniscial nucleus   MPL
medial parietal association cortex   MPtA
medial preoptic nucleus   MPO
medial preoptic nucleus, central part   MPOC
medial preoptic nucleus, lateral part   MPOL
medial preoptic nucleus, medial part   MPOM
medial pretectal nucleus   MPT
medial septal nucleus   MS
medial superior olive   MSO
medial terminal nucleus of the accessory optic tract   MT
medial tuberal nucleus   MTu
medial vestibular nucleus   MVe
medial vestibular nucleus, magnocellular part   MVeMC
medial vestibular nucleus, parvicellular part   MVePC
median accessory nucleus of the medulla   MnA
median eminence   ME
median preoptic nucleus   MnPO
mediodorsal thalamic nucleus   MD
mediodorsal thalamic nucleus, central part   MDC
mediodorsal thalamic nucleus, lateral part   MDL
mediodorsal thalamic nucleus, medial part   MDM
mediodorsal thalamic nucleus, paralaminar part   MDPL
medioventral periolivary nucleus   MVPO
medullary reticular nucleus, dorsal part   MdD
medullary reticular nucleus, ventral part   MdV
mesencephalic trigeminal nucleus   Me5
mesencephalic trigeminal tract   me5
microcellular tegmental nucleus   MiTg
middle cerebellar peduncle   mcp
middle cerebral artery   mcer
mitral cell layer of the accessory olfactory bulb   MiA
mitral cell layer of the olfactory bulb   Mi
molecular layer of the cerebellum   MoCb
molecular layer of the dentate gyrus   MoDG
molecular layer of the subiculum   MoS
motor root of the trigeminal nerve   m5
motor trigeminal nucleus   5N
motor trigeminal nucleus, accessory subnucleus   5Acs
motor trigeminal nucleus, masseter part   5Ma
motor trigeminal nucleus, mylohyoid part   5MHy
motor trigeminal nucleus, temporalis part   5Te
motor trigeminal nucleus, ventromedial part   5VM

N

navicular nucleus of the basal forebrain   Nv
nervus intermedius component of the facial nerve   7ni
nigrostriatal bundle   ns
nucleus of Darkschewitsch   Dk
nucleus of origin of efferents of the vestibular nerve   EVe
nucleus of Roller   Ro
nucleus of the ansa lenticularis   AL
nucleus of the brachium of the inferior colliculus   BIC
nucleus of the central acoustic tract   CAT
nucleus of the commissural stria terminalis   CST
nucleus of the dorsal hipp commissure   DHC
nucleus of the fields of Forel   F
nucleus of the horizontal limb of the diagonal band   HDB
nucleus of the lateral olfactory tract   LOT
nucleus of the lateral olfactory tract, dorsal part   LOTD
nucleus of the lateral olfactory tract, layer 1   LOT1
nucleus of the optic tract   OT
nucleus of the posterior commissure   PCom
nucleus of the solitary tract   Sol
nucleus of the solitary tract, central part   SolCe
nucleus of the solitary tract, commissural part   SolC
nucleus of the solitary tract, gelatinous part   SolG
nucleus of the solitary tract, intermediate part   SolIM
nucleus of the solitary tract, interstitial part   SolI
nucleus of the solitary tract, lateral part   SolL
nucleus of the solitary tract, rostrolateral part   SolRL
nucleus of the solitary tract, ventrolateral part   SolVL
nucleus of the stria medullaris   SM
nucleus of the trapezoid body   Tz
nucleus of the vertical limb of the diagonal band   VDB
nucleus X   X
nucleus Y   Y
nucleus Z   Z

O

obex   Obex
oculomotor nerve   3n
oculomotor nucleus, parvicellular part   3PC

olfactory bulb   OB
olfactory nerve layer   ON
olfactory tubercle   Tu
olfactory tubercle layer 1   Tu1
olfactory ventricle (olfactory part of lateral ventricle)   OV
olivary pretectal nucleus   OPT
olivocerebellar tract   oc
olivocochlear bundle   ocb
optic chiasm   och
optic nerve   2n
optic nerve layer of the superior colliculus   Op
optic tract   opt
oriens layer of the hippocampus   Or
oval paracentral thalamic nucleus   OPC

P

paraabducens nucleus   Pa6
parabigeminal nucleus   PBG
parabrachial pigmented nucleus of the VTA   PBP
paracentral thalamic nucleus   PC
paracochlear glial substance   PCGS
paracommissural nucleus of the posterior commissure   PaC
parafascicular thalamic nucleus   PF
parafloccular sulcus   pfs
paraflocculus   PFl
parainterfascicular nucleus of the VTA   PIF
paralambdoid septal nucleus   PLd
paralemniscal nucleus   PL
paramedian lobule   PM
paramedian raphe nucleus   PMnR
paramedian reticular nucleus   PMn
paramedian sulcus   pms
paranigral nucleus of the VTA   PN
parapyramidal nucleus   PPy
pararubral nucleus   PaR
parasolitary nucleus   PSol
parastrial nucleus   PS
parasubiculum   PaS
parasubthalamic nucleus   PSTh
paratenial thalamic nucleus   PT
paraterete nucleus   PTe
paratrigeminal nucleus   Pa5
paratrochlear nucleus   Pa4
paraventricular hypoth nucleus   Pa
paraventricular hypothalamic nucleus, anterior parvicellular part   PaAP

paraventricular hypothalamic nucleus, dorsal cap   PaDC
paraventricular hypothalamic nucleus, lateral magnocellular part   PaLM
paraventricular hypothalamic nucleus, medial magnocellular part   PaMM
paraventricular hypothalamic nucleus, medial parvicellular part   PaMP
paraventricular hypothalamic nucleus, posterior part   PaPo
paraventricular hypothalamic nucleus, ventral part   PaV
paraventricular thalamic nucleus   PV
paraventricular thalamic nucleus, anterior part   PVA
paraventricular thalamic nucleus, posterior part   PVP
paraxiphoid nucleus of thalamus   PaXi
parietal cortex, posterior area, caudal part   PtPC
parietal cortex, posterior area, dorsal part   PtPD
parietal cortex, posterior area, rostral part   PtPR
parvicellular motor trigeminal nucleus   PC5
parvicellular reticular nucleus   PCRt
parvicellular reticular nucleus, alpha part   PCRtA
peduncular part of lateral hypothalamus   PLH
pedunculopontine tegmental nucleus   PPTg
periaqueductal gray   PAG
perifacial zone   P7
periformical nucleus   PeF
periformical part of lateral hypothalamus   PeFLH
perilemniscal nucleus, ventral part   PLV
periolivary horn   POH
peripeduncular nucleus   PP
perirhinal cortex   PRh
peritrigeminal zone   P5
periventricular gray   PVG
periventricular hypothalamic nucleus   Pe
periventricular hypothalamic nucleus, anterior parvicellular part   PeAP
pineal gland   Pi
pineal stalk   PiSt
piriform cortex   Pir
piriform cortex, layer 1   Pir1
piriform cortex, layer 1a   Pir1a
plioglial part of periaqueductai gray   PlPAG
polymorph layer of the dentate gyrus   PoDG
pontine nuclei   Pn
pontine raphe nucleus   PnR
pontine reticular nucleus, caudal part   PnC

pontine reticular nucleus, oral part   PnO
posterior cerebral artery   pcer
posterior commissure   pc
posterior hypothalamic area   PHA
posterior hypothalamic area, dorsal part   PHD
posterior hypothalamic nucleus   PH
posterior intralaminar thalamic nucleus   PIL
posterior limitans thalamic nucleus   PLi
posterior lobe of pituitary   PPit
posterior pretectal nucleus   PPT
posterior superior fissure   psf
posterior thalamic nuclear group   Po
posterior thalamic nuclear group, triangular part   PoT
posterodorsal preoptic nucleus   PDPO
posterodorsal raphe nucleus   PDR
posterodorsal tegmental nucleus   PDTg
posterolateral cortical amygdaloid nucleus   PLCo
posterolateral cortical amygdaloid nucleus, layer 1   PLCo1
posterolateral fissure   plf
posteromedial cortical amygdaloid nucleus   PMCo
posteromedian thalamic nucleus   PoMn
postsubiculum   Post
pre-Botzinger complex   PrBo
precommissural fornix   pcf
precommissural nucleus   PrC
preculminate fissure   pcuf
prelimbic cortex   PrL
premammillary nucleus, dorsal part   PMD
premammillary nucleus, ventral part   PMV
preoptic recess of the 3rd ventricle   P3V
prepyramidal fissure   ppf
prerubral field   PR
presubiculum   PrS
primary auditory cortex   Au1
primary fissure   prf
primary motor cortex   M1
primary somatosensory cortex   S1
primary somatosensory cortex, barrel field   S1BF
primary somatosensory cortex, dysgranular zone   S1DZ
primary somatosensory cortex, forelimb region   S1FL
primary somatosensory cortex, hindlimb region   S1HL
primary somatosensory cortex, jaw region   S1J
primary somatosensory cortex, jaw region, oral surface   S1JO

primary somatosensory cortex, oral dysgranular zone   S1DZO
primary somatosensory cortex, shoulder region   S1Sh
primary somatosensory cortex, shoulder/neck region   S1ShNc
primary somatosensory cortex, trunk region   S1Tr
primary somatosensory cortex, upper lip region   S1ULp
primary visual cortex   V1
primary visual cortex, binocular area   V1B
primary visual cortex, monocular area   V1M
principal mammillary tract   pm
principal sensory trigeminal nucleus   Pr5
principal sensory trigeminal nucleus, dorsomedial part   Pr5DM
principal sensory trigeminal nucleus, ventrolateral part   Pr5VL
Purkinje cell layer of the cerebellum   Pk
pyramidal cell layer of the hippocampus   Py
pyramidal decussation   pyd
pyramidal tract   py

R

radiatum layer of the hippocampus   Rad
raphe interpositus nucleus   RIP
raphe magnus nucleus   RMg
raphe obscurus nucleus   ROb
raphe pallidus nucleus   RPa
recess of the inferior colliculus   ReIC
red nucleus   R
red nucleus, magnocellular part   RMC
red nucleus, parvicellular part   RPC
region where VA and VL overlap   VA/VL
reticluostrial nucleus   RtSt
reticular thalamic nucleus   Rt
reticulotegmental nucleus of the pons   RtTg
reticulotegmental nucleus of the pons, lateral part   RtTgL
retroambiguus nucleus   RAmb
retrochiasmatic area, lateral part   RChL
retroethmoid nucleus   REth
retrolemniscal nucleus   RL
retroparafascicular nucleus   RPF
retrorubral field   RRF
retrorubral nucleus   RR
retrosplenial dysgranular cortex   RSD
retrosplenial granular cortex   RSG

retrosplenial granular cortex, c region   RSGc
retrouniens area   RRe
reuniens thalamic nucleus   Re
rhabdoid nucleus   Rbd
rhinal fissure   rf
rhinal incisure   ri
rhomboid thalamic nucleus   Rh
root of abducens nerve   6n
root of accessory nerve   11n
root of hypoglossal nerve   12n
rostral interstitial nucleus of medial longitudinal fasciculus   RI
rostral ventral respiratory group   RVRG
rostroventrolateral reticular nucleus   RVL
rostrum of the corpus callosum   rcc
rubrospinal tract   rs

S

scaphoid thalamic nucleus   Sc
secondary auditory cortex, dorsal area   AuD
secondary auditory cortex, ventral area   AuV
secondary fissure   sf
secondary motor cortex   M2
secondary somatosensory cortex   S2
secondary visual cortex, lateral area   V2L
secondary visual cortex, mediolateral area   V2ML
secondary visual cortex, mediomedial area   V2MM
sensory root of the trigeminal nerve   s5
septofimbrial nucleus   SFi
septohippocampal nucleus   SHi
septohypothalamic nucleus   SHy
simple lobule   Sim
simple lobule A   SimA
simple lobule B   SimB
simplex fissure   simf
solitary nucleus, dorsolateral part   SolDL
solitary nucleus, ventral part   SolV
solitary tract   sol
sphenoid nucleus   Sph
spinal trigeminal nucleus, caudal part   Sp5C
spinal trigeminal nucleus, interpolar part   Sp5I
spinal trigeminal nucleus, oral part   Sp5O
spinal trigeminal tract   sp5
spinal vestibular nucleus   SpVe
splenium of the corpus callosum   scc
stigmoid hypothalamic nucleus   Stg
stratum lucidum of the hippocampus   SLu
stria medullaris of the thalamus   sm

stria terminalis   st
strial part of the preoptic area   StA
striohypothalamic nucleus   StHy
subbrachial nucleus   SubB
subcoeruleus nucleus, alpha part   SubCA
subcoeruleus nucleus, dorsal part   SubCD
subcoeruleus nucleus, ventral part   SubCV
subcommissural organ   SCO
subfornical organ   SFO
subgeniculate nucleus   SubG
subiculum   S
subiculum, transition area   STr
subincertal nucleus   SubI
sublenticular extended amygdala   EA
sublenticular extended amygdala, central part   EAC
sublenticular extended amygdala, medial part   EAM
submammillothalamic nucleus   SMT
submedius thalamic nucleus   Sub
submedius thalamic nucleus, dorsal part   SubD
submedius thalamic nucleus, ventral part   SubV
subparafascicular thalamic nucleus, parvicellular part   SPFPC
subparaventricular zone of the hypothalamus   SPa
subpeduncular tegmental nucleus   SPTg
subpostrema area   SubP
substantia innominata   SI
substantia innominata, basal part   SIB
substantia innominata, dorsal part   SID
substantia innominata, ventral part   SIV
substantia nigra, compact part, dorsal tier   SNCD
substantia nigra, compact part, medial tier   SNCM
substantia nigra, lateral part   SNL
substantia nigra, reticular part   SNR
subthalamic nucleus   STh
superficial gray layer of the superior colliculus   SuG
superior cerebellar peduncle (brachium conjunctivum)   scp
superior cerebellar peduncle, descending limb   scpd
superior medullary velum   SMV
superior paraolivary nucleus   SPO
superior salivatory nucleus   SuS
superior thalamic radiation   str
superior vestibular nucleus   SuVe
suprachiasmatic nucleus   SCh
suprachiasmatic nucleus, ventromedial part   SChVM
suprachiasmatic nucleus, dorsolateral part   SChDL
suprageniculate thalamic nucleus   SG
supragenual nucleus   SGe
supramammillary decussation   sumd

supramammillary nucleus   SuM
supramammillary nucleus, lateral part   SuML
supramammillary nucleus, medial part   SuMM
supraoculomotor cap   Su3C
supraoculomotor periaqueductal gray   Su3
supraoptic decussation   sod
supraoptic nucleus   SO
supraoptic nucleus, retrochiasmatic part   SOR
supratrigeminal nucleus   Su5

T

tectospinal tract   ts
temporal associatin cortex   TeA
terete hypothalamic nucleus   Te
transverse fibers of the pons   tfp
trapezoid body   tz
triangular nucleus   Tr
triangular septal nucleus   TS
trigeminal ganglion   5Gn
trigeminal transition zone   5Tr
trigeminal-solitary transition zone   5Sol
trigeminothalamic tract   tth
trochlear nerve   4n
trochlear nerve decussation   4d
trochlear nucleus   4N
trochlear nucleus shell region   4Sh
tuberal region of lateral hypothalamus   TuLH

U

uncinate fasciculus decussation   und
uncinate fasciculus of the cerebellum   un

V

vagus nerve   10n
vascular organ of the lamina terminalis   VOLT
vein   v
ventral anterior thalamic nucleus   VA
ventral cochlear nucleus, anterior part   VCA
ventral cochlear nucleus, capsular part   VCCap
ventral cochlear nucleus, granule cell layer   VCAGr
ventral cochlear nucleus, posterior part   VCP
ventral cochlear nucleus, posterior part, octopus cell area   VCPO
ventral endopiriform nucleus   VEn
ventral hippocampal commissure   vhc
ventral intermediate entorhinal cortex   VIEnt
ventral lateral geniculate nucleus   VLG
ventral lateral geniculate nucleus, layer 1   VLG1
ventral linear nucleus of the thalamus   VLi
ventral nucleus of the lat lemniscus   VLL
ventral orbital cortex   VO
ventral pallidum   VP
ventral part of claustrum   VCl

ventral posterior nucleus of the thalamus, parvicellular part   VPPC
ventral posterolateral thalamic nucleus   VPL
ventral posteromedial thalamic nucleus   VPM
ventral reuniens thalamic nucleus   VRe
ventral spinocerebellar tract   vsc
ventral spinocerebellar tract decussation   vscd
ventral subiculum   VS
ventral tegmental area   VTA
ventral tegmental area, rostral part   VTAR
ventral tegmental decussation   vtgd
ventral tegmental nucleus   VTg
ventral tenia tecta   VTT
ventral tenia tecta, layer 1   VTT1
ventral tuberomammillary nucleus   VTM
ventrolateral preoptic nucleus   VLPO
ventrolateral thalamic nucleus   VL
ventromedial hypothalamic nucleus   VMH
ventromedial hypothalamic nucleus, anterior part   VMHA
ventromedial hypothalamic nucleus, central part   VMHC
ventromedial hypothalamic nucleus, dorsomedial part   VMHDM
ventromedial hypothalamic nucleus, ventrolateral part   VMHVL
ventromedial nucleus of the hypothalamus shell   VMHSh

ventromedial preoptic nucleus   VMPO
ventromedial thalamic nucleus   VM
vertebral artery   vert
vestibular root of the vestibulocochlear nerve   8vn
vestibulocerebellar nucleus   VeCb
vestibulocochlear ganglion   8Gn
vestibulocochlear nerve   8n
vestibulomesencephalic tract   veme
vestibulospinal tract   vesp
vomeronasal nerve   vn
vomeronasal nerve layer   VN
ventrolateral hypothalamic tract   vlh

X

xiphoid thalamic nucleus   Xi

Z

zona incerta   ZI
zona incerta, caudal part   ZIC
zona incerta, dorsal part   ZID
zona incerta, rostral part   ZIR
zona incerta, ventral part   ZIV
zona layer of the superior colliculus   Zo
zona limitans   ZL
zonal layer of the superior colliculus   GI

Index of Abbreviations

The abbreviations are listed in alphabetical order. Each abbreviation is followed by the structure name and the number of the figures on which the abbreviation appears.

1 layer 1 of cortex   9-21, 30-34, 52, 54-55, 59-75, 94, 99-109
1b layer 1b of cortex   8
1Cb 1st cerebellar lobule (lingula)   112-128
2 layer 2 of cortex   7-75, 94, 99-109
2/3Cb 2nd and 3rd cerebellar lobules   106-109
2bCb 2b cerebellar lobule   110
2Cb 2nd cerebellar lobule   103-120
3 layer 3 of cortex   8-75, 99-109
3/4 layers 3 and 4 of cortex   94
3/4Cb 3rd and 4th cerebellar lobules   127-129
3Cb 3rd cerebellar lobule   106-120, 122, 124-126, 134
3N oculomotor nucleus   85-91
3n oculomotor nerve   79-82
3PC oculomotor nucleus, parvicellular part   84-91
3V 3rd ventricle   29-73
4 layer 4 of cortex   8-10, 30-33, 52
4/5Cb 4th and 5th cerebellar lobules   106-109
4Cb 4th cerebellar lobule   108-126
4N trochlear nucleus   92-95
4n trochlear nerve   94-112
4Sh trochlear nucleus shell region   92-95
4V 4th ventricle   104-146
5 layer 5 of cortex   52
5a layer 5a of cortex   52, 94
5Acs motor trigeminal nucleus, accessory subnucleus   113-116
5b layer 5b of cortex   52, 94
5Cb 5th cerebellar lobule   109-120, 122-131
5Ma motor trigeminal nucleus, masseter part   109-115
5MHy motor trigeminal nucleus, mylohyoid part   109-110
5N motor trigeminal nucleus   108
5Sol trigeminal-solitary transition zone   122-148
5Te motor trigeminal nucleus, temporalis part   109-115

5Tr trigeminal transition zone   113-117
5VM motor trigeminal nucleus, ventromedial part   111-113
6 layer 6 of cortex   94
6a layer 6a of cortex   52
6aCb 6a cerebellar lobule   121-127, 129-133
6Acs accessory abducens nucleus   117-120
6b layer 6b of cortex   52
6bCb 6b cerebellar lobule   134-138
6Cb 6th cerebellar lobule   128, 133
6cCb 6c cerebellar lobule   134-137, 139-143
6N abducens nucleus   117-119
6n root of abducens nerve   115-116
7Acs facial motor nucleus, accessory part   120-125, 128-132
7Cb 7th cerebellar lobule   137-150
7DI facial nucleus, dorsal intermediate subnucleus   123-131
7DL facial nucleus, dorsolateral subnucleus   122-132
7DM facial nucleus, dorsomedial subnucleus   121-131
7L facial nucleus, lateral subnucleus   122-134
7n facial nerve   111-119
7ni nervus intermedius component of the facial nerve   121-123
7VI facial nucleus, ventral intermediate subnucleus   121-132
7VM facial nucleus, ventromedial subnucleus   121-130
8Cb 8th cerebellar lobule   135-155
8cn cochlear root of the vestibulocochlear nerve   112-117, 121-123, 126-127
8vn vestibular root of the vestibulocochlear nerve   115-125
9a,bCb 9th cerebellar lobule, a and b   140-159
9Cb 9th cerebellar lobule   131-133, 135-139
9cCb 9th cerebellar lobule, c   140-158
9n glossopharyngeal nerve   130-132
10Cb 10th cerebellar lobule (nodule)   129-149
10N dorsal motor nucleus of vagus   136-156
10n vagus nerve   132-133, 135-137, 139, 141-143, 147, 149-150
11N accessory nerve nucleus   157-161
12GH hypoglossal nucleus, geniohyoid part   147-156
12N hypoglossal nucleus   138-157
12n root of hypoglossal nerve   137, 141, 144, 149-155

A

artery   8-9, 38-39, 41, 43-45, 47-49, 52-63, 65-74, 76-79, 81, 83-85, 87, 91-105, 107, 109, 113-114, 118, 120-126, 128, 130-131, 148-149
A1 A1 noradrenaline cells   152-161
A1/C1 A1 noradrenaline cells/C1 adrenaline cells   151
A11 A11dopamine cells   59-70
A13 A13 dopamine cells   50-57
A2 A2 noradrenaline cells   155-161
A5 A5 noradrenaline cells   108-113, 115-125, 127
A7 A7 noradrenaline cells   102-108
AA anterior amygdaloid area   35-48
ac anterior commissure   34-37
aca anterior commissure, anterior part   8-33
AcbC accumbens nucleus, core   11-29
AcbR accumbens nucleus, rostral pole   10
AcbSh accumbens nucleus, shell   11-29
acer anterior cerebral artery   13-20, 29-30, 35-37
aci anterior commissure, intrabulbar part   2-7
ACo anterior cortical amygdaloid nucleus   37-55
acp anterior commissure, posterior part   34-43
AD anterodorsal thalamic nucleus   43-52
af amygdaloid fissure   46-47, 50, 54, 68-80
AHA anterior hypothalamic area, anterior part   41-45
AHC anterior hypothalamic area, central part   46-50
AHiAL amygdalohippocampal area, anterolateral part   59-66
AHiPL amygdalohippocampal area, posterolateral   67-70
AHiPM amygdalohippocampal area, posteromedial part   67-81
AHP anterior hypothalamic area, posterior part   47-52
AID agranular insular cortex, dorsal part   8-32
AIP agranular insular cortex, posterior part   33-57
AIV agranular insular cortex, ventral part   8-32
Al alar nucleus   33-34
alv alveus of the hippocampus   47-96
AM anteromedial thalamic nucleus   42-53
AmbC ambiguus nucleus, compact part   133-140
AmbL ambiguus nucleus, loose part   146-151
AmbSC ambiguus nucleus, subcompact part   141-145
AMV anteromedial thalamic nucleus, ventral part   45-49

AngT angular thalamic nucleus   53-55
ANS accessory neurosecretory nuclei   43-50
AOD anterior olfactory nucleus, dorsal part   4-7
AOE anterior olfactory nucleus, external part   3-5
AOL anterior olfactory nucleus, lateral part   3-8
AOM anterior olfactory nucleus, medial part   4-7
AOP anterior olfactory nucleus, posterior part   9-11
aot accessory olfactory tract   46
AOV anterior olfactory nucleus, ventral part   4-5
AOVP anterior olfactory nucleus, ventroposterior part   6-9
AP area postrema   147-152
APF anterior perifornical nucleus   39-41
APir amygdalopiriform transition area   64-88
apmf ansoparamedian fissure   139-148, 150
APT anterior pretectal nucleus   80-82
APTD anterior pretectal nucleus, dorsal part   69-79
APTV anterior pretectal nucleus, ventral part   70-79
Aq aqueduct   74-103
Arc arcuate hypothalamic nucleus   47-48
ArcD arcuate hypothalamic nucleus, dorsal part   49-61
ArcL arcuate hypothalamic nucleus, lateral part   49-61
ArcLP arcuate hypothalamic nucleus, lateroposterior part   62-68
ArcM arcuate hypothalamic nucleus, medial part   49-61
ArcMP arcuate hypothalamic nucleus, medial posterior part   62-69
asc7 ascending fibers of the facial nerve   122-126
asp anterior spinal artery   137, 139-142, 144-145
ASt amygdalostriatal transition area   46-50, 54-61
ATg anterior tegmental nucleus   96-98
Au1 primary auditory cortex   60-90
AuD secondary auditory cortex, dorsal area   58-90
AuV secondary auditory cortex, ventral area   60-90
AV anteroventral thalamic nucleus   42
AVDM anterovent thalamic nucleus, dorsomedial part   43-53
AVPe anteroventral periventricular nucleus   32-34
AVVL anteroventral thalamic nucleus, ventrolateral part   43-51
azac azygous anterior cerebral artery   10-12
azp azygous pericallosal artery   10-18

B

B   basal nucleus (Meynert)   36-55, 57-59

B9   B9 serotonin cells   90-96, 100

BAC   bed nucleus of the anterior commissure   37-40

BAOT   bed nucleus of the accessory olfactory tract   46-51

Bar   Barrington's nucleus   108-112

bas   basilar artery   85-96, 98-135

BIC   nucleus of the brachium of the inferior colliculus   88-98

bic   brachium of the inferior colliculus   82-102

BL   basolateral amygdaloid nucleus   74-75

BLA   basolateral amygdaloid nucleus, anterior part   46-61

BLP   basolateral amygdaloid nucleus, posterior part   53-73

BLV   basolateral amygdaloid nucleus, ventral part   49-61

BMA   basomedial amygdaloid nucleus, anterior part   43-56

BMP   basomedial amygdaloid nucleus, posterior part   54-67

Bo   Botzinger complex   133-137

bsc   brachium of the superior colliculus   69-71, 74-86

C

C1   C1 adrenaline cells   131-146

C1/A1   C1 adrenaline cells and A1 noradrenaline cells   147-150

C2   C2 adrenaline cells   138-139, 141-143

C3   C3 adrenaline cells   135-137

CA1   field CA1 of the hippocampus   52-66, 68-90

CA2   field CA2 of the hippocampus   52-61, 63-77

CA3   field CA3 of the hippocampus   47-81, 83-84

CAT   nucleus of the central acoustic tract   105-109

CB   cell bridges of the ventral striatum   24-34

cbc   cerebellar commissure   120-122, 124, 126

cbw   cerebellar white matter   108-156

CC   central canal   147-161

cc   corpus callosum   21-74

CeC   central amygdaloid nucleus, capsular part   46-61

CeCv   central cervical nucleus of the spinal cord   149-161

CeL   central amygdaloid nucleus, lateral division   49-60

CeM   central amygdaloid nucleus, medial division   45-57

CEnt   caudomedial entothinal cortex   89-112

CG   central gray   106-108

cg   cingulum   11-83

Cg1   cingulate cortex, area 1   8-46

Cg2   cingulate cortex, area 2   14-46

CGA   central gray, alpha part   109-118

CGB   central gray, beta part   109-115

CGG   central gray, gamma part   115-118

CGO   central gray, nucleus O   110-114

CGPn   central gray of the pons   113-115

chp   choroid plexus   34-41, 43-76, 118-119, 124-143, 145-146

CI   caudal interstitial nucleus of the medial longitudinal fasciculus   122-128

CIC   central nucleus of the inferior colliculus   98-109

cic   commissure of the inferior colliculus   98-102

Cir   circular nucleus   45

CL   centrolateral thalamic nucleus   51-69

Cl   claustrum   8-14

CLi   caudal linear nucleus of the raphe   85-94

cll   commissure of the lateral lemniscus   100-103

CM   central medial thalamic nucleus   44-66

CnFD   cuneiform nucleus, dorsal part   101-106

CnFI   cuneiform nucleus, intermediate part   98-105

CnFV   cuneiform nucleus, ventral part   98-105

Com   commissural nucleus of the inferior colliculus   99-102

Cop   copula of the pyramis   135-154

cp   cerebral peduncle   61-93

CPO   caudal periolivary nucleus   121

CPu   caudate putamen (striatum)   12-66

Crus1   crus 1 of the ansiform lobule   117-141

Crus2   crus 2 of the ansiform lobule   132-150

csc   commissure of the superior colliculus   76- 86

CST   nucleus of the commissural stria terminalis   35

cst   commissural stria terminalis   39, 41-64

Ct   conterminal nucleus   152-155

Cu   cuneate nucleus   140-161

cu   cuneate fasciculus   143-162

CuR   cuneate nucleus, rotundus part   147-154

CVL   caudoventrolateral reticular nucleus   138-145, 150-151

Cx   cerebral cortex   112-113

CxA1   cortex-amygdala transition zone, layer 1   34-47

D

D   dorsal nucleus (Clarke)   162c-d

D3V   dorsal 3rd ventricle   42-76

DA   dorsal hypothalamic area   52-60

das   dorsal acoustic stria   126-129

DCDp   dorsal cochlear nucleus, deep core   121-130

DCFu   dorsal cochlear nucleus, fusiform layer   121-131

DCGr   dorsal cochlear nucleus, granular layer   121

DCIC   dorsal cortex of the inferior colliculus   101-110

DCl   dorsal part of claustrum   15-49

DCMo   dorsal cochlear nucleus, molecular layer   121-131

dcs   dorsal corticospinal tract   160-162

dcw   deep cerebral white matter   55-98

DEn   dorsal endopiriform nucleus   8-70

df   dorsal fornix   40-68

DG   dentate gyrus   66

dhc   dorsal hippocampal commissure   49-92

DI   dysgranular insular cortex   10-57

DIEnt   dorsal intermediate entorhinal cortex   85-95

Dk   nucleus of Darkschewitsch   72-82

DLEnt   dorsolateral entorhinal cortex   85-103

DLG   dorsal lateral geniculate nucleus   60-78

DLL   dorsal nucleus of the lateral lemniscus   100-106

DLO   dorsolateral orbital cortex   6-7

dlo   dorsal lateral olfactory tract   3-4

DLPAG   dorsolateral periaqueductal gray   82-101

DM   dorsomedial hypothalamic nucleus   63-64

DMC   dorsomedial hypothalamic nucleus, compact part   58-62

DMD   dorsomedial hypothalamic nucleus, dorsal part   52-62

DMPAG   dorsomedial periaqueductal gray   77-106

DMSp5   dorsomedial spinal trigeminal nucleus   120-140

DMTg   dorsomedial tegmental area   105-115

DMV   dorsomedial hypothalamic nucleus, ventral part   59-62

DP   dorsal peduncular cortex   9-15

DpG   deep gray layer of the superior colliculus   78-100

DPGi   dorsal paragigantocellular nucleus   120-132

DpMe   deep mesencephalic nucleus   81-100

DPO   dorsal periolivary region   113-116, 118-119

DPPn   dorsal peduncular pontine nucleus   102

DpWh   deep white layer of the superior colliculus   79-100

DR   dorsal raphe nucleus   90-91, 111

DRC   dorsal raphe nucleus, caudal part   104-110

DRD   dorsal raphe nucleus, dorsal part   92-103

DRL   dorsal raphe nucleus, lateral part   93-100

DRV   dorsal raphe nucleus, ventral part   93-103

DS   dorsal subiculum   74-90

dsc   dorsal spinocerebellar tract   150-161

dsc/oc   dorsal spinocerebellar fibres and olivocerebellar fibres   137-149

dscp   decussation of the superior cerebellar peduncle   96-98

DT   dorsal terminal nucleus of the accessory optic tract   83-87

DTgC   dorsal tegmental nucleus, central part   107-110

dtgd   dorsal tegmental decussation   84-89

DTgP   dorsal tegmental nucleus, pericentral part   104-110

DTM   dorsal tuberomammillary nucleus   63-65

DTr   dorsal transition zone   8-9

DTT   dorsal tenia tecta   9-14

DTT1   dorsal tenia tecta layer 1   8

E

E   ependyma and subependymal layer   1-5, 13-39

E/OV   ependymal and subendymal layer/olfactory ventricle   6-12

E5   ectotrigeminal nucleus   143

EA   sublenticular extended amygdala   47-48

EAC   sublenticular extended amygdala, central part   39-46

EAM   sublenticular extended amygdala, medial part   43-46

ec   external capsule   14-66

ECIC   external cortex of the inferior colliculus   91-110

Ect   ectorhinal cortex   58-111

ECu   external cuneate nucleus   135-151

EF   epifascicular nucleus   132-134

EGP   external globus pallidus   35-59

ELm   epilemniscal nucleus   82

eml   external medullary lamina   60-65

EPl   external plexiform layer of the olfactory bulb   1-5

EPlA   external plexiform layer of the accessory olfactory bulb   3-4

TS triangular septal nucleus 35-45
ts tectospinal tract 85-161
tth trigeminothalamic tract 83-117, 120-128
Tu olfactory tubercle 10-35
TuLH tuberal region of lateral hypothalamus 46-62
Tz nucleus of the trapezoid body 107-120
tz trapezoid body 105-125
tzd decussation of the trapezoid body 118

U

un uncinate fasciculus of the cerebellum 115-122
und uncinate fasciculus decussation 123-128

V

v vein 92-93, 95, 98, 101-102, 105-107
V1 primary visual cortex 70-81
V1B primary visual cortex, binocular area 82-111
V1M primary visual cortex, monocular area 82-111
V2L secondary visual cortex, lateral area 79-111
V2ML secondary visual cortex, mediolateral area 68-98
V2MM secondary visual cortex, mediomedial area 68-107
VA ventral anterior thalamic nucleus 46, 48-52
VA/VL region where VA and VL overlap 47

VCA ventral cochlear nucleus, anterior part 109-121
VCAGr ventral cochlear nucleus, granule cell layer 115-118, 120
VCCap ventral cochlear nucleus, capsular part 122-125
VCl ventral part of claustrum 15-49
VCP ventral cochlear nucleus, posterior part 120-122, 125-127
VCPO ventral cochlear nucleus, posterior part, octopus cell area 123-125
VDB nucleus of the vertical limb of the diagonal band 18-29
VeCb vestibulocerebellar nucleus 122-126
veme vestibulomesencephalic tract 116-122
VEn ventral endopiriform nucleus 40-58
vert vertebral artery 137, 139-145, 147-161
vesp vestibulospinal tract 123-126
vhc ventral hippocampal commissure 38-48
VIEnt ventral intermediate entorhinal cortex 85-100
VL ventrolateral thalamic nucleus 48-59
VLG ventral lateral geniculate nucleus 60-63, 65-71, 73-77
VLG1 ventral lateral geniculate nucleus, layer 1 64, 72
VLH ventrolateral hypothalamic nucleus 39-46
vlh ventrolateral hypothalamic tract 47-56
VLi ventral linear nucleus of the thalamus 69-74

VLL ventral nucleus of the lat lemniscus 97-108
VLPAG ventrolateral periaqueductal gray 88-106
VLPO ventrolateral preoptic nucleus 33-38
VM ventromedial thalamic nucleus 48-64
VMH ventromedial hypothalamic nucleus 48-49, 60-61
VMHA ventromedial hypothalamic nucleus, anterior part 47
VMHC ventromedial hypothalamic nucleus, central part 50-59
VMHDM ventromedial hypothalamic nucleus, dorsomedial part 50-59
VMHSh ventromedial nucleus of the hypothalamus shell 47-61
VMHVL ventromedial hypothalamic nucleus, ventrolateral part 50-59
VMPO ventromedial preoptic nucleus 31-36, 107-109
VO ventral orbital cortex 5-11
VOLT vascular organ of the lamina terminalis 28-33
VP ventral pallidum 11-42
VPL ventral posterolateral thalamic nucleus 51-70
VPM ventral posteromedial thalamic nucleus 54-73
VPPC ventral posterior nucleus of the thalamus, parvicellular part 62-69
VRe ventral reuniens thalamic nucleus 43-61
VS ventral subiculum 71-90

vsc ventral spinocerebellar tract 105-161
vscd ventral spinocerebellar tract decussation 123
VTA ventral tegmental area 89-90
VTAR ventral tegmental area, rostral part 72-75
VTg ventral tegmental nucleus 100-105
vtgd ventral tegmental decussation 78-86
VTM ventral tuberomammillary nucleus 64-74
VTT ventral tenia tecta 7, 10
VTT1 ventral tenia tecta, layer 1 8-9

X

X nucleus X 125-139
Xi xiphoid thalamic nucleus 45-48, 51-53

Y

Y nucleus Y 124-127

Z

Z nucleus Z 140-142
ZI zona incerta 54-55
ZIC zona incerta, caudal part 70-76
ZID zona incerta, dorsal part 56-69
ZIR zona incerta, rostral part 49-53
ZIV zona incerta, ventral part 56-69
ZL zona limitans 30-32
Zo zona layer of the superior colliculus 77-99

Figures

Coronal sections of the brain   *Figures 1-161*

Transverse sections of the spinal cord   *Figures 162a, b, c, d, e*

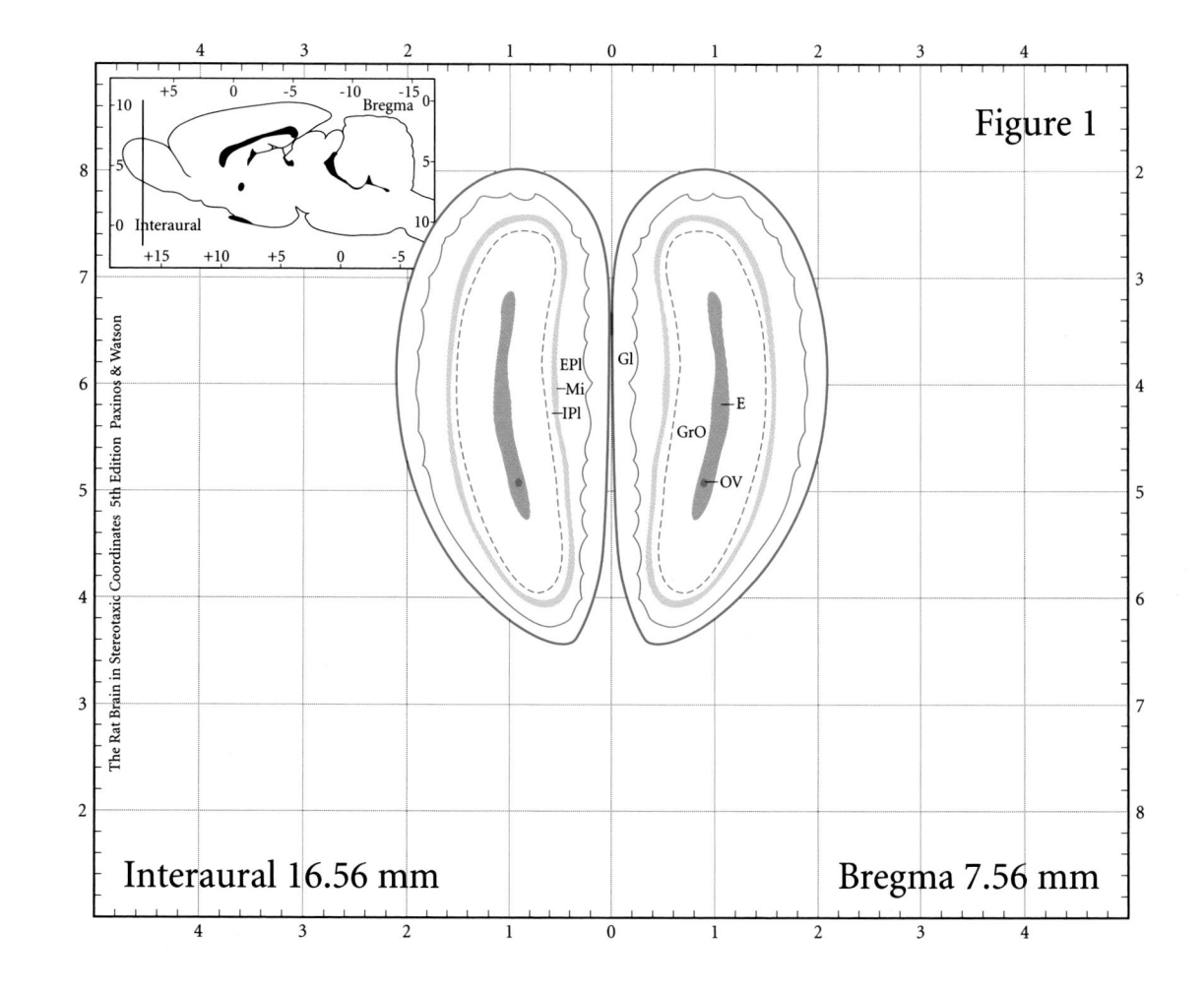

Figure 1

The Rat Brain in Stereotaxic Coordinates 5th Edition Paxinos & Watson

Interaural 16.56 mm

Bregma 7.56 mm

E ependyma/subepen
EPl ext plex olf bulf
Gl glomerular olf b
GrO granular olf bulb
IPl int plexi olf bulb
Mi mitral olf bulb
OV olfact ventricle

aci ac intrabulbar
E ependyma/subepen
EPl ext plex olf bulf
Gl glomerular olf b
GrA gran acc olf bulb

GrO granular olf bulb
IPl int plexi olf bulb
Mi mitral olf bulb
OV olfact ventricle

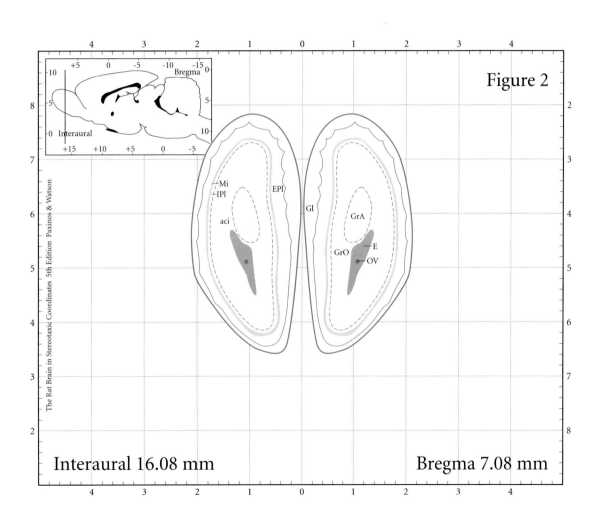

Figure 2

The Rat Brain in Stereotaxic Coordinates 5th Edition Paxinos & Watson

Interaural 16.08 mm

Bregma 7.08 mm

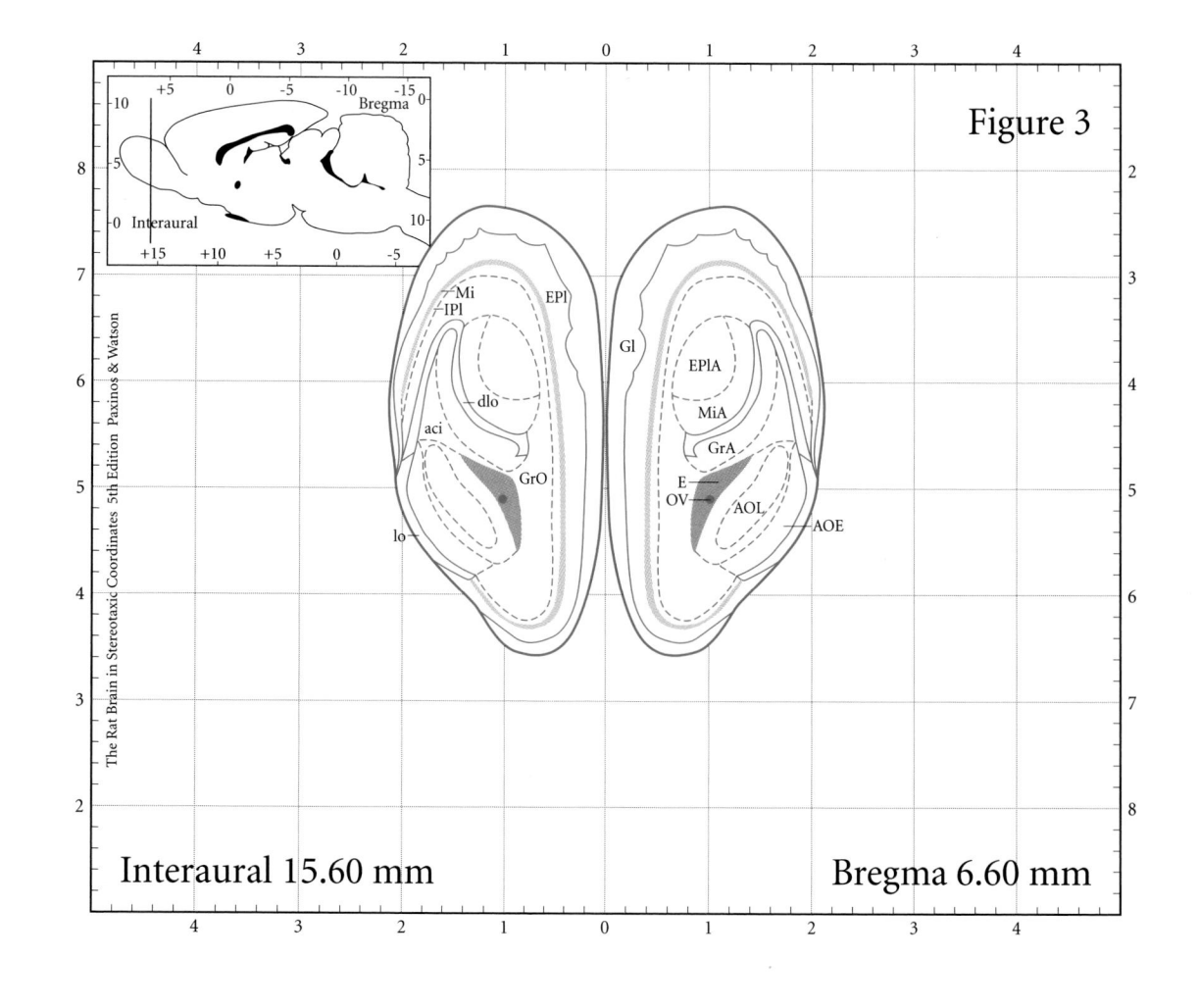

Figure 3

Interaural 15.60 mm

Bregma 6.60 mm

aci ac intrabulbar
AOE ant olfact ext
AOL ant olfact lat
dlo dorsal lat olf tr
E ependyma/subepen
EPl ext plex olf bulf
EPlA ext plex acc olf b
Gl glomerular olf b

GrA gran acc olf bulb
GrO granular olf bulb
IPl int plexi olf bulb
lo lat olfactory tr
Mi mitral olf bulb
MiA mitral acc olf bulb
OV olfact ventricle

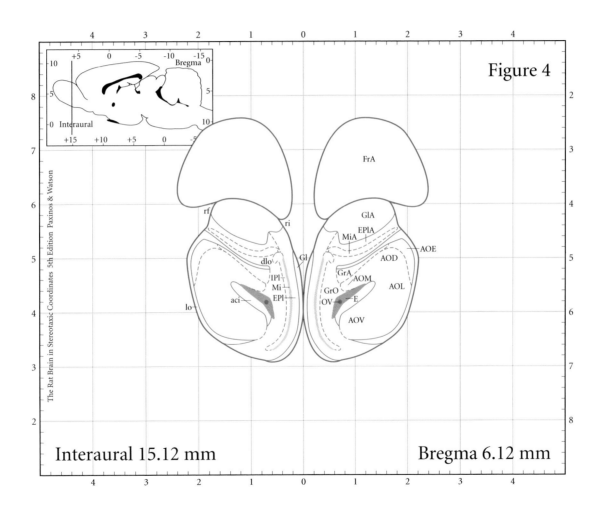

Figure 4

The Rat Brain in Stereotaxic Coordinates 5th Edition Paxinos & Watson

Interaural 15.12 mm

Bregma 6.12 mm

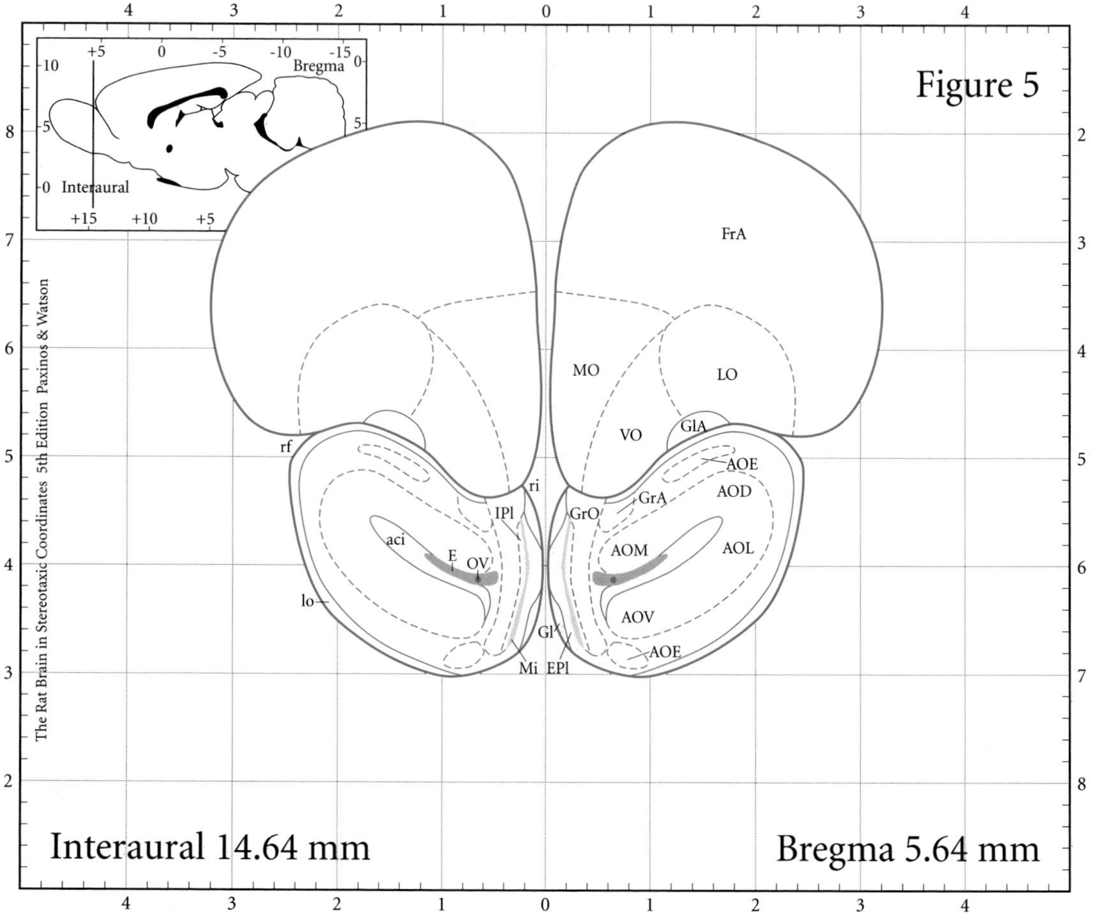

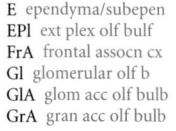

Figure 5

The Rat Brain in Stereotaxic Coordinates 5th Edition Paxinos & Watson

Bregma

Interaural

Interaural 14.64 mm

Bregma 5.64 mm

FrA

MO   LO

VO   GlA

rf   AOE

ri   GrA   AOD

IPl   GrO

aci   AOM   AOL

E   OV

lo   AOV

Gl   AOE

Mi   EPl

aci ac intrabulbar
AOD ant olf dorsal
AOE ant olfact ext
AOL ant olfact lat
AOM ant olfact medial
AOV ant olfact vent

E ependyma/subepen
EPl ext plex olf bulf
FrA frontal assocn cx
Gl glomerular olf b
GlA glom acc olf bulb
GrA gran acc olf bulb

GrO granular olf bulb
IPl int plexi olf bulb
LO lat orbital cx
lo lat olfactory tr
Mi mitral olf bulb
MO medial orbital cx

OV olfact ventricle
rf rhinal fissure
ri rhinal incisure
VO ventral orbital cx

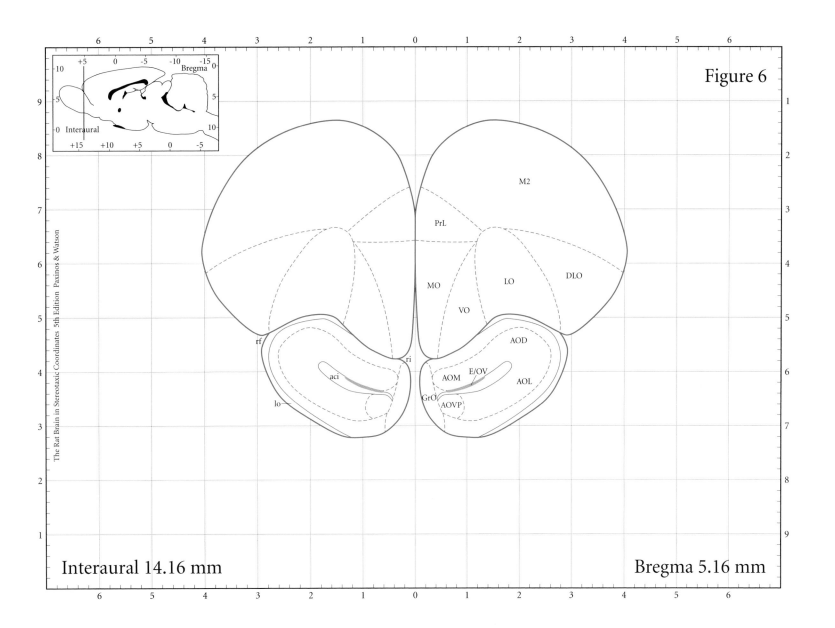

aci ac intrabulbar	**AOVP** ant olf ventropost	**LO** lat orbital cx	**PrL** prelimbic cx
AOD ant olf dorsal	**DLO** dorsolat orbital cx	**lo** lat olfactory tr	**rf** rhinal fissure
AOL ant olfact lat	**E/OV** epend/olf ventr	**M2** 2ary motor cx	**ri** rhinal incisure
AOM ant olfact medial	**GrO** granular olf bulb	**MO** medial orbital cx	**VO** ventral orbital cx

Figure 6

Interaural 14.16 mm

Bregma 5.16 mm

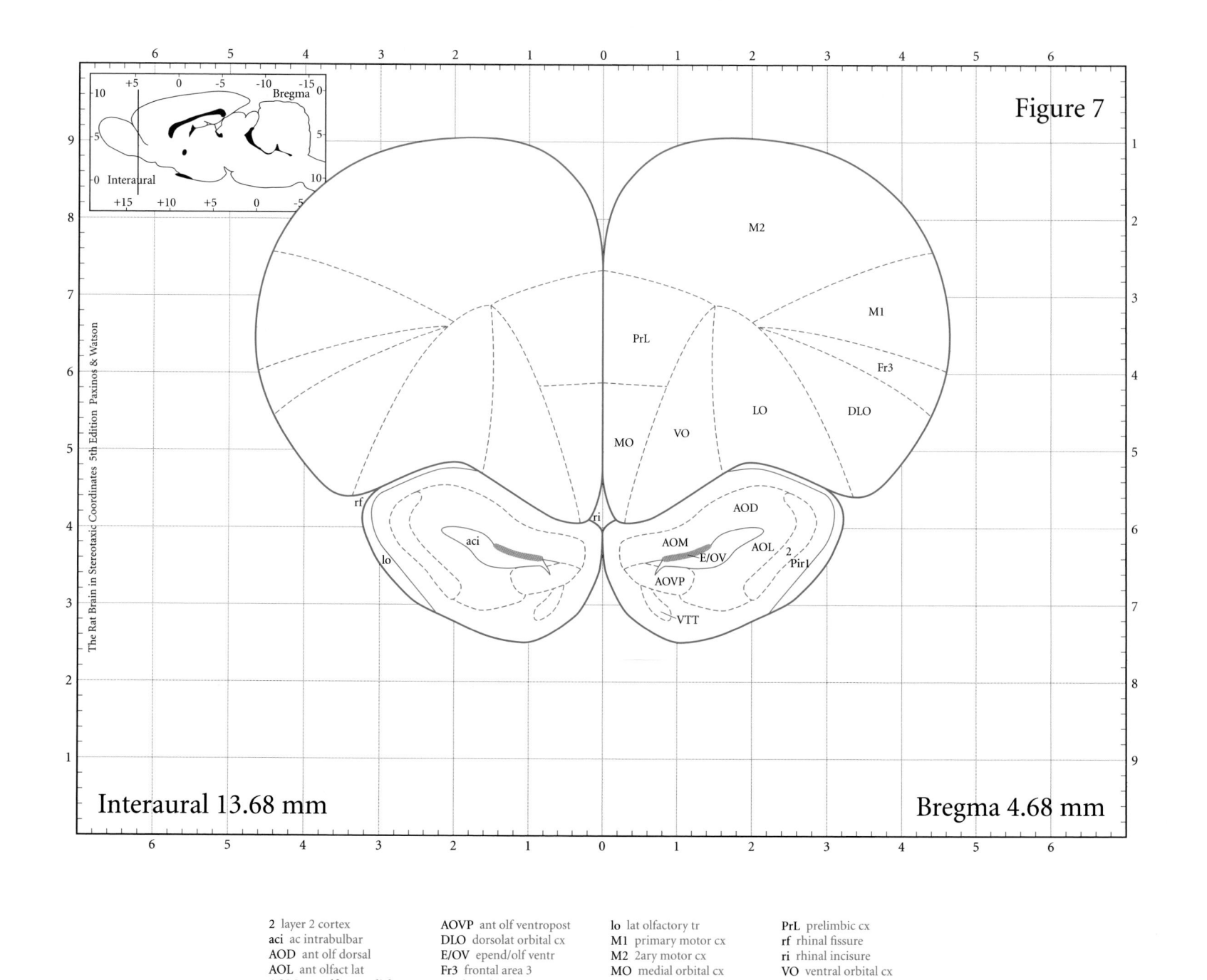

Figure 7

The Rat Brain in Stereotaxic Coordinates 5th Edition Paxinos & Watson

Interaural 13.68 mm

Bregma 4.68 mm

2 layer 2 cortex	AOVP ant olf ventropost	lo lat olfactory tr	PrL prelimbic cx
aci ac intrabulbar	DLO dorsolat orbital cx	M1 primary motor cx	rf rhinal fissure
AOD ant olf dorsal	E/OV epend/olf ventr	M2 2ary motor cx	ri rhinal incisure
AOL ant olfact lat	Fr3 frontal area 3	MO medial orbital cx	VO ventral orbital cx
AOM ant olfact medial	LO lat orbital cx	Pir1 piriform layer 1	VTT ventral tenia tecta

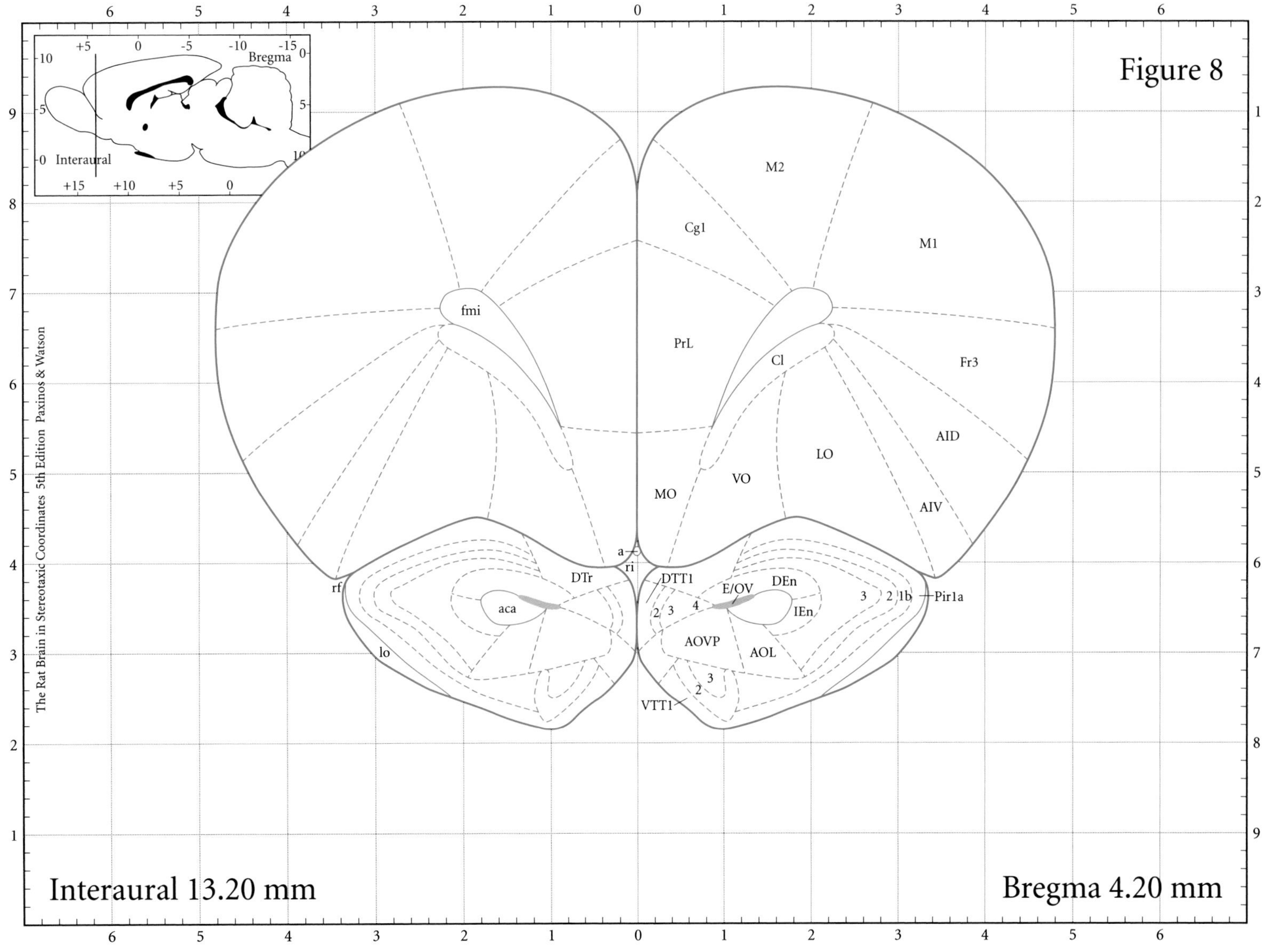

Figure 8

Interaural 13.20 mm

Bregma 4.20 mm

1b layer 1b cortex
2 layer 2 cortex
3 layer 3 cortex
4 layer 4 cortex
a artery
aca ant comm, ant
AID ant insular dorsal
AIV ant insular ventral

AOL ant olfact lat
AOVP ant olf ventropost
Cg1 cingulate area 1
Cl claustrum
DEn dorsal endopirif
DTr dorsal trans zone
DTT1 dors TT layer 1
E/OV epend/olf ventr

fmi forceps minor
Fr3 frontal area 3
IEn intermed endopir
LO lat orbital cx
lo lat olfactory tr
M1 primary motor cx
M2 2ary motor cx
MO medial orbital cx

Pir1 piriform layer 1
PrL prelimbic cx
ri rhinal incisure
VO ventral orbital cx
VTT1 vent TT layer 1

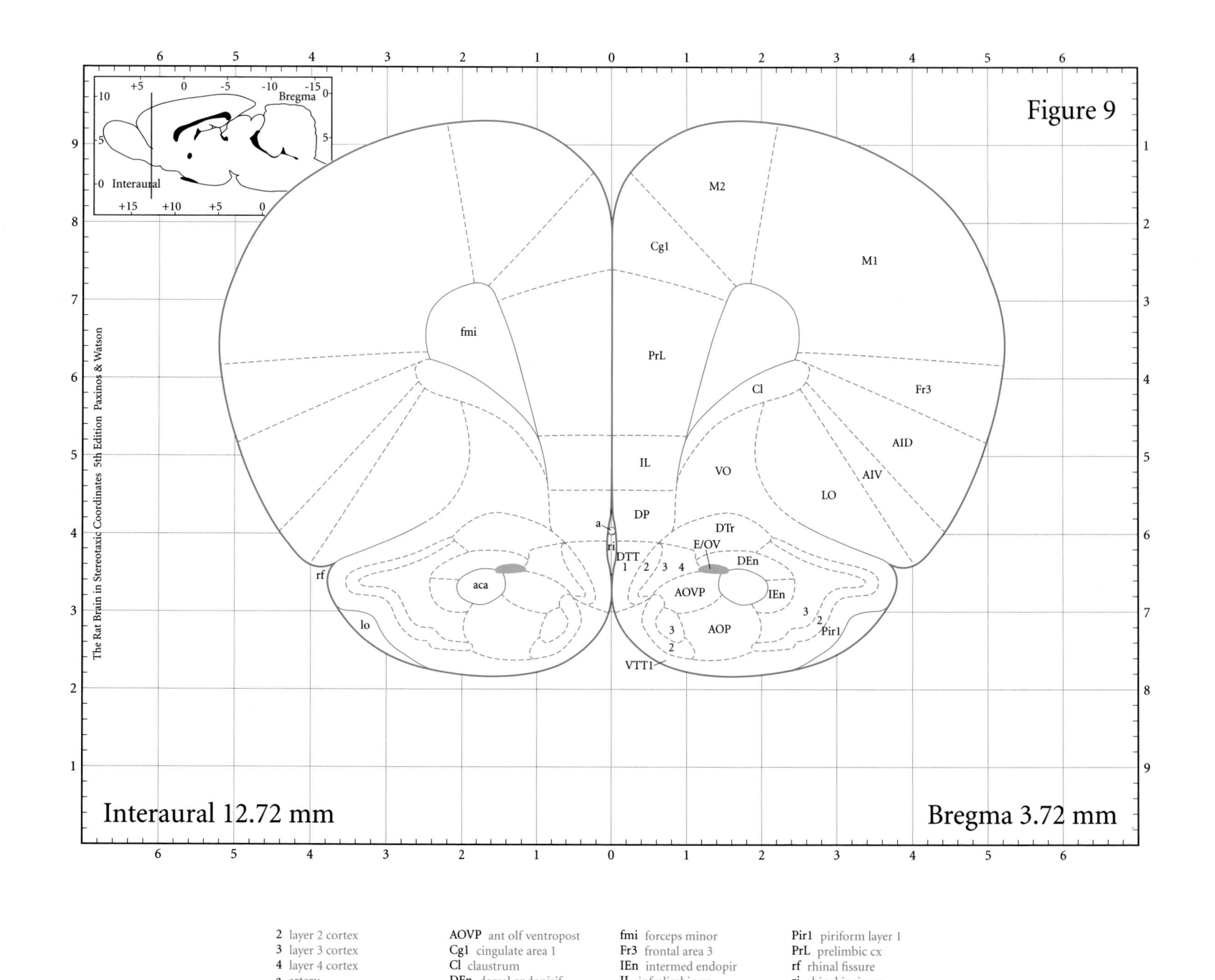

Figure 9

Interaural 12.72 mm

Bregma 3.72 mm

The Rat Brain in Stereotaxic Coordinates 5th Edition Paxinos & Watson

2 layer 2 cortex
3 layer 3 cortex
4 layer 4 cortex
a artery
aca ant comm, ant
AID ant insular dorsal
AIV ant insular ventral
AOP ant olfact post

AOVP ant olf ventropost
Cg1 cingulate area 1
Cl claustrum
DEn dorsal endopirif
DP dorsal pedunc cx
DTr dorsal trans zone
DTT dorsal tenia tecta
E/OV epend/olf ventr

fmi forceps minor
Fr3 frontal area 3
IEn intermed endopir
IL infralimbic cx
LO lat orbital cx
lo lat olfactory tr
M1 primary motor cx
M2 2ary motor cx

Pir1 piriform layer 1
PrL prelimbic cx
rf rhinal fissure
ri rhinal incisure
VO ventral orbital cx
VTT1 vent TT layer 1

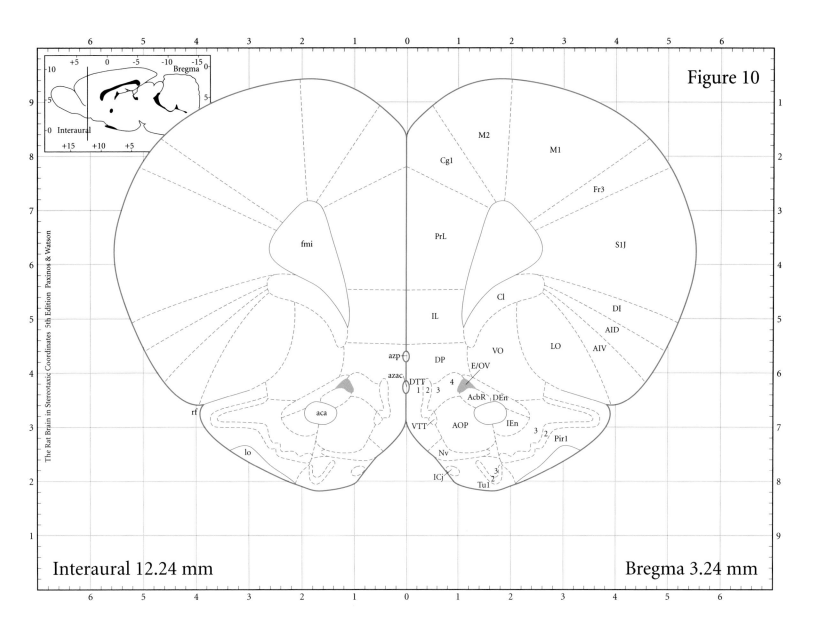

Figure 10

The Rat Brain in Stereotaxic Coordinates 5th Edition Paxinos & Watson

Interaural 12.24 mm

Bregma 3.24 mm

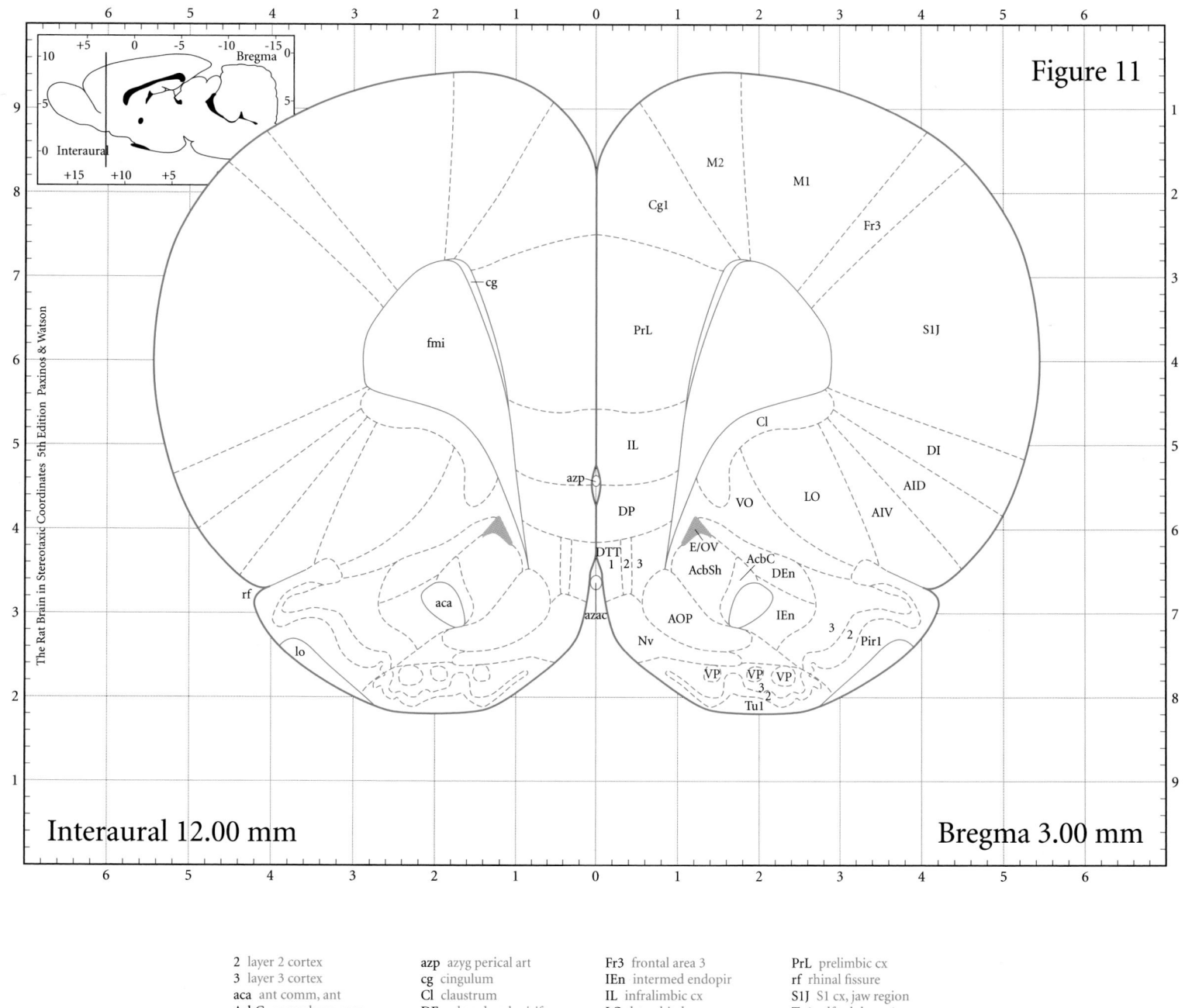

Figure 11

Interaural 12.00 mm

Bregma 3.00 mm

The Rat Brain in Stereotaxic Coordinates 5th Edition Paxinos & Watson

2 layer 2 cortex	azp azyg perical art	Fr3 frontal area 3	PrL prelimbic cx
3 layer 3 cortex	cg cingulum	IEn intermed endopir	rf rhinal fissure
aca ant comm, ant	Cl claustrum	IL infralimbic cx	S1J S1 cx, jaw region
AcbC accumbens core	DEn dorsal endopirif	LO lat orbital cx	Tu1 olf tub layer 1
AcbSh accumbens shell	DI dysgran insular	lo lat olfactory tr	VO ventral orbital cx
AID ant insular dorsal	DP dorsal pedunc cx	M1 primary motor cx	VP ventral pallidum
AIV ant insular ventral	DTT dorsal tenia tecta	M2 2ary motor cx	
AOP ant olfact post	E/OV epend/olf ventr	Nv navicular nu	
azac azyg ant cer art	fmi forceps minor	Pir1 piriform layer 1	

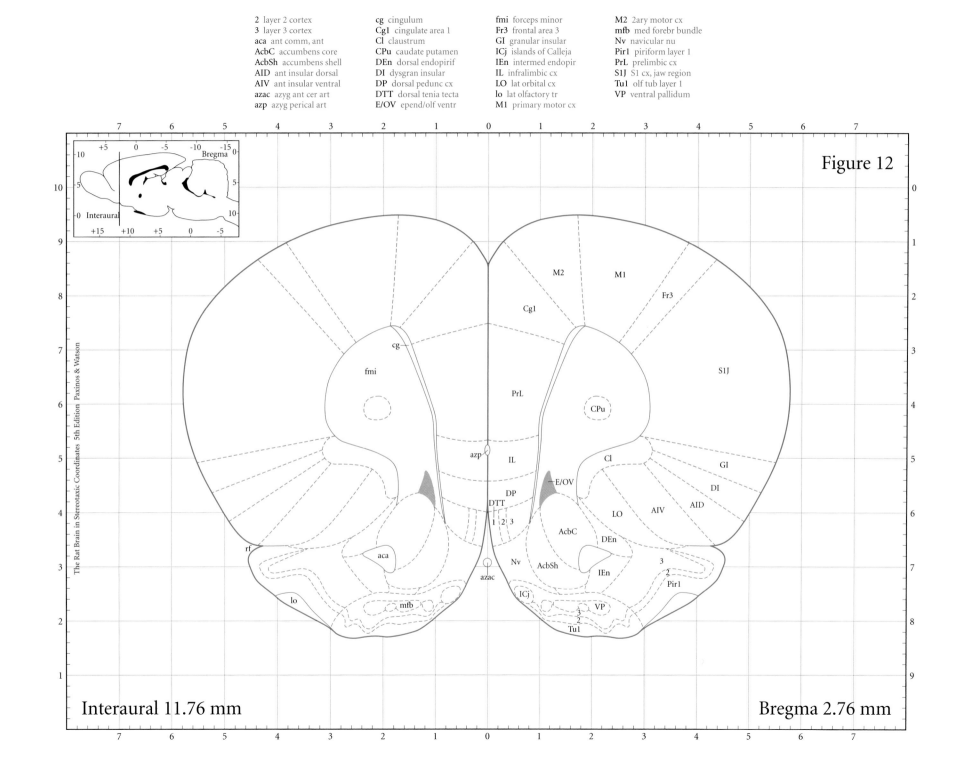

Figure 12

Interaural 11.76 mm

Bregma 2.76 mm

2 layer 2 cortex
3 layer 3 cortex
aca ant comm, ant
AcbC accumbens core
AcbSh accumbens shell
AID ant insular dorsal
AIV ant insular ventral
azac azyg ant cer art
azp azyg perical art

cg cingulum
Cg1 cingulate area 1
Cl claustrum
CPu caudate putamen
DEn dorsal endopirif
DI dysgran insular
DP dorsal pedunc cx
DTT dorsal tenia tecta
E/OV epend/olf ventr

fmi forceps minor
Fr3 frontal area 3
GI granular insular
ICj islands of Calleja
IEn intermed endopir
IL infralimbic cx
LO lat orbital cx
lo lat olfactory tr
M1 primary motor cx

M2 2ary motor cx
mfb med forebr bundle
Nv navicular nu
Pir1 piriform layer 1
PrL prelimbic cx
S1J S1 cx, jaw region
Tu1 olf tub layer 1
VP ventral pallidum

The Rat Brain in Stereotaxic Coordinates 5th Edition Paxinos & Watson

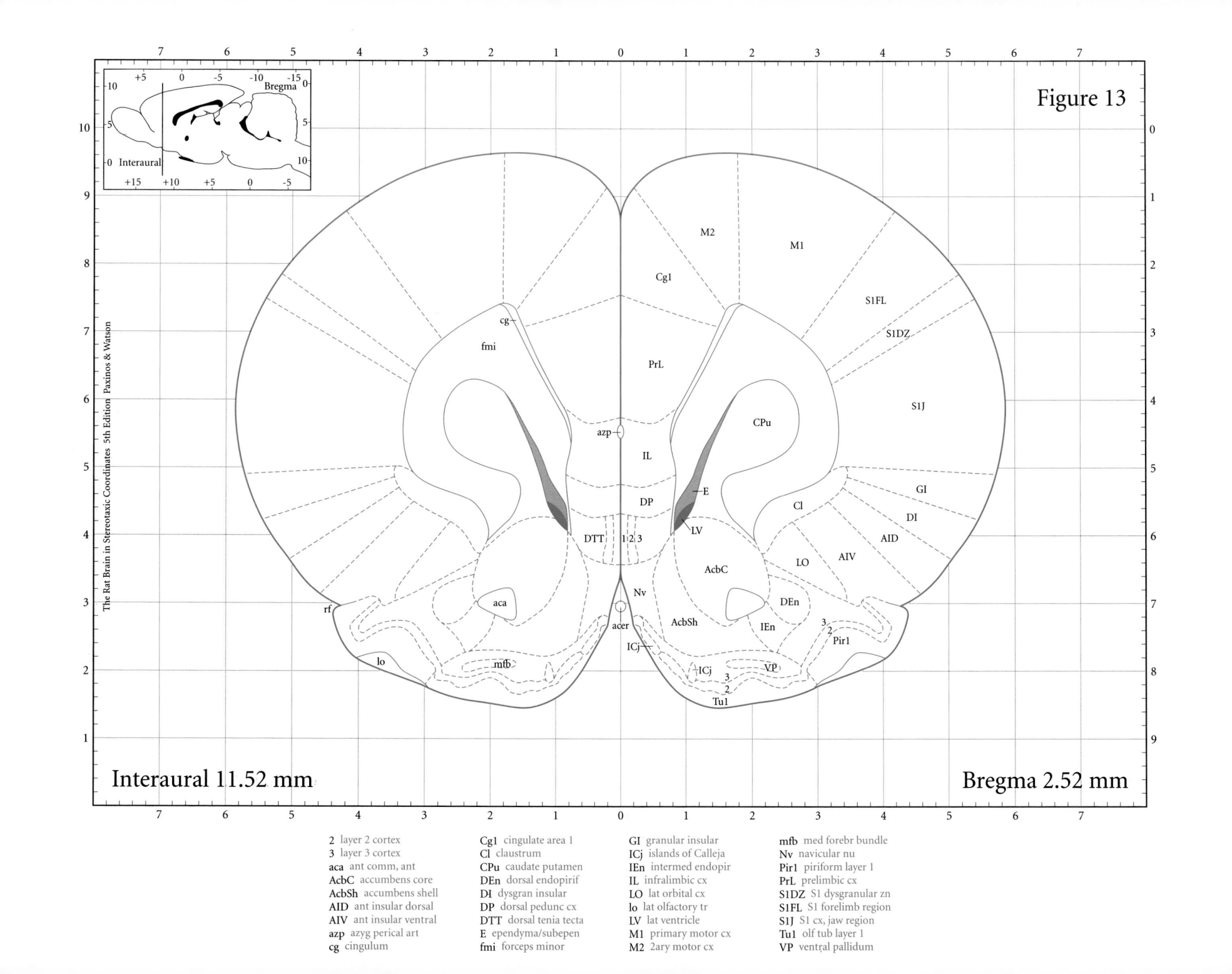

Figure 13

Interaural 11.52 mm

Bregma 2.52 mm

2   layer 2 cortex
3   layer 3 cortex
aca   ant comm, ant
AcbC   accumbens core
AcbSh   accumbens shell
AID   ant insular dorsal
AIV   ant insular ventral
azp   azyg perical art
cg   cingulum

Cg1   cingulate area 1
Cl   claustrum
CPu   caudate putamen
DEn   dorsal endopirif
DI   dysgran insular
DP   dorsal pedunc cx
DTT   dorsal tenia tecta
E   ependyma/subepen
fmi   forceps minor

GI   granular insular
ICj   islands of Calleja
IEn   intermed endopir
IL   infralimbic cx
LO   lat orbital cx
lo   lat olfactory tr
LV   lat ventricle
M1   primary motor cx
M2   2ary motor cx

mfb   med forebr bundle
Nv   navicular nu
Pir1   piriform layer 1
PrL   prelimbic cx
S1DZ   S1 dysgranular zn
S1FL   S1 forelimb region
S1J   S1 cx, jaw region
Tu1   olf tub layer 1
VP   ventral pallidum

The Rat Brain in Stereotaxic Coordinates 5th Edition Paxinos & Watson

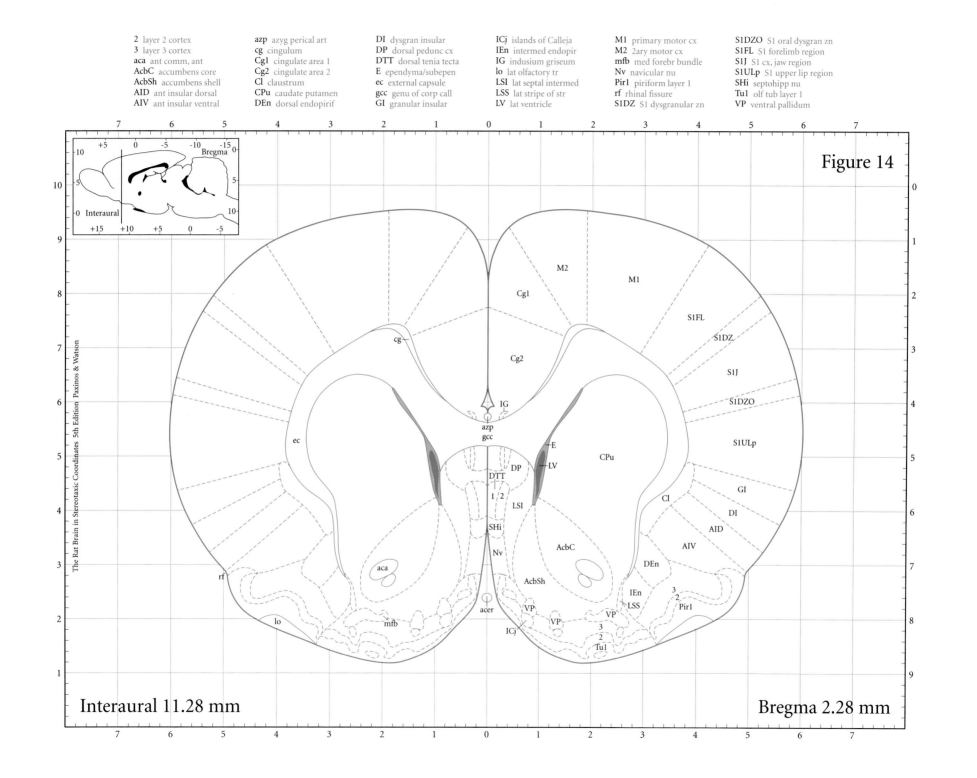

Figure 14

Interaural 11.28 mm

Bregma 2.28 mm

2 layer 2 cortex
3 layer 3 cortex
aca ant comm, ant
AcbC accumbens core
AcbSh accumbens shell
AID ant insular dorsal
AIV ant insular ventral

azp azyg perical art
cg cingulum
Cg1 cingulate area 1
Cg2 cingulate area 2
Cl claustrum
CPu caudate putamen
DEn dorsal endopirif

DI dysgran insular
DP dorsal pedunc cx
DTT dorsal tenia tecta
E ependyma/subepen
ec external capsule
gcc genu of corp call
GI granular insular

ICj islands of Calleja
IEn intermed endopir
IG indusium griseum
lo lat olfactory tr
LSI lat septal intermed
LSS lat stripe of str
LV lat ventricle

M1 primary motor cx
M2 2ary motor cx
mfb med forebr bundle
Nv navicular nu
Pir1 piriform layer 1
rf rhinal fissure
S1DZ S1 dysgranular zn

S1DZO S1 oral dysgran zn
S1FL S1 forelimb region
S1J S1 cx, jaw region
S1ULp S1 upper lip region
SHi septohipp nu
Tu1 olf tub layer 1
VP ventral pallidum

The Rat Brain in Stereotaxic Coordinates 5th Edition Paxinos & Watson

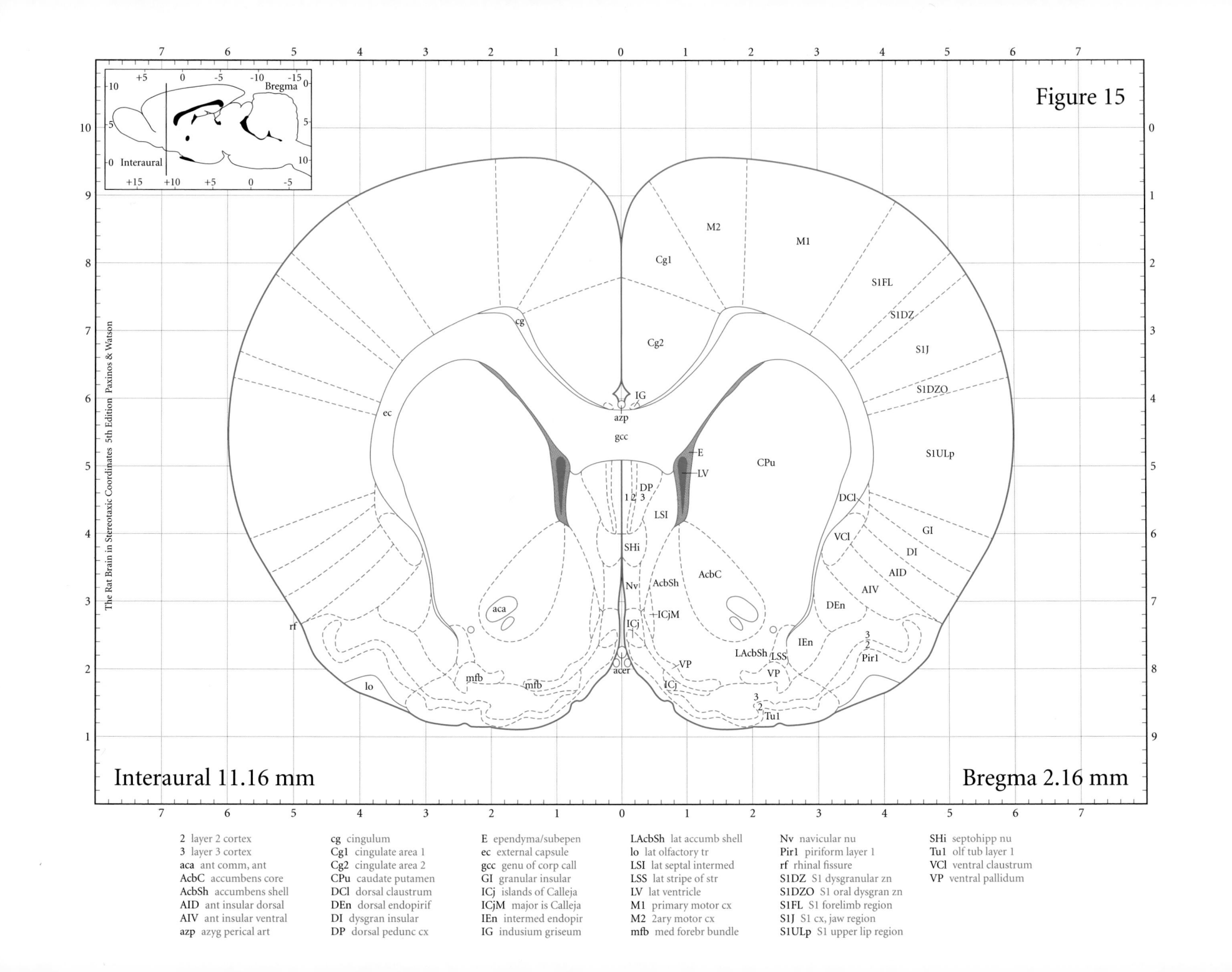

Figure 15

Interaural 11.16 mm

Bregma 2.16 mm

2 layer 2 cortex
3 layer 3 cortex
aca ant comm, ant
AcbC accumbens core
AcbSh accumbens shell
AID ant insular dorsal
AIV ant insular ventral
azp azyg perical art

cg cingulum
Cg1 cingulate area 1
Cg2 cingulate area 2
CPu caudate putamen
DCl dorsal claustrum
DEn dorsal endopirif
DI dysgran insular
DP dorsal pedunc cx

E ependyma/subepen
ec external capsule
gcc genu of corp call
GI granular insular
ICj islands of Calleja
ICjM major is Calleja
IEn intermed endopir
IG indusium griseum

LAcbSh lat accumb shell
lo lat olfactory tr
LSI lat septal intermed
LSS lat stripe of str
LV lat ventricle
M1 primary motor cx
M2 2ary motor cx
mfb med forebr bundle

Nv navicular nu
Pir1 piriform layer 1
rf rhinal fissure
S1DZ S1 dysgranular zn
S1DZO S1 oral dysgran zn
S1FL S1 forelimb region
S1J S1 cx, jaw region
S1ULp S1 upper lip region

SHi septohipp nu
Tu1 olf tub layer 1
VCl ventral claustrum
VP ventral pallidum

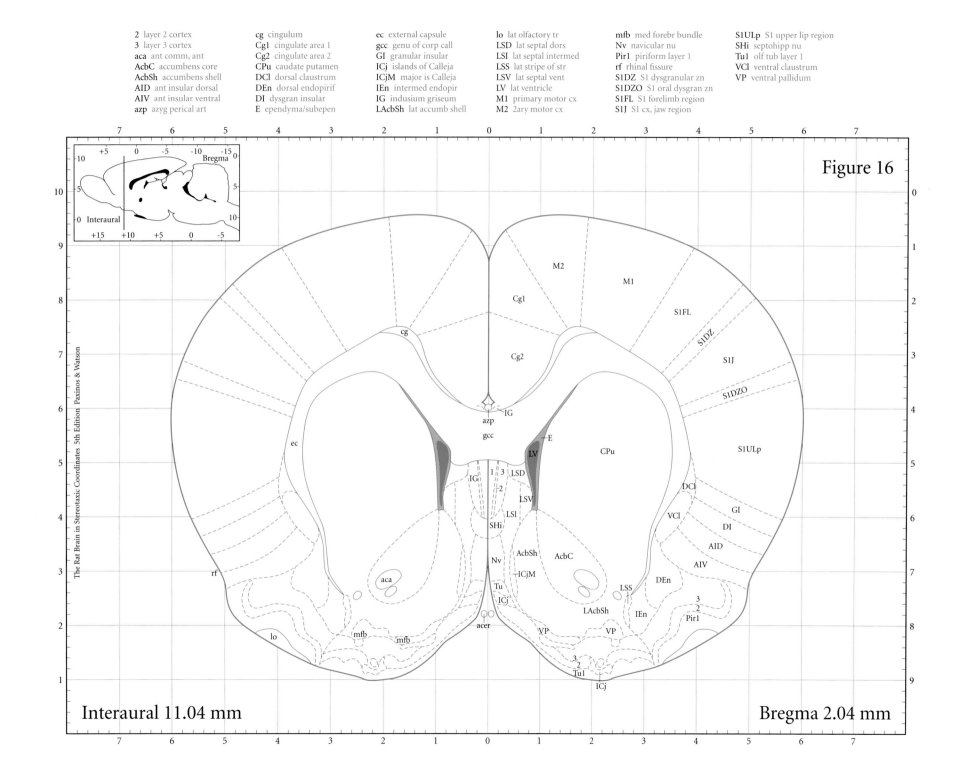

Figure 16

Glossary:

2 layer 2 cortex
3 layer 3 cortex
aca ant comm, ant
AcbC accumbens core
AcbSh accumbens shell
AID ant insular dorsal
AIV ant insular ventral
azp azyg perical art

cg cingulum
Cg1 cingulate area 1
Cg2 cingulate area 2
CPu caudate putamen
DCl dorsal claustrum
DEn dorsal endopirif
DI dysgran insular
E ependyma/subepen

ec external capsule
gcc genu of corp call
GI granular insular
ICj islands of Calleja
ICjM major is Calleja
IEn intermed endopir
IG indusium griseum
LAcbSh lat accumb shell

lo lat olfactory tr
LSD lat septal dors
LSI lat septal intermed
LSS lat stripe of str
LSV lat septal vent
LV lat ventricle
M1 primary motor cx
M2 2ary motor cx

mfb med forebr bundle
Nv navicular nu
Pir1 piriform layer 1
rf rhinal fissure
S1DZ S1 dysgranular zn
S1DZO S1 oral dysgran zn
S1FL S1 forelimb region
S1J S1 cx, jaw region

S1ULp S1 upper lip region
SHi septohipp nu
Tu1 olf tub layer 1
VCl ventral claustrum
VP ventral pallidum

Interaural 11.04 mm

Bregma 2.04 mm

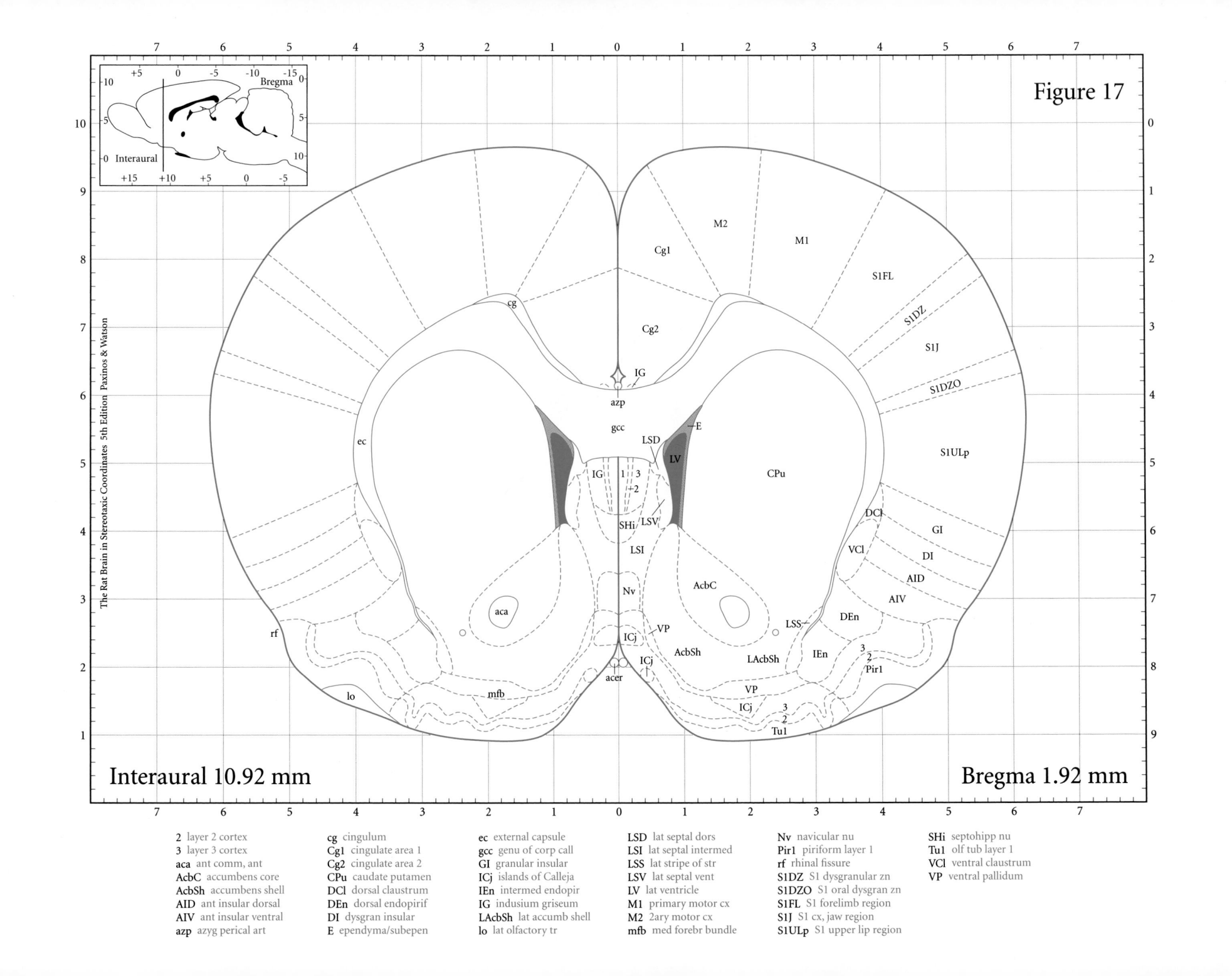

Figure 17

Interaural 10.92 mm

Bregma 1.92 mm

The Rat Brain in Stereotaxic Coordinates 5th Edition Paxinos & Watson

2 layer 2 cortex
3 layer 3 cortex
aca ant comm, ant
AcbC accumbens core
AcbSh accumbens shell
AID ant insular dorsal
AIV ant insular ventral
azp azyg perical art

cg cingulum
Cg1 cingulate area 1
Cg2 cingulate area 2
CPu caudate putamen
DCl dorsal claustrum
DEn dorsal endopirif
DI dysgran insular
E ependyma/subepen

ec external capsule
gcc genu of corp call
GI granular insular
ICj islands of Calleja
IEn intermed endopir
IG indusium griseum
LAcbSh lat accumb shell
lo lat olfactory tr

LSD lat septal dors
LSI lat septal intermed
LSS lat stripe of str
LSV lat septal vent
LV lat ventricle
M1 primary motor cx
M2 2ary motor cx
mfb med forebr bundle

Nv navicular nu
Pir1 piriform layer 1
rf rhinal fissure
S1DZ S1 dysgranular zn
S1DZO S1 oral dysgran zn
S1FL S1 forelimb region
S1J S1 cx, jaw region
S1ULp S1 upper lip region

SHi septohipp nu
Tu1 olf tub layer 1
VCl ventral claustrum
VP ventral pallidum

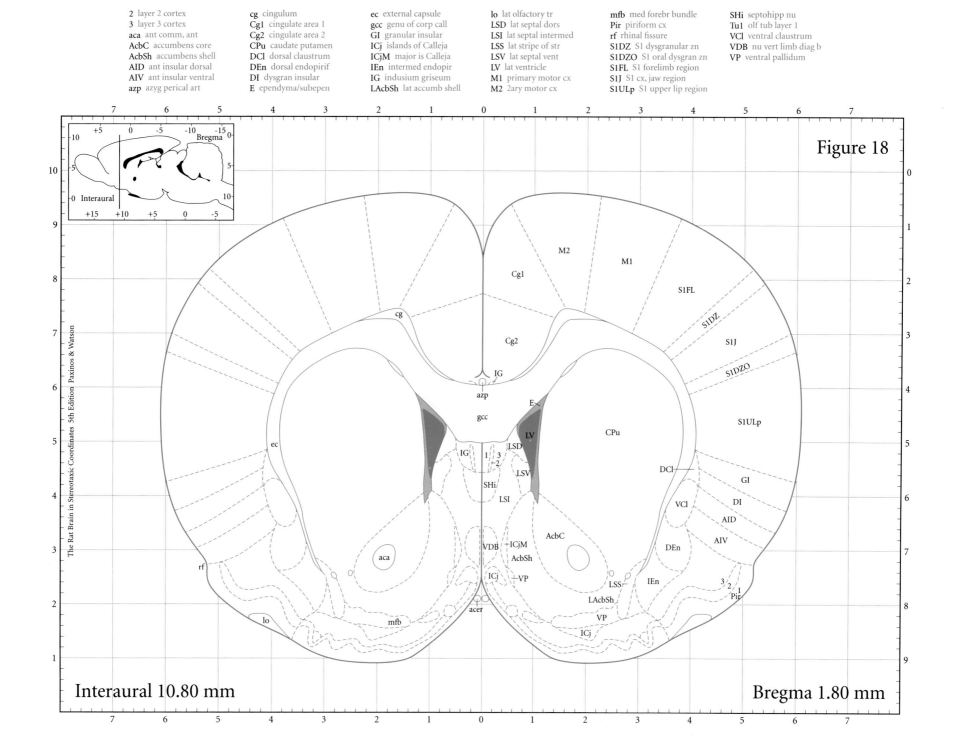

2 layer 2 cortex
3 layer 3 cortex
aca ant comm, ant
AcbC accumbens core
AcbSh accumbens shell
AID ant insular dorsal
AIV ant insular ventral
azp azyg perical art

cg cingulum
Cg1 cingulate area 1
Cg2 cingulate area 2
CPu caudate putamen
DCl dorsal claustrum
DEn dorsal endopirif
DI dysgran insular
E ependyma/subepen

ec external capsule
gcc genu of corp call
GI granular insular
ICj islands of Calleja
ICjM major is Calleja
IEn intermed endopir
IG indusium griseum
LAcbSh lat accumb shell

lo lat olfactory tr
LSD lat septal dors
LSI lat septal intermed
LSS lat stripe of str
LSV lat septal vent
LV lat ventricle
M1 primary motor cx
M2 2ary motor cx

mfb med forebr bundle
Pir piriform cx
rf rhinal fissure
S1DZ S1 dysgranular zn
S1DZO S1 oral dysgran zn
S1FL S1 forelimb region
S1J S1 cx, jaw region
S1ULp S1 upper lip region

SHi septohipp nu
Tu1 olf tub layer 1
VCl ventral claustrum
VDB nu vert limb diag b
VP ventral pallidum

Figure 18

Interaural 10.80 mm

Bregma 1.80 mm

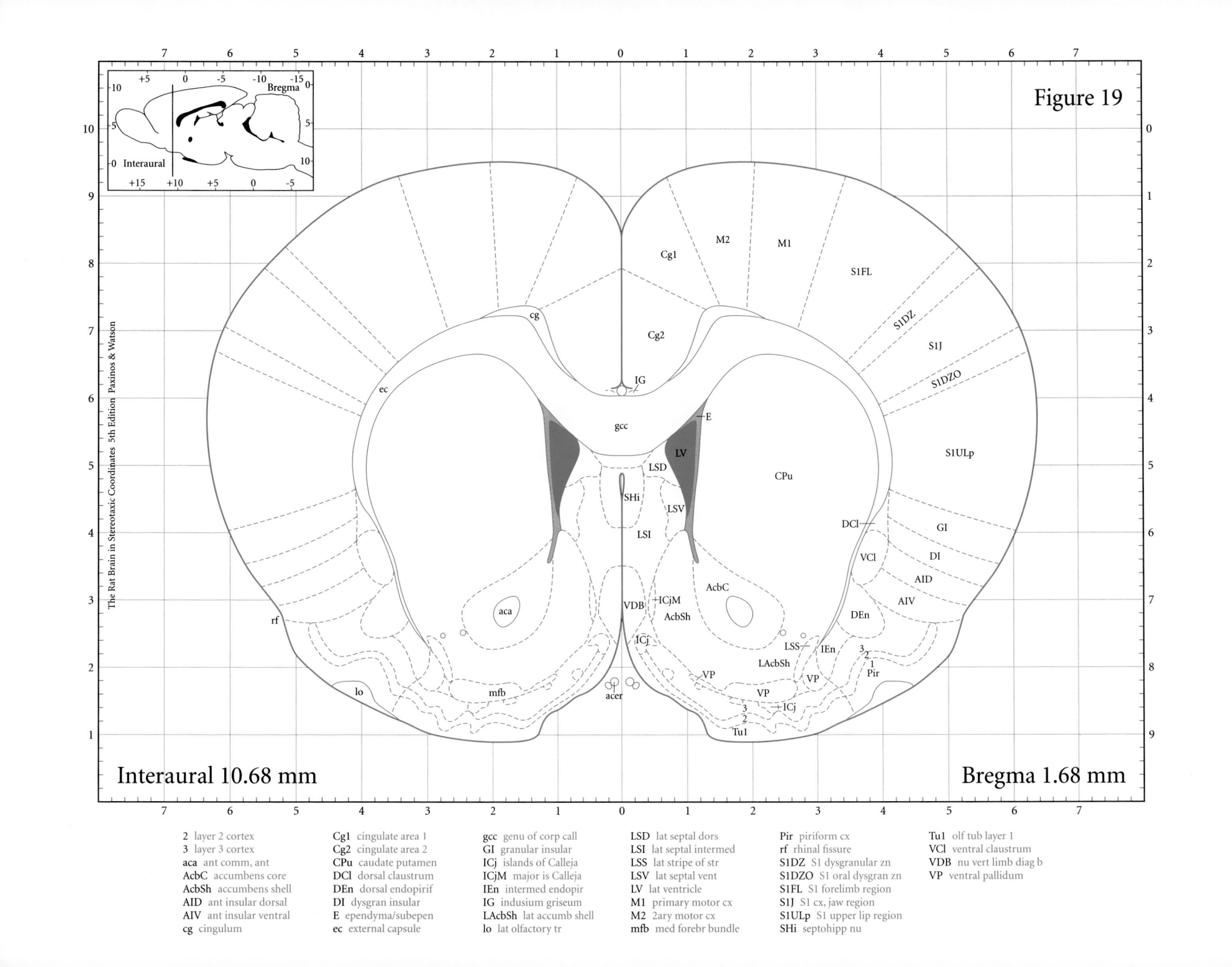

Figure 19

Interaural 10.68 mm

Bregma 1.68 mm

2 layer 2 cortex	Cg1 cingulate area 1	gcc genu of corp call	LSD lat septal dors	Pir piriform cx	Tu1 olf tub layer 1
3 layer 3 cortex	Cg2 cingulate area 2	GI granular insular	LSI lat septal intermed	rf rhinal fissure	VCl ventral claustrum
aca ant comm, ant	CPu caudate putamen	ICj islands of Calleja	LSS lat stripe of str	S1DZ S1 dysgranular zn	VDB nu vert limb diag b
AcbC accumbens core	DCl dorsal claustrum	ICjM major is Calleja	LSV lat septal vent	S1DZO S1 oral dysgran zn	VP ventral pallidum
AcbSh accumbens shell	DEn dorsal endopirif	IEn intermed endopir	LV lat ventricle	S1FL S1 forelimb region	
AID ant insular dorsal	DI dysgran insular	IG indusium griseum	M1 primary motor cx	S1J S1 cx, jaw region	
AIV ant insular ventral	E ependyma/subepen	LAcbSh lat accumb shell	M2 2ary motor cx	S1ULp S1 upper lip region	
cg cingulum	ec external capsule	lo lat olfactory tr	mfb med forebr bundle	SHi septohipp nu	

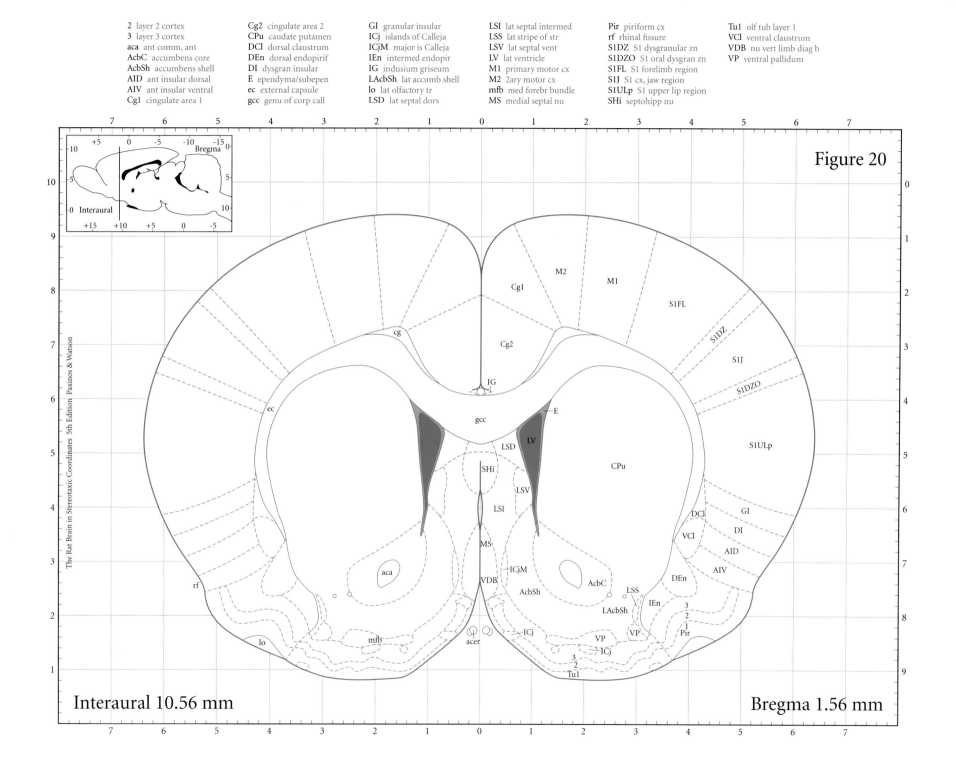

Figure 20

Interaural 10.56 mm

Bregma 1.56 mm

2 layer 2 cortex
3 layer 3 cortex
aca ant comm, ant
AcbC accumbens core
AcbSh accumbens shell
AID ant insular dorsal
AIV ant insular ventral
Cg1 cingulate area 1

Cg2 cingulate area 2
CPu caudate putamen
DCl dorsal claustrum
DEn dorsal endopirif
DI dysgran insular
E ependyma/subepen
ec external capsule
gcc genu of corp call

GI granular insular
ICj islands of Calleja
ICjM major is Calleja
IEn intermed endopir
IG indusium griseum
LAcbSh lat accumb shell
lo lat olfactory tr
LSD lat septal dors

LSI lat septal intermed
LSS lat stripe of str
LSV lat septal vent
LV lat ventricle
M1 primary motor cx
M2 2ary motor cx
mfb med forebr bundle
MS medial septal nu

Pir piriform cx
rf rhinal fissure
S1DZ S1 dysgranular zn
S1DZO S1 oral dysgran zn
S1FL S1 forelimb region
S1J S1 cx, jaw region
S1ULp S1 upper lip region
SHi septohipp nu

Tu1 olf tub layer 1
VCl ventral claustrum
VDB nu vert limb diag b
VP ventral pallidum

The Rat Brain in Stereotaxic Coordinates 5th Edition Paxinos & Watson

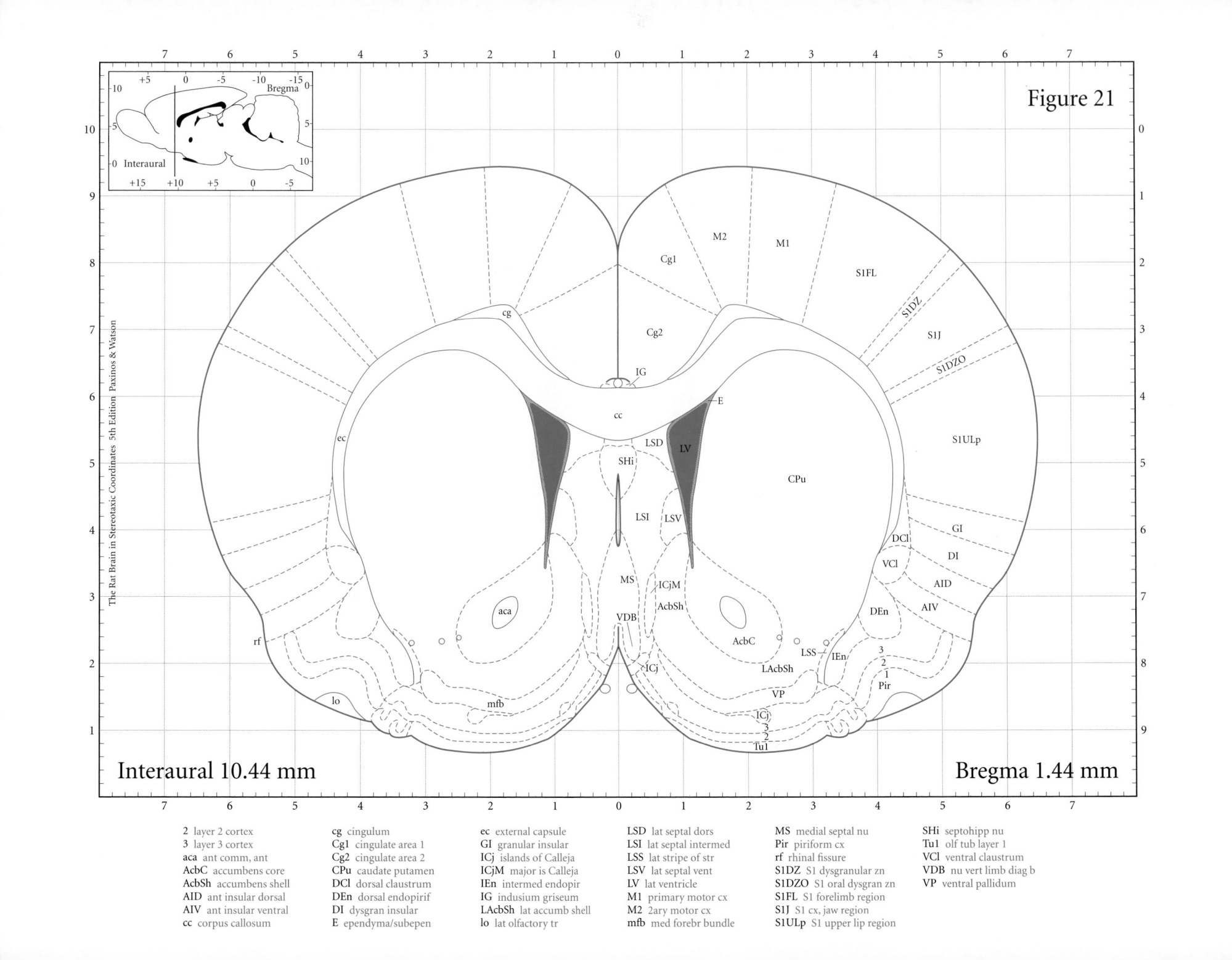

Figure 21

Interaural 10.44 mm

Bregma 1.44 mm

The Rat Brain in Stereotaxic Coordinates 5th Edition Paxinos & Watson

2 layer 2 cortex	cg cingulum	ec external capsule	LSD lat septal dors	MS medial septal nu	SHi septohipp nu
3 layer 3 cortex	Cg1 cingulate area 1	GI granular insular	LSI lat septal intermed	Pir piriform cx	Tu1 olf tub layer 1
aca ant comm, ant	Cg2 cingulate area 2	ICj islands of Calleja	LSS lat stripe of str	rf rhinal fissure	VCl ventral claustrum
AcbC accumbens core	CPu caudate putamen	ICjM major is Calleja	LSV lat septal vent	S1DZ S1 dysgranular zn	VDB nu vert limb diag b
AcbSh accumbens shell	DCl dorsal claustrum	IEn intermed endopir	LV lat ventricle	S1DZO S1 oral dysgran zn	VP ventral pallidum
AID ant insular dorsal	DEn dorsal endopirif	IG indusium griseum	M1 primary motor cx	S1FL S1 forelimb region	
AIV ant insular ventral	DI dysgran insular	LAcbSh lat accumb shell	M2 2ary motor cx	S1J S1 cx, jaw region	
cc corpus callosum	E ependyma/subepen	lo lat olfactory tr	mfb med forebr bundle	S1ULp S1 upper lip region	

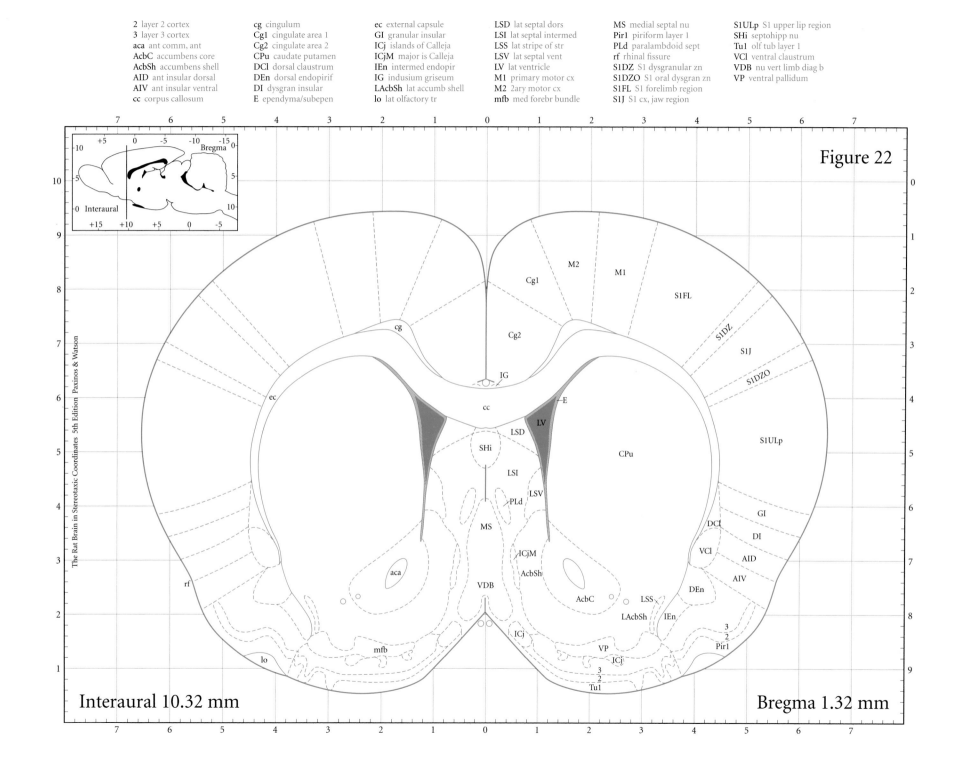

Figure 22

Interaural 10.32 mm

Bregma 1.32 mm

2 layer 2 cortex
3 layer 3 cortex
aca ant comm, ant
AcbC accumbens core
AcbSh accumbens shell
AID ant insular dorsal
AIV ant insular ventral
cc corpus callosum

cg cingulum
Cg1 cingulate area 1
Cg2 cingulate area 2
CPu caudate putamen
DCl dorsal claustrum
DEn dorsal endopirif
DI dysgran insular
E ependyma/subepen

ec external capsule
GI granular insular
ICj islands of Calleja
ICjM major is Calleja
IEn intermed endopir
IG indusium griseum
LAcbSh lat accumb shell
lo lat olfactory tr

LSD lat septal dors
LSI lat septal intermed
LSS lat stripe of str
LSV lat septal vent
LV lat ventricle
M1 primary motor cx
M2 2ary motor cx
mfb med forebr bundle

MS medial septal nu
Pir1 piriform layer 1
PLd paralambdoid sept
rf rhinal fissure
S1DZ S1 dysgranular zn
S1DZO S1 oral dysgran zn
S1FL S1 forelimb region
S1J S1 cx, jaw region

S1ULp S1 upper lip region
SHi septohipp nu
Tu1 olf tub layer 1
VCl ventral claustrum
VDB nu vert limb diag b
VP ventral pallidum

The Rat Brain in Stereotaxic Coordinates 5th Edition Paxinos & Watson

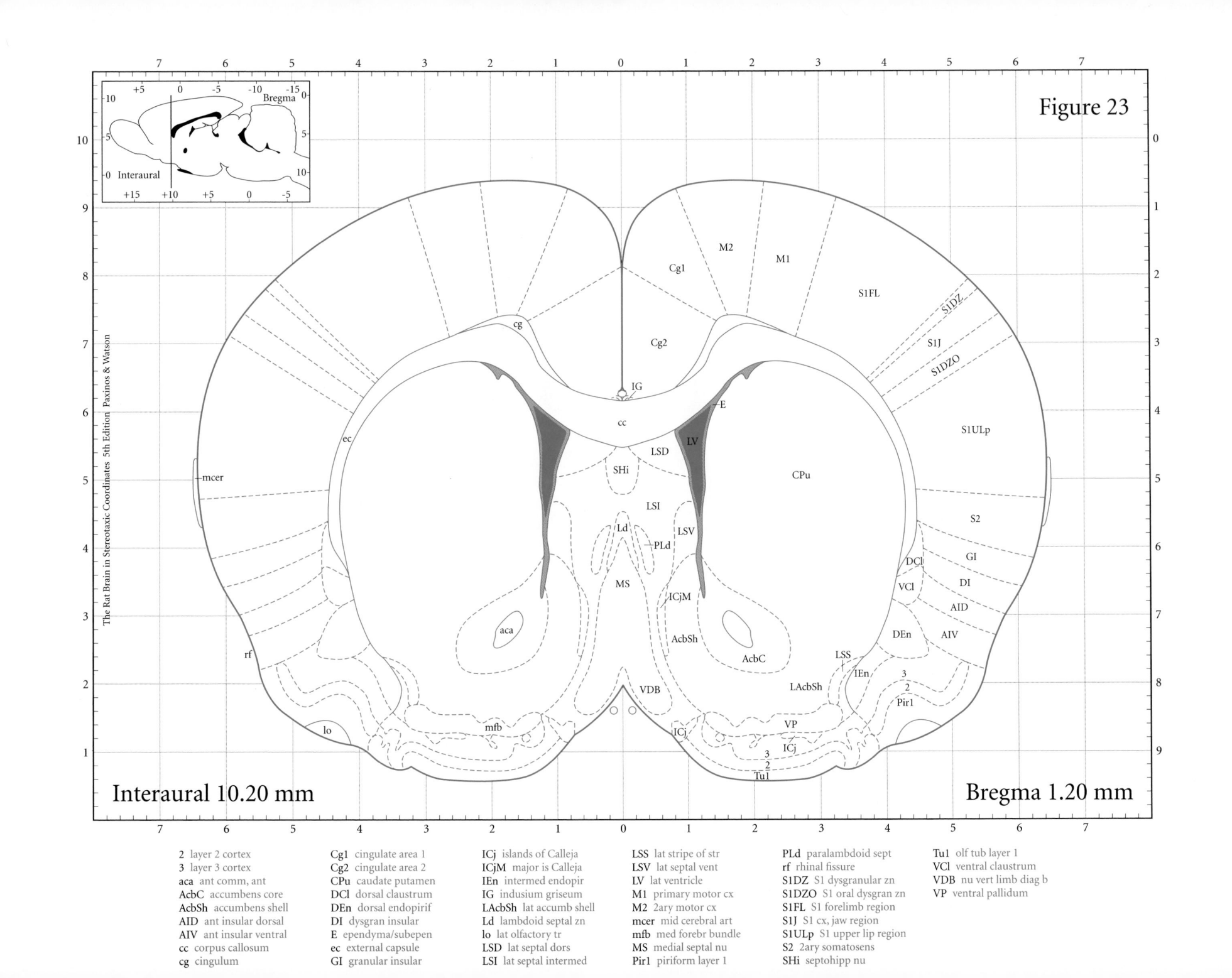

Figure 23

Interaural 10.20 mm

Bregma 1.20 mm

The Rat Brain in Stereotaxic Coordinates 5th Edition Paxinos & Watson

2 layer 2 cortex	Cg1 cingulate area 1	ICj islands of Calleja	LSS lat stripe of str	PLd paralambdoid sept	Tu1 olf tub layer 1
3 layer 3 cortex	Cg2 cingulate area 2	ICjM major is Calleja	LSV lat septal vent	rf rhinal fissure	VCl ventral claustrum
aca ant comm, ant	CPu caudate putamen	IEn intermed endopir	LV lat ventricle	S1DZ S1 dysgranular zn	VDB nu vert limb diag b
AcbC accumbens core	DCl dorsal claustrum	IG indusium griseum	M1 primary motor cx	S1DZO S1 oral dysgran zn	VP ventral pallidum
AcbSh accumbens shell	DEn dorsal endopirif	LAcbSh lat accumb shell	M2 2ary motor cx	S1FL S1 forelimb region	
AID ant insular dorsal	DI dysgran insular	Ld lambdoid septal zn	mcer mid cerebral art	S1J S1 cx, jaw region	
AIV ant insular ventral	E ependyma/subepen	lo lat olfactory tr	mfb med forebr bundle	S1ULp S1 upper lip region	
cc corpus callosum	ec external capsule	LSD lat septal dors	MS medial septal nu	S2 2ary somatosens	
cg cingulum	GI granular insular	LSI lat septal intermed	Pir1 piriform layer 1	SHi septohipp nu	

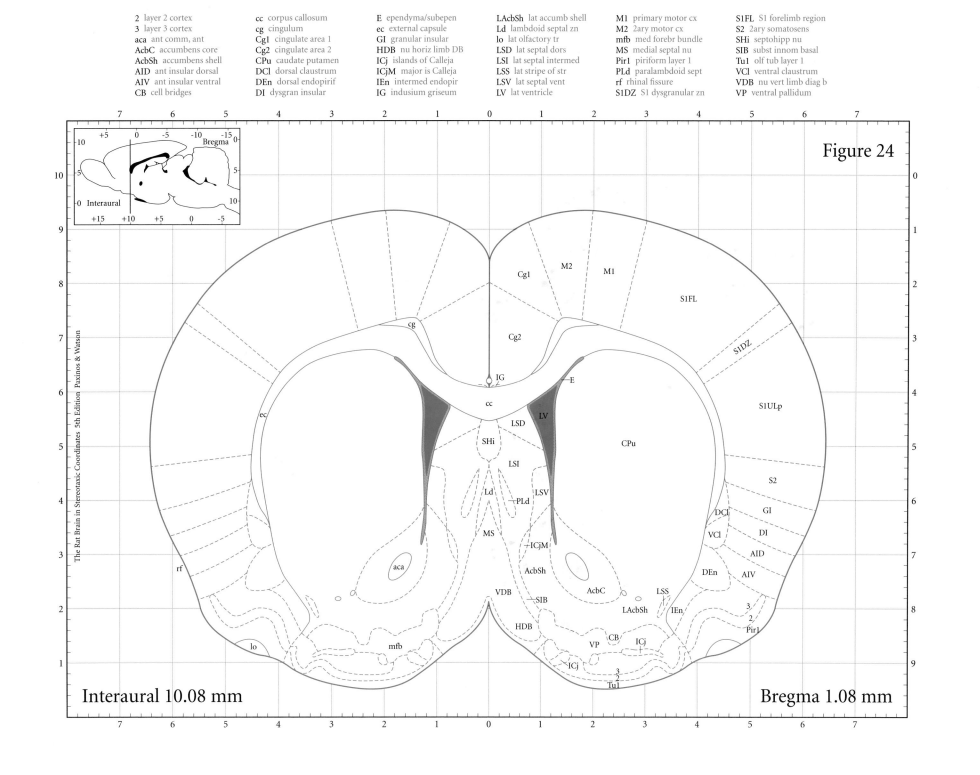

Figure 24

Interaural 10.08 mm

Bregma 1.08 mm

2 layer 2 cortex
3 layer 3 cortex
aca ant comm, ant
AcbC accumbens core
AcbSh accumbens shell
AID ant insular dorsal
AIV ant insular ventral
CB cell bridges

cc corpus callosum
cg cingulum
Cg1 cingulate area 1
Cg2 cingulate area 2
CPu caudate putamen
DCl dorsal claustrum
DEn dorsal endopirif
DI dysgran insular

E ependyma/subepen
ec external capsule
GI granular insular
HDB nu horiz limb DB
ICj islands of Calleja
ICjM major is Calleja
IEn intermed endopir
IG indusium griseum

LAcbSh lat accumb shell
Ld lambdoid septal zn
lo lat olfactory tr
LSD lat septal dors
LSI lat septal intermed
LSS lat stripe of str
LSV lat septal vent
LV lat ventricle

M1 primary motor cx
M2 2ary motor cx
mfb med forebr bundle
MS medial septal nu
Pir1 piriform layer 1
PLd paralambdoid sept
rf rhinal fissure
S1DZ S1 dysgranular zn

S1FL S1 forelimb region
S2 2ary somatosens
SHi septohipp nu
SIB subst innom basal
Tu1 olf tub layer 1
VCl ventral claustrum
VDB nu vert limb diag b
VP ventral pallidum

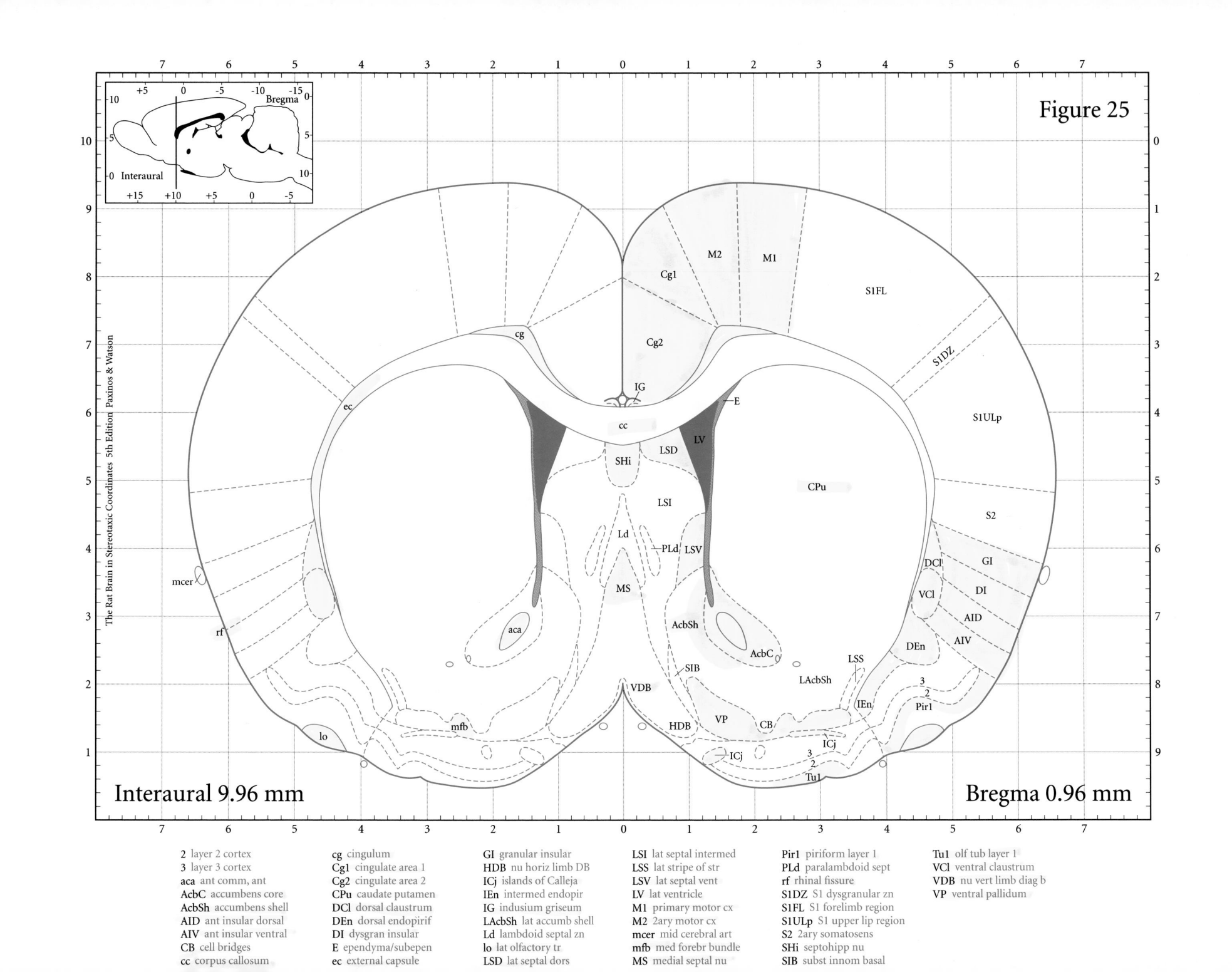

Figure 25

The Rat Brain in Stereotaxic Coordinates 5th Edition Paxinos & Watson

Interaural 9.96 mm

Bregma 0.96 mm

2 layer 2 cortex	cg cingulum	GI granular insular
3 layer 3 cortex	Cg1 cingulate area 1	HDB nu horiz limb DB
aca ant comm, ant	Cg2 cingulate area 2	ICj islands of Calleja
AcbC accumbens core	CPu caudate putamen	IEn intermed endopir
AcbSh accumbens shell	DCl dorsal claustrum	IG indusium griseum
AID ant insular dorsal	DEn dorsal endopirif	LAcbSh lat accumb shell
AIV ant insular ventral	DI dysgran insular	Ld lambdoid septal zn
CB cell bridges	E ependyma/subepen	lo lat olfactory tr
cc corpus callosum	ec external capsule	LSD lat septal dors

LSI lat septal intermed	Pir1 piriform layer 1	Tu1 olf tub layer 1
LSS lat stripe of str	PLd paralambdoid sept	VCl ventral claustrum
LSV lat septal vent	rf rhinal fissure	VDB nu vert limb diag b
LV lat ventricle	S1DZ S1 dysgranular zn	VP ventral pallidum
M1 primary motor cx	S1FL S1 forelimb region	
M2 2ary motor cx	S1ULp S1 upper lip region	
mcer mid cerebral art	S2 2ary somatosens	
mfb med forebr bundle	SHi septohipp nu	
MS medial septal nu	SIB subst innom basal	

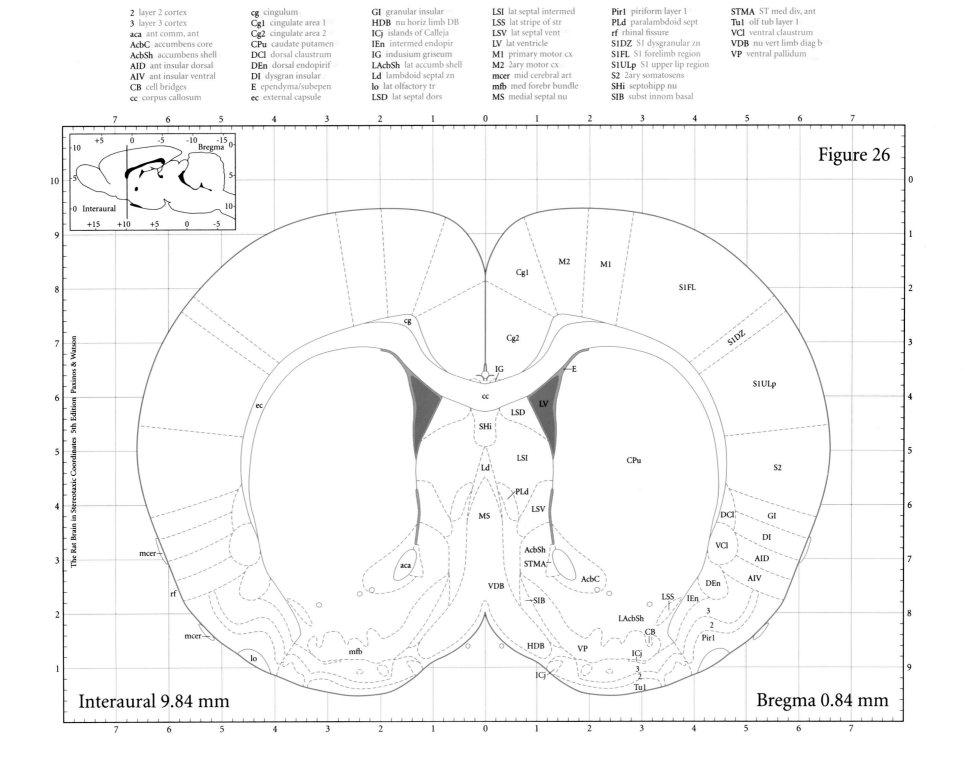

Figure 26

Interaural 9.84 mm

Bregma 0.84 mm

2 layer 2 cortex
3 layer 3 cortex
aca ant comm, ant
AcbC accumbens core
AcbSh accumbens shell
AID ant insular dorsal
AIV ant insular ventral
CB cell bridges
cc corpus callosum

cg cingulum
Cg1 cingulate area 1
Cg2 cingulate area 2
CPu caudate putamen
DCl dorsal claustrum
DEn dorsal endopirif
DI dysgran insular
E ependyma/subepen
ec external capsule

GI granular insular
HDB nu horiz limb DB
ICj islands of Calleja
IEn intermed endopir
IG indusium griseum
LAcbSh lat accumb shell
Ld lambdoid septal zn
lo lat olfactory tr
LSD lat septal dors

LSI lat septal intermed
LSS lat stripe of str
LSV lat septal vent
LV lat ventricle
M1 primary motor cx
M2 2ary motor cx
mcer mid cerebral art
mfb med forebr bundle
MS medial septal nu

Pir1 piriform layer 1
PLd paralambdoid sept
rf rhinal fissure
S1DZ S1 dysgranular zn
S1FL S1 forelimb region
S1ULp S1 upper lip region
S2 2ary somatosens
SHi septohipp nu
SIB subst innom basal

STMA ST med div, ant
Tu1 olf tub layer 1
VCl ventral claustrum
VDB nu vert limb diag b
VP ventral pallidum

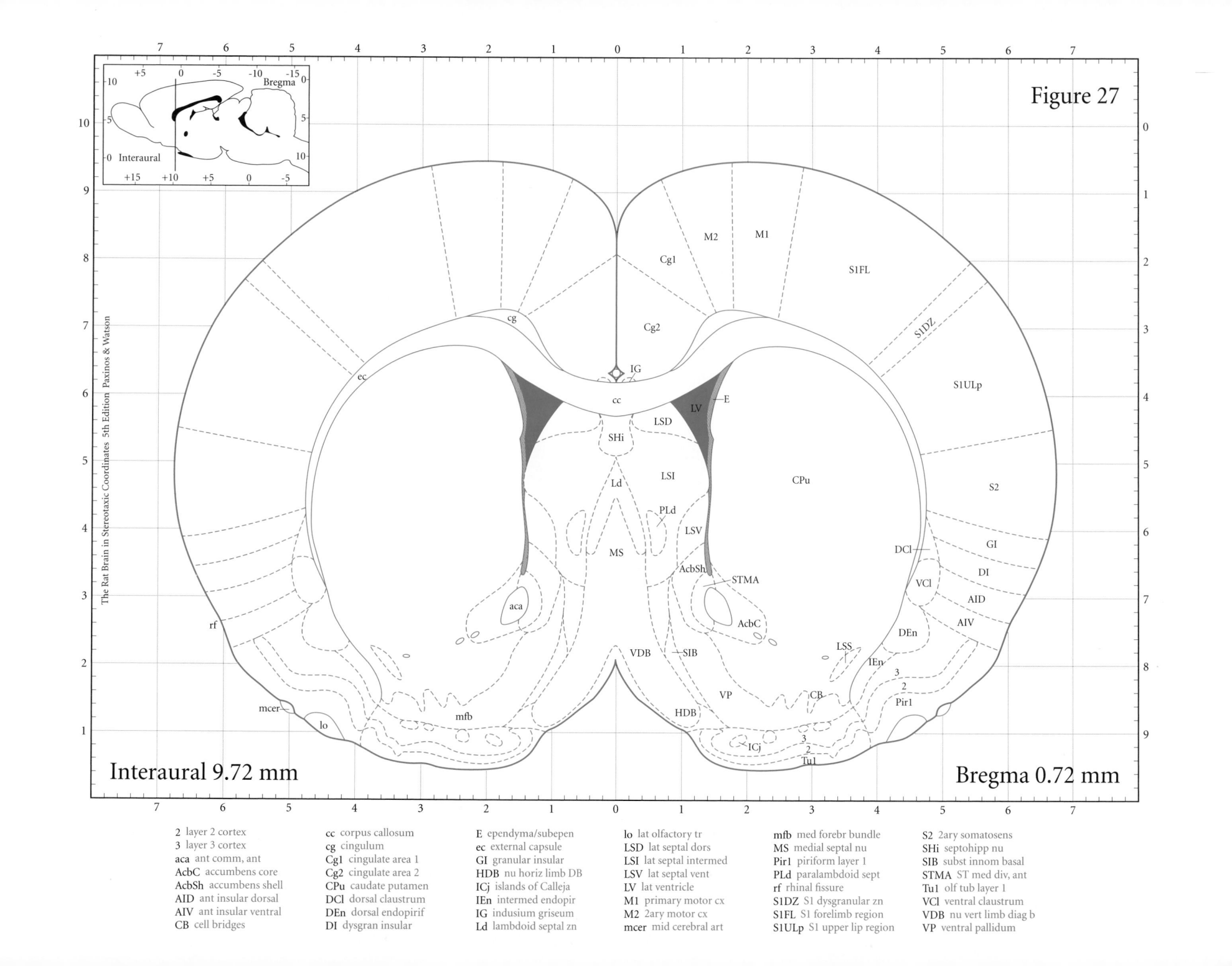

Figure 27

Interaural 9.72 mm

Bregma 0.72 mm

The Rat Brain in Stereotaxic Coordinates 5th Edition Paxinos & Watson

2 layer 2 cortex
3 layer 3 cortex
aca ant comm, ant
AcbC accumbens core
AcbSh accumbens shell
AID ant insular dorsal
AIV ant insular ventral
CB cell bridges

cc corpus callosum
cg cingulum
Cg1 cingulate area 1
Cg2 cingulate area 2
CPu caudate putamen
DCl dorsal claustrum
DEn dorsal endopirif
DI dysgran insular

E ependyma/subepen
ec external capsule
GI granular insular
HDB nu horiz limb DB
ICj islands of Calleja
IEn intermed endopir
IG indusium griseum
Ld lambdoid septal zn

lo lat olfactory tr
LSD lat septal dors
LSI lat septal intermed
LSV lat septal vent
LV lat ventricle
M1 primary motor cx
M2 2ary motor cx
mcer mid cerebral art

mfb med forebr bundle
MS medial septal nu
Pir1 piriform layer 1
PLd paralambdoid sept
rf rhinal fissure
S1DZ S1 dysgranular zn
S1FL S1 forelimb region
S1ULp S1 upper lip region

S2 2ary somatosens
SHi septohipp nu
SIB subst innom basal
STMA ST med div, ant
Tu1 olf tub layer 1
VCl ventral claustrum
VDB nu vert limb diag b
VP ventral pallidum

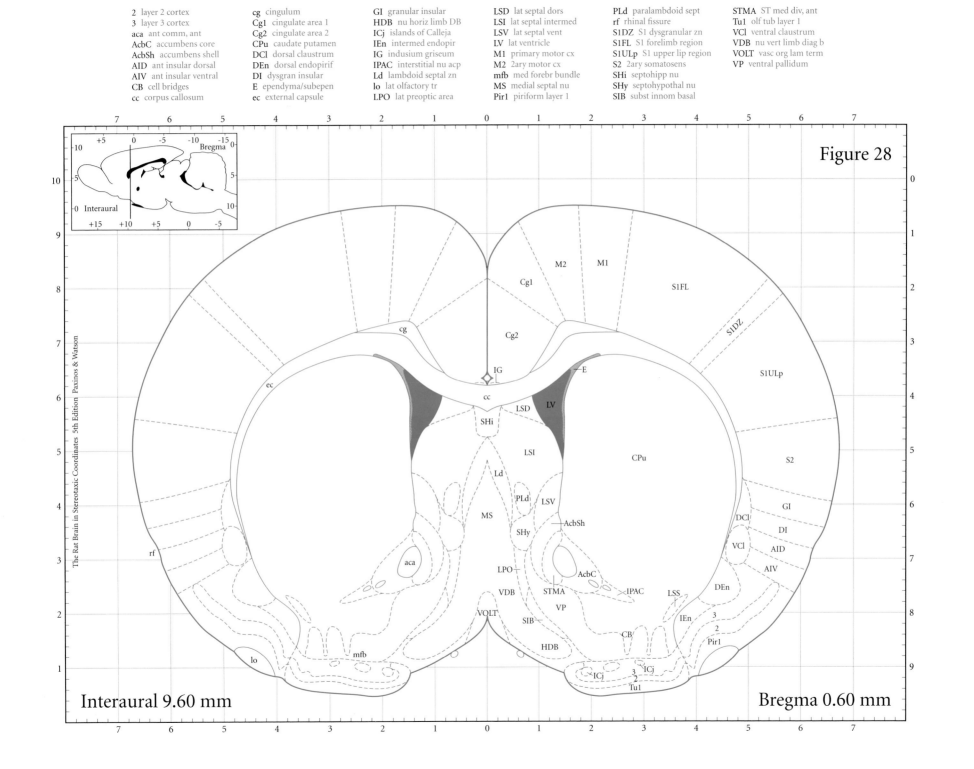

The Rat Brain in Stereotaxic Coordinates 5th Edition Paxinos & Watson

Figure 28

2	layer 2 cortex	cg	cingulum	GI	granular insular
3	layer 3 cortex	Cg1	cingulate area 1	HDB	nu horiz limb DB
aca	ant comm, ant	Cg2	cingulate area 2	ICj	islands of Calleja
AcbC	accumbens core	CPu	caudate putamen	IEn	intermed endopir
AcbSh	accumbens shell	DCl	dorsal claustrum	IG	indusium griseum
AID	ant insular dorsal	DEn	dorsal endopirif	IPAC	interstitial nu acp
AIV	ant insular ventral	DI	dysgran insular	Ld	lambdoid septal zn
CB	cell bridges	E	ependyma/subepen	lo	lat olfactory tr
cc	corpus callosum	ec	external capsule	LPO	lat preoptic area

LSD	lat septal dors	PLd	paralambdoid sept
LSI	lat septal intermed	rf	rhinal fissure
LSV	lat septal vent	S1DZ	S1 dysgranular zn
LV	lat ventricle	S1FL	S1 forelimb region
M1	primary motor cx	S1ULp	S1 upper lip region
M2	2ary motor cx	S2	2ary somatosens
mfb	med forebr bundle	SHi	septohipp nu
MS	medial septal nu	SHy	septohypothal nu
		SIB	subst innom basal

STMA	ST med div, ant
Tu1	olf tub layer 1
VCl	ventral claustrum
VDB	nu vert limb diag b
VOLT	vasc org lam term
VP	ventral pallidum

Interaural 9.60 mm

Bregma 0.60 mm

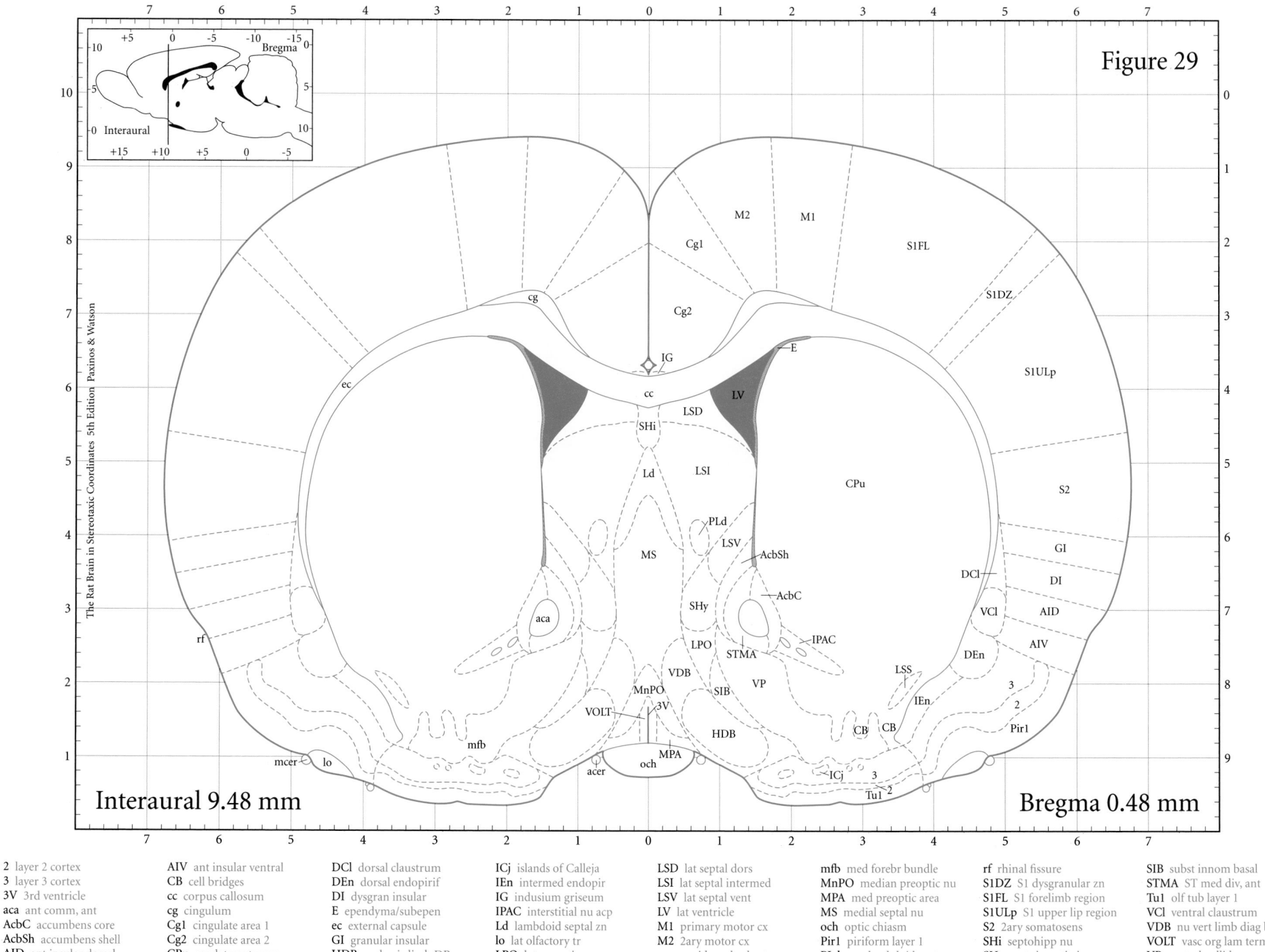

Figure 29

The Rat Brain in Stereotaxic Coordinates 5th Edition Paxinos & Watson

Interaural 9.48 mm

Bregma 0.48 mm

2 layer 2 cortex
3 layer 3 cortex
3V 3rd ventricle
aca ant comm, ant
AcbC accumbens core
AcbSh accumbens shell
AID ant insular dorsal

AIV ant insular ventral
CB cell bridges
cc corpus callosum
cg cingulum
Cg1 cingulate area 1
Cg2 cingulate area 2
CPu caudate putamen

DCl dorsal claustrum
DEn dorsal endopirif
DI dysgran insular
E ependyma/subepen
ec external capsule
GI granular insular
HDB nu horiz limb DB

ICj islands of Calleja
IEn intermed endopir
IG indusium griseum
IPAC interstitial nu acp
Ld lambdoid septal zn
lo lat olfactory tr
LPO lat preoptic area

LSD lat septal dors
LSI lat septal intermed
LSV lat septal vent
LV lat ventricle
M1 primary motor cx
M2 2ary motor cx
mcer mid cerebral art

mfb med forebr bundle
MnPO median preoptic nu
MPA med preoptic area
MS medial septal nu
och optic chiasm
Pir1 piriform layer 1
PLd paralambdoid sept

rf rhinal fissure
S1DZ S1 dysgranular zn
S1FL S1 forelimb region
S1ULp S1 upper lip region
S2 2ary somatosens
SHi septohipp nu
SHy septohypothal nu

SIB subst innom basal
STMA ST med div, ant
Tu1 olf tub layer 1
VCl ventral claustrum
VDB nu vert limb diag b
VOLT vasc org lam term
VP ventral pallidum

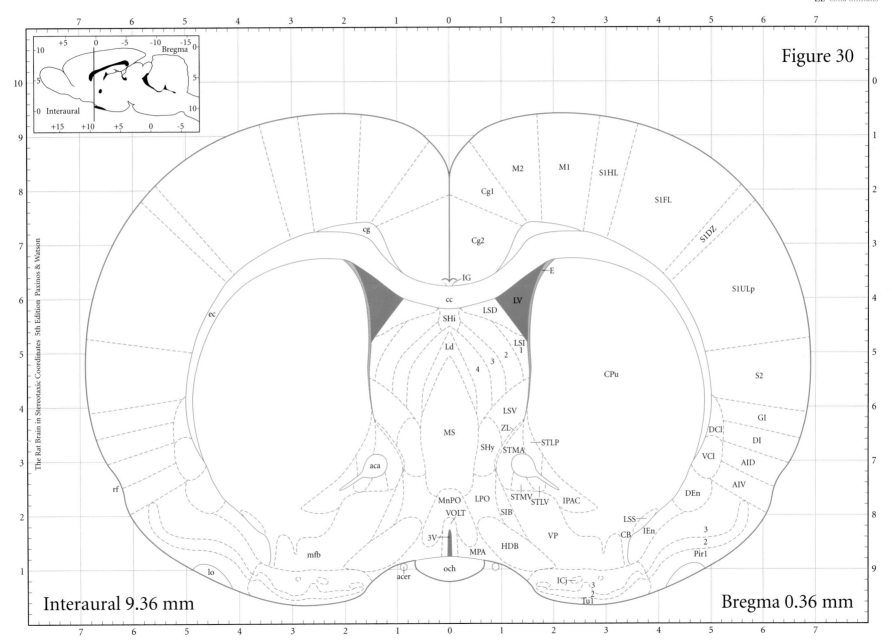

Figure 30

The Rat Brain in Stereotaxic Coordinates 5th Edition Paxinos & Watson

Interaural 9.36 mm

Bregma 0.36 mm

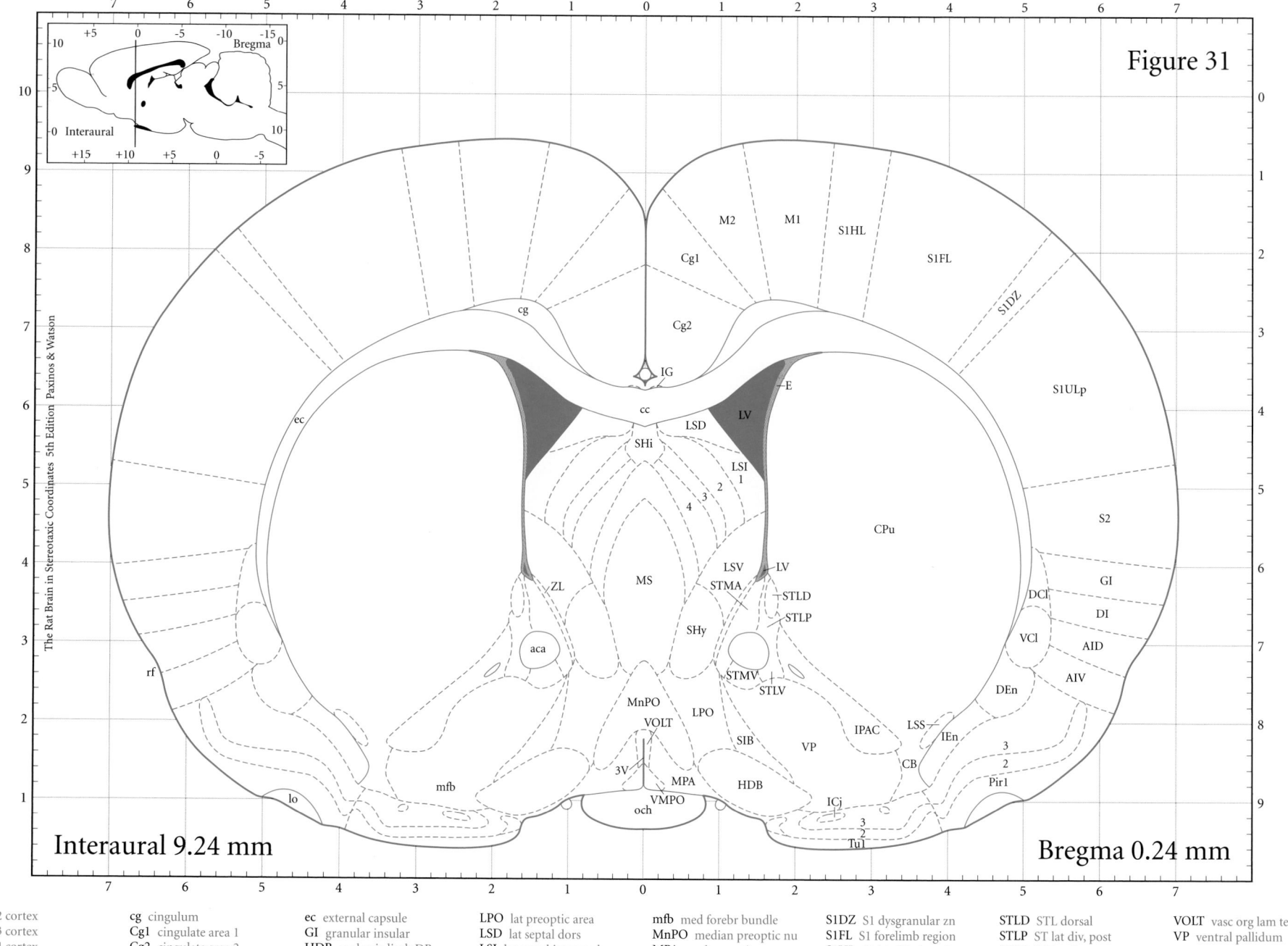

Figure 31

The Rat Brain in Stereotaxic Coordinates 5th Edition Paxinos & Watson

Interaural 9.24 mm

Bregma 0.24 mm

2 layer 2 cortex	cg cingulum	ec external capsule	LPO lat preoptic area	mfb med forebr bundle	S1DZ S1 dysgranular zn	STLD STL dorsal	VOLT vasc org lam term
3 layer 3 cortex	Cg1 cingulate area 1	GI granular insular	LSD lat septal dors	MnPO median preoptic nu	S1FL S1 forelimb region	STLP ST lat div, post	VP ventral pallidum
4 layer 4 cortex	Cg2 cingulate area 2	HDB nu horiz limb DB	LSI lat septal intermed	MPA med preoptic area	S1HL S1 hindlimb region	STLV ST lat div, vent	
aca ant comm, ant	CPu caudate putamen	ICj islands of Calleja	LSS lat stripe of str	MS medial septal nu	S1ULp S1 upper lip region	STMA ST med div, ant	
AID ant insular dorsal	DCl dorsal claustrum	IEn intermed endopir	LSV lat septal vent	och optic chiasm	S2 2ary somatosens	STMV ST med div, vent	
AIV ant insular ventral	DEn dorsal endopirif	IG indusium griseum	LV lat ventricle	P3V preopt recess 3V	SHi septohipp nu	Tu1 olf tub layer 1	
CB cell bridges	DI dysgran insular	IPAC interstitial nu acp	M1 primary motor cx	Pir1 piriform layer 1	SHy septohypothal nu	VCl ventral claustrum	
cc corpus callosum	E ependyma/subepen	lo lat olfactory tr	M2 2ary motor cx	rf rhinal fissure	SIB subst innom basal	VMPO ventromed preopt	

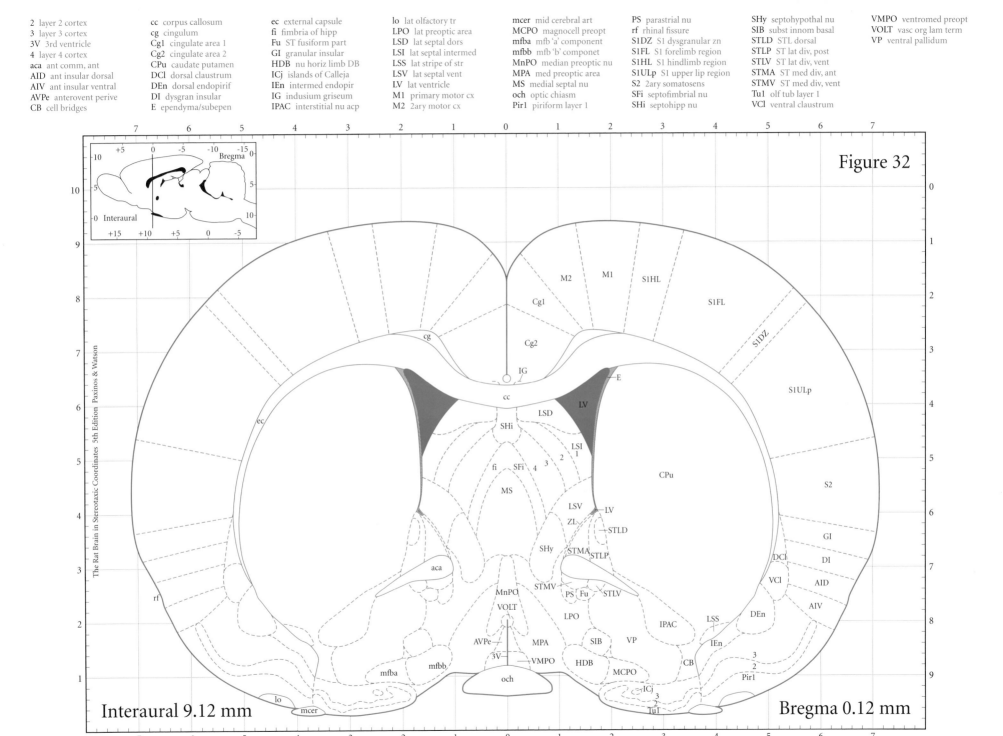

Figure 32

Interaural 9.12 mm

Bregma 0.12 mm

Key:
2 layer 2 cortex
3 layer 3 cortex
3V 3rd ventricle
4 layer 4 cortex
aca ant comm, ant
AID ant insular dorsal
AIV ant insular ventral
AVPe anterovent perive
CB cell bridges

cc corpus callosum
cg cingulum
Cg1 cingulate area 1
Cg2 cingulate area 2
CPu caudate putamen
DCl dorsal claustrum
DEn dorsal endopirif
DI dysgran insular
E ependyma/subepen

ec external capsule
fi fimbria of hipp
Fu ST fusiform part
GI granular insular
HDB nu horiz limb DB
ICj islands of Calleja
IEn intermed endopir
IG indusium griseum
IPAC interstitial nu acp

lo lat olfactory tr
LPO lat preoptic area
LSD lat septal dors
LSI lat septal intermed
LSS lat stripe of str
LSV lat septal vent
LV lat ventricle
M1 primary motor cx
M2 2ary motor cx

mcer mid cerebral art
MCPO magnocell preopt
mfba mfb 'a' component
mfbb mfb 'b' componet
MnPO median preoptic nu
MPA med preoptic area
MS medial septal nu
och optic chiasm
Pir1 piriform layer 1

PS parastrial nu
rf rhinal fissure
S1DZ S1 dysgranular zn
S1FL S1 forelimb region
S1HL S1 hindlimb region
S1ULp S1 upper lip region
S2 2ary somatosens
SFi septofimbrial nu
SHi septohipp nu

SHy septohypothal nu
SIB subst innom basal
STLD STL dorsal
STLP ST lat div, post
STLV ST lat div, vent
STMA ST med div, ant
STMV ST med div, vent
Tu1 olf tub layer 1
VCl ventral claustrum

VMPO ventromed preopt
VOLT vasc org lam term
VP ventral pallidum

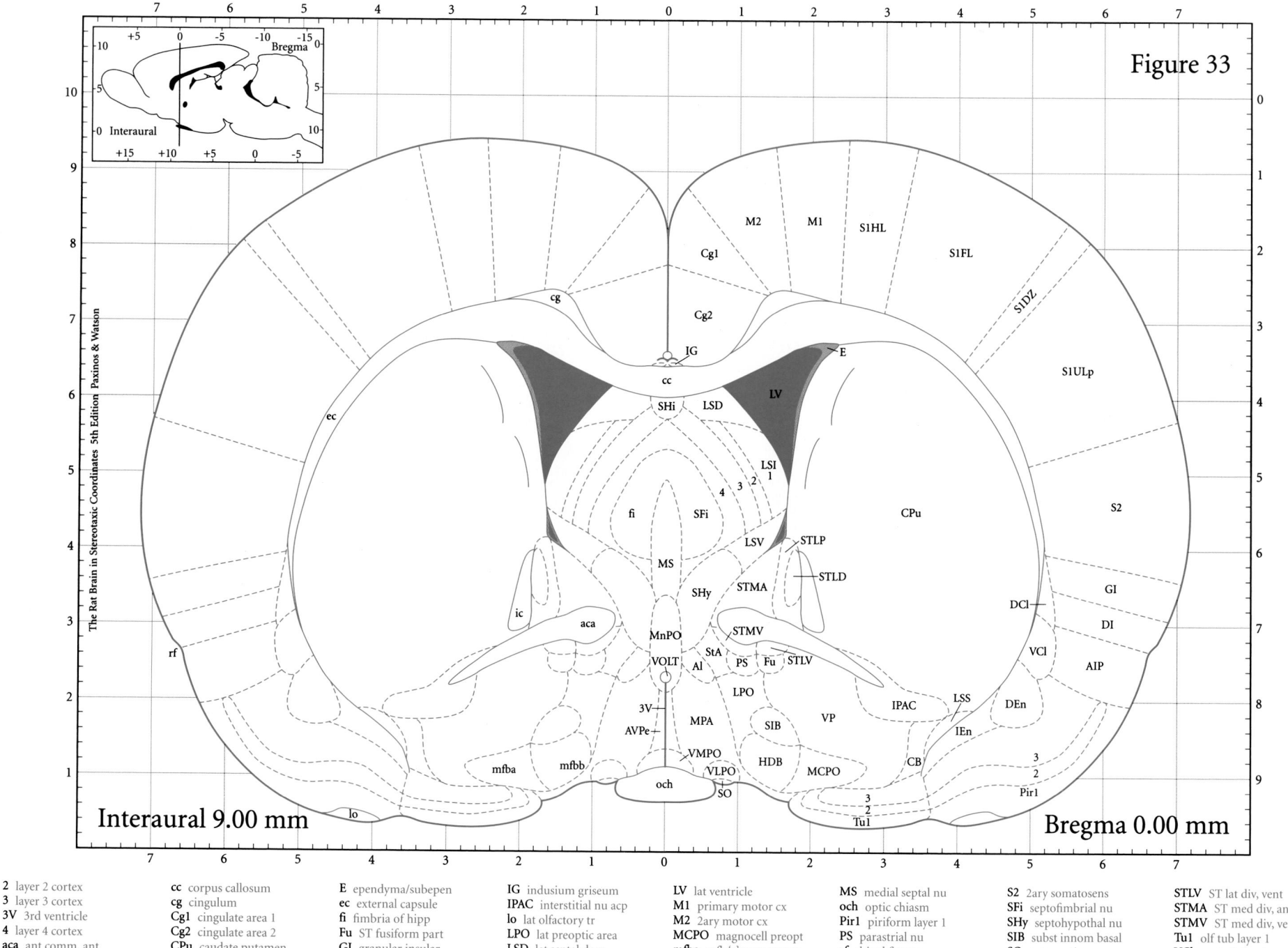

Figure 33

The Rat Brain in Stereotaxic Coordinates 5th Edition Paxinos & Watson

Interaural 9.00 mm

Bregma 0.00 mm

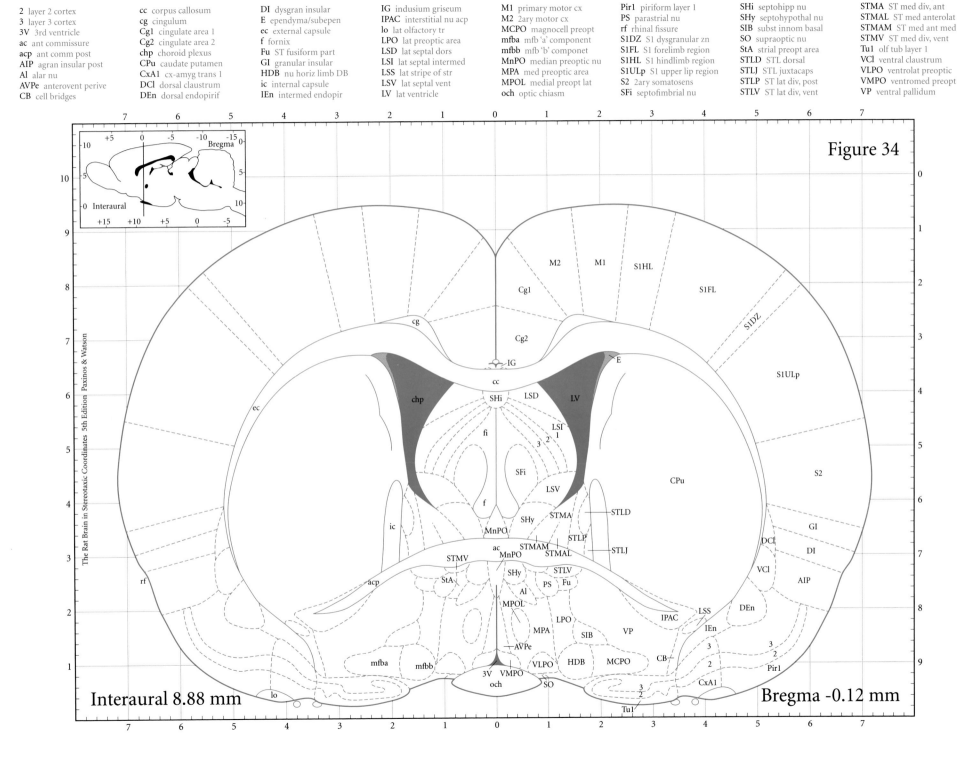

Figure 34

Interaural 8.88 mm

Bregma -0.12 mm

The Rat Brain in Stereotaxic Coordinates 5th Edition Paxinos & Watson

2 layer 2 cortex
3 layer 3 cortex
3V 3rd ventricle
ac ant commissure
acp ant comm post
AIP agran insular post
Al alar nu
AVPe anterovent perive
CB cell bridges

cc corpus callosum
cg cingulum
Cg1 cingulate area 1
Cg2 cingulate area 2
chp choroid plexus
CPu caudate putamen
CxA1 cx-amyg trans 1
DCl dorsal claustrum
DEn dorsal endopirif

DI dysgran insular
E ependyma/subepen
ec external capsule
f fornix
Fu ST fusiform part
GI granular insular
HDB nu horiz limb DB
ic internal capsule
IEn intermed endopir

IG indusium griseum
IPAC interstitial nu acp
lo lat olfactory tr
LPO lat preoptic area
LSD lat septal dors
LSI lat septal intermed
LSS lat stripe of str
LSV lat septal vent
LV lat ventricle

M1 primary motor cx
M2 2ary motor cx
MCPO magnocell preopt
mfba mfb 'a' component
mfbb mfb 'b' componet
MnPO median preoptic nu
MPA med preoptic area
MPOL medial preopt lat
och optic chiasm

Pir1 piriform layer 1
PS parastrial nu
rf rhinal fissure
S1DZ S1 dysgranular zn
S1FL S1 forelimb region
S1HL S1 hindlimb region
S1ULp S1 upper lip region
S2 2ary somatosens
SFi septofimbrial nu

SHi septohipp nu
SHy septohypothal nu
SIB subst innom basal
SO supraoptic nu
StA strial preopt area
STLD STL dorsal
STLJ STL juxtacaps
STLP ST lat div, post
STLV ST lat div, vent

STMA ST med div, ant
STMAL ST med anterolat
STMAM ST med ant med
STMV ST med div, vent
Tu1 olf tub layer 1
VCl ventral claustrum
VLPO ventrolat preoptic
VMPO ventromed preopt
VP ventral pallidum

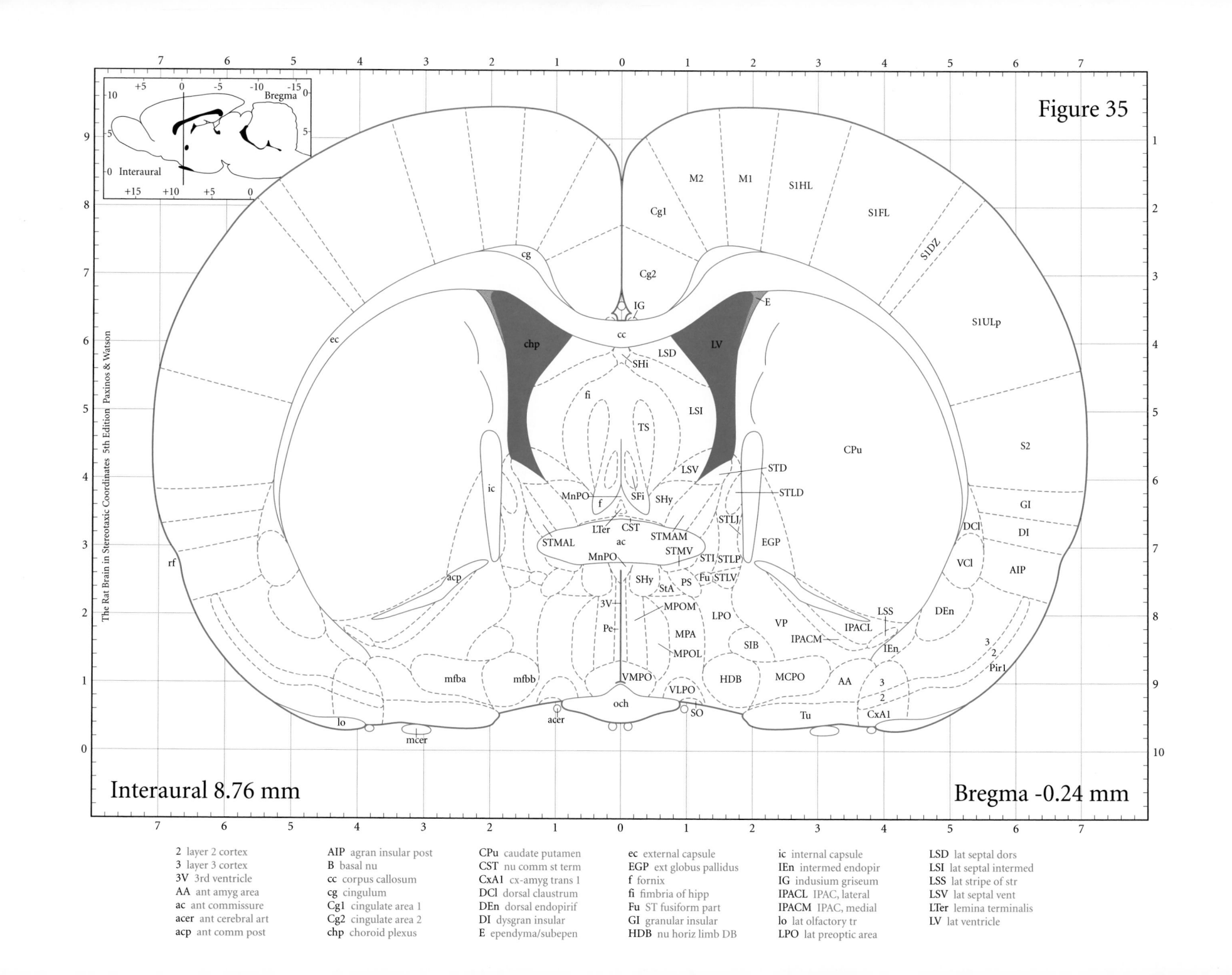

Figure 35

Interaural 8.76 mm

Bregma -0.24 mm

2 layer 2 cortex
3 layer 3 cortex
3V 3rd ventricle
AA ant amyg area
ac ant commissure
acer ant cerebral art
acp ant comm post

AIP agran insular post
B basal nu
cc corpus callosum
cg cingulum
Cg1 cingulate area 1
Cg2 cingulate area 2
chp choroid plexus

CPu caudate putamen
CST nu comm st term
CxA1 cx-amyg trans 1
DCl dorsal claustrum
DEn dorsal endopirif
DI dysgran insular
E ependyma/subepen

ec external capsule
EGP ext globus pallidus
f fornix
fi fimbria of hipp
Fu ST fusiform part
GI granular insular
HDB nu horiz limb DB

ic internal capsule
IEn intermed endopir
IG indusium griseum
IPACL IPAC, lateral
IPACM IPAC, medial
lo lat olfactory tr
LPO lat preoptic area

LSD lat septal dors
LSI lat septal intermed
LSS lat stripe of str
LSV lat septal vent
LTer lemina terminalis
LV lat ventricle

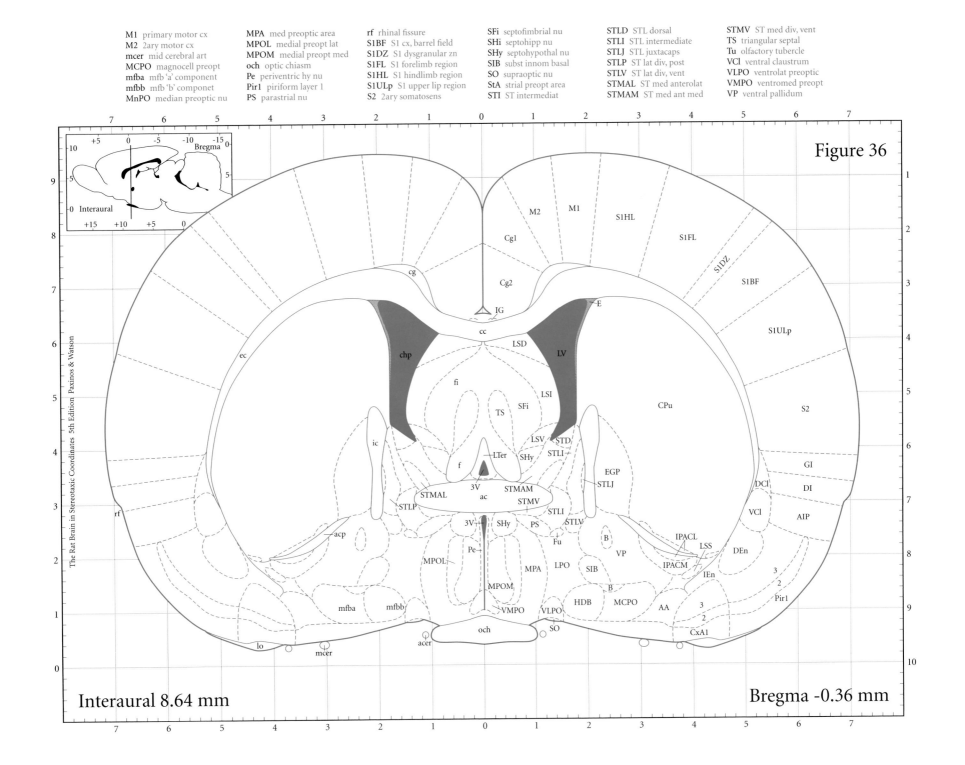

M1   primary motor cx
M2   2ary motor cx
mcer   mid cerebral art
MCPO   magnocell preopt
mfba   mfb 'a' component
mfbb   mfb 'b' componet
MnPO   median preoptic nu

MPA   med preoptic area
MPOL   medial preopt lat
MPOM   medial preopt med
och   optic chiasm
Pe   periventric hy nu
Pir1   piriform layer 1
PS   parastrial nu

rf   rhinal fissure
S1BF   S1 cx, barrel field
S1DZ   S1 dysgranular zn
S1FL   S1 forelimb region
S1HL   S1 hindlimb region
S1ULp   S1 upper lip region
S2   2ary somatosens

SFi   septofimbrial nu
SHi   septohipp nu
SHy   septohypothal nu
SIB   subst innom basal
SO   supraoptic nu
StA   strial preopt area
STI   ST intermediat

STLD   STL dorsal
STLI   STL intermediate
STLJ   STL juxtacaps
STLP   ST lat div, post
STLV   ST lat div, vent
STMAL   ST med anterolat
STMAM   ST med ant med

STMV   ST med div, vent
TS   triangular septal
Tu   olfactory tubercle
VCl   ventral claustrum
VLPO   ventrolat preoptic
VMPO   ventromed preopt
VP   ventral pallidum

Figure 36

Interaural 8.64 mm

Bregma -0.36 mm

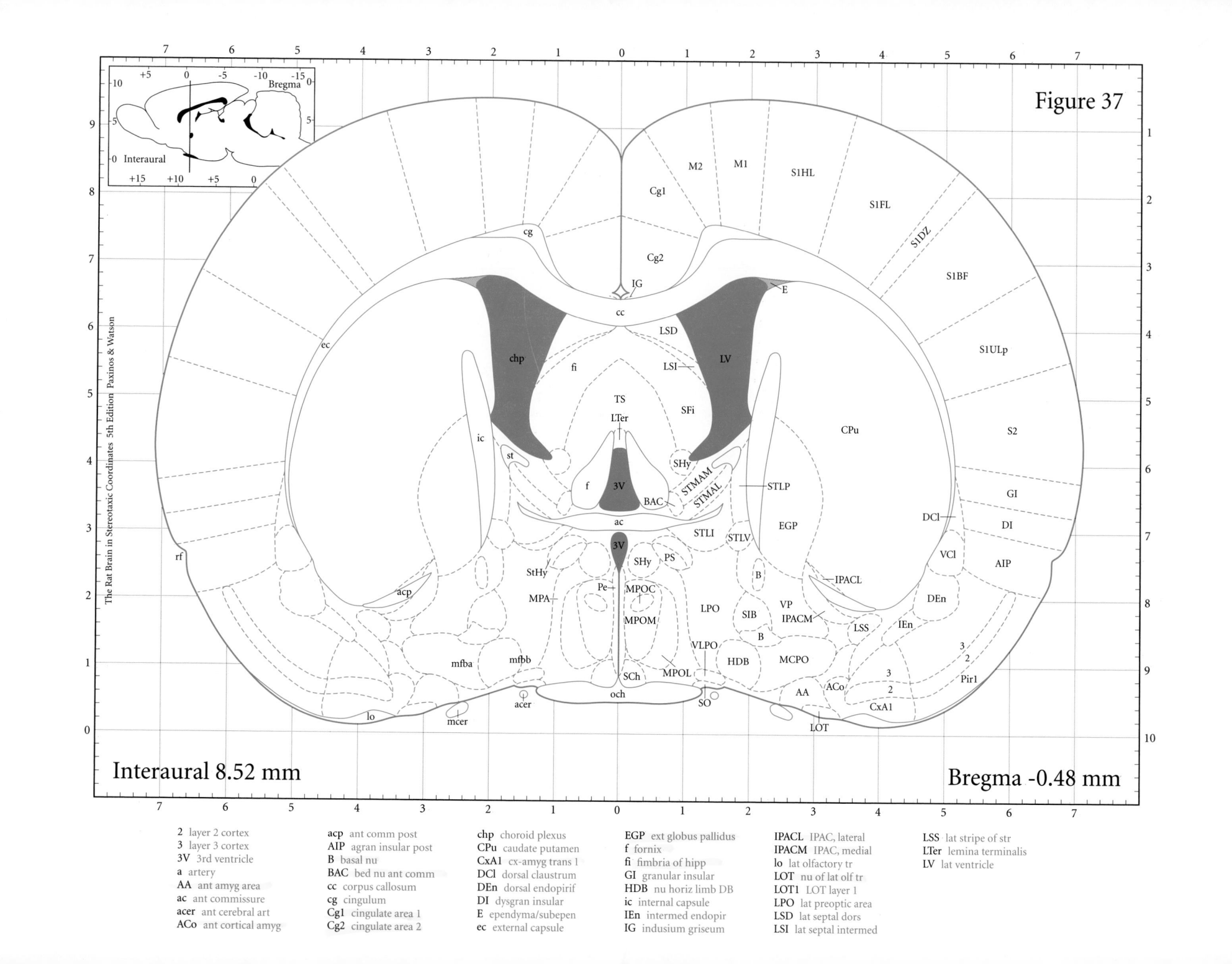

Figure 37

Interaural 8.52 mm

Bregma -0.48 mm

The Rat Brain in Stereotaxic Coordinates 5th Edition Paxinos & Watson

2   layer 2 cortex
3   layer 3 cortex
3V   3rd ventricle
a   artery
AA   ant amyg area
ac   ant commissure
acer ant cerebral art
ACo ant cortical amyg

acp ant comm post
AIP agran insular post
B   basal nu
BAC bed nu ant comm
cc   corpus callosum
cg   cingulum
Cg1 cingulate area 1
Cg2 cingulate area 2

chp choroid plexus
CPu caudate putamen
CxA1 cx-amyg trans 1
DCl dorsal claustrum
DEn dorsal endopirif
DI   dysgran insular
E   ependyma/subepen
ec   external capsule

EGP ext globus pallidus
f   fornix
fi   fimbria of hipp
GI   granular insular
HDB nu horiz limb DB
ic   internal capsule
IEn intermed endopir
IG   indusium griseum

IPACL IPAC, lateral
IPACM IPAC, medial
lo   lat olfactory tr
LOT nu of lat olf tr
LOT1 LOT layer 1
LPO lat preoptic area
LSD lat septal dors
LSI lat septal intermed

LSS lat stripe of str
LTer lemina terminalis
LV   lat ventricle

Figure 38

M1	primary motor cx
M2	2ary motor cx
mcer	mid cerebral art
mch	med corthypo tr
MCPO	magnocell preopt
mfba	mfb 'a' component
mfbb	mfb 'b' componet
MPA	med preoptic area

MPOC	medial preopt cent
MPOL	medial preopt lat
MPOM	medial preopt med
och	optic chiasm
PDPO	postdors preopt
Pe	periventric hy nu
PeAP	periv ant parvic
Pir1	piriform layer 1

PS	parastrial nu
rf	rhinal fissure
S1BF	S1 cx, barrel field
S1DZ	S1 dysgranular zn
S1FL	S1 forelimb region
S1HL	S1 hindlimb region
S1ULp	S1 upper lip region
S2	2ary somatosens

SCh	suprachiasmatic nu
SFi	septofimbrial nu
SFO	subfornical organ
SHy	septohypothal nu
SIB	subst innom basal
SO	supraoptic nu
sod	supraoptic decussn

st	stria terminalis
StHy	striohypothal nu
STLI	STL intermediate
STLP	ST lat div, post
STLV	ST lat div, vent
sm	stria medullaris
STMAL	ST med anterolat
STMAM	ST med ant med
STMPI	STM postintermed

STMPL	STM posterolat
STMPM	STM posteromed
TS	triangular septal
VCl	ventral claustrum
vhc	ventral hipp comm
VLPO	ventrolat preoptic
VP	ventral pallidum

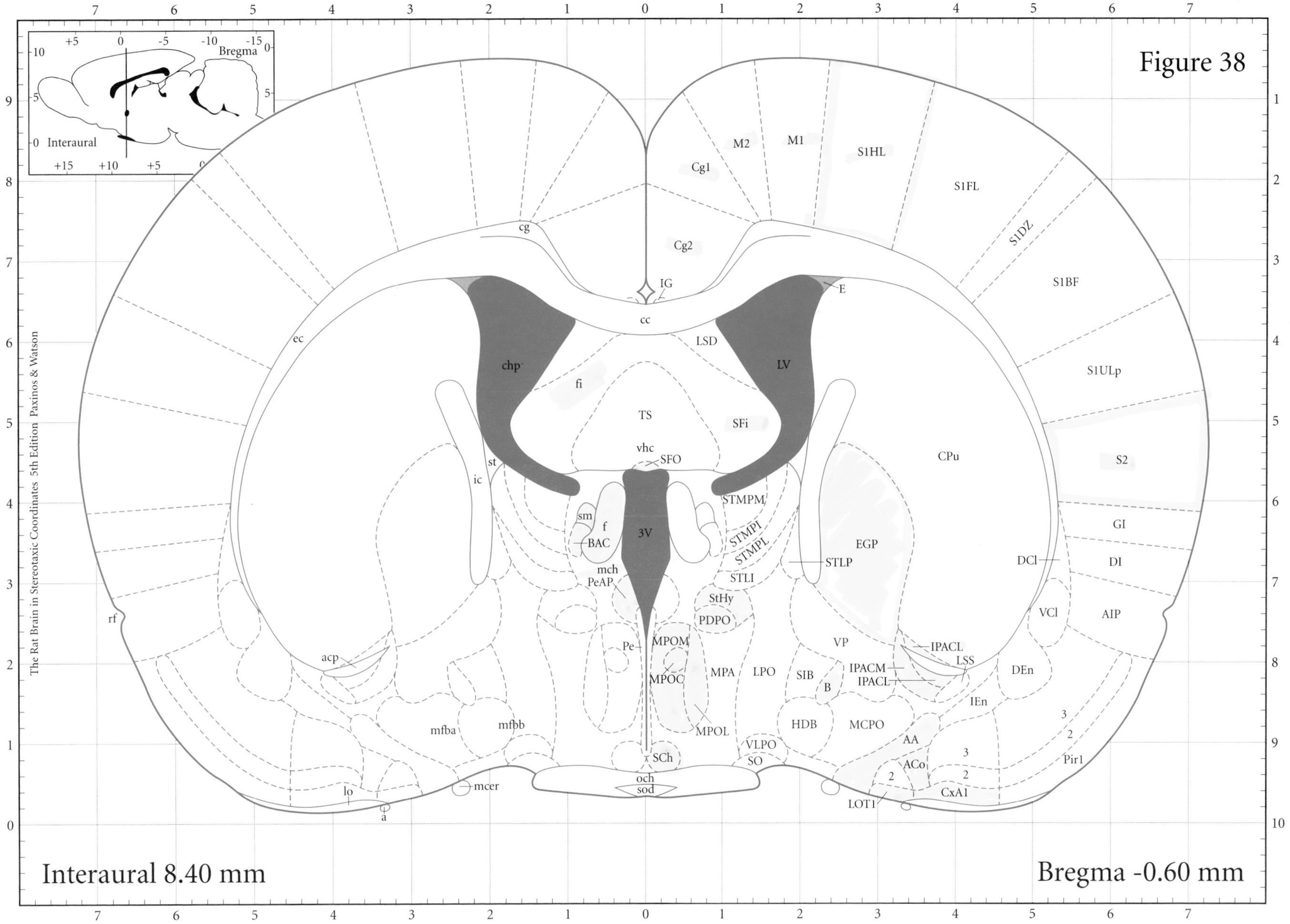

Interaural 8.40 mm

Bregma -0.60 mm

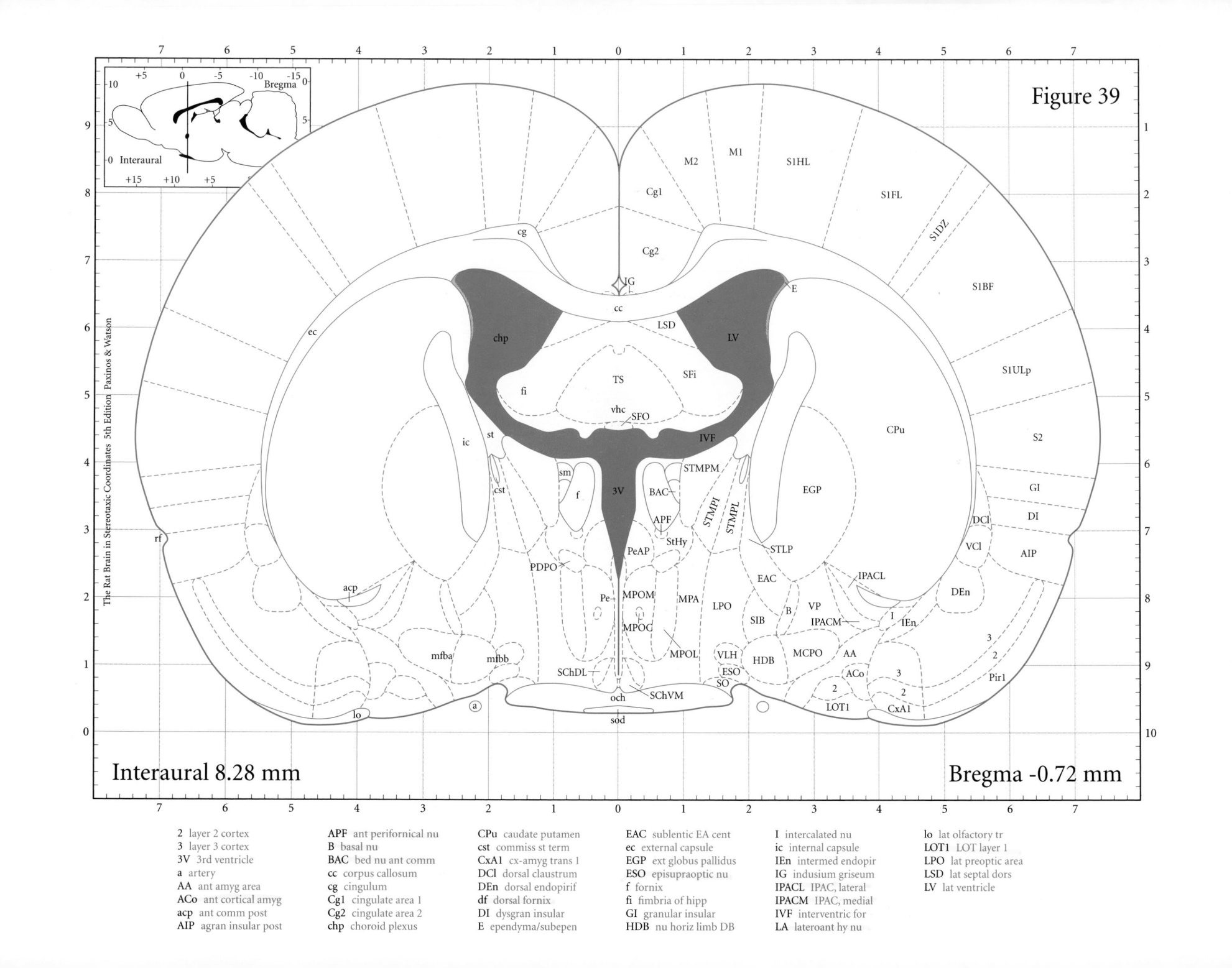

Figure 39

Interaural 8.28 mm

Bregma -0.72 mm

The Rat Brain in Stereotaxic Coordinates 5th Edition Paxinos & Watson

2   layer 2 cortex
3   layer 3 cortex
3V 3rd ventricle
a   artery
AA ant amyg area
ACo ant cortical amyg
acp ant comm post
AIP agran insular post

APF ant perifornical nu
B basal nu
BAC bed nu ant comm
cc corpus callosum
cg cingulum
Cg1 cingulate area 1
Cg2 cingulate area 2
chp choroid plexus

CPu caudate putamen
cst commiss st term
CxA1 cx-amyg trans 1
DCl dorsal claustrum
DEn dorsal endopirif
df dorsal fornix
DI dysgran insular
E ependyma/subepen

EAC sublentic EA cent
ec external capsule
EGP ext globus pallidus
ESO episupraoptic nu
f fornix
fi fimbria of hipp
GI granular insular
HDB nu horiz limb DB

I intercalated nu
ic internal capsule
IEn intermed endopir
IG indusium griseum
IPACL IPAC, lateral
IPACM IPAC, medial
IVF interventric for
LA lateroant hy nu

lo lat olfactory tr
LOT1 LOT layer 1
LPO lat preoptic area
LSD lat septal dors
LV lat ventricle

Figure 40

The Rat Brain in Stereotaxic Coordinates 5th Edition Paxinos & Watson

M1 primary motor cx
M2 2ary motor cx
mch med corthypo tr
MCPO magnocell preopt
mfb med forebr bundle
mfba mfb 'a' component
mfbb mfb 'b' componet
MPA med preoptic area

MPOC medial preopt cent
MPOL medial preopt lat
MPOM medial preopt med
och optic chiasm
PaAP Pa ant parvicell
Pe periventric hy nu
PeAP periv ant parvic
Pir1 piriform layer 1

PVA paraventric th ant
rf rhinal fissure
SChDL suprach dorsolat
SChVM suprach ventromed
S1BF S1 cx, barrel field
S1DZ S1 dysgranular zn
S1FL S1 forelimb region
S1HL S1 hindlimb region

S1ULp S1 upper lip region
S2 2ary somatosens
SFi septofimbrial nu
SFO subfornical organ
SIB subst innom basal
sm stria medullaris
SO supraoptic nu
sod supraoptic decussn

st stria terminalis
StHy striohypothal nu
STLP ST lat div, post
STMPI STM postintermed
STMPL STM posterolat
STMPM STM posteromed
TS triangular septal
VCl ventral claustrum

VEn ventral endopir
vhc ventral hipp comm
VLH ventrolat hy nu
VP ventral pallidum

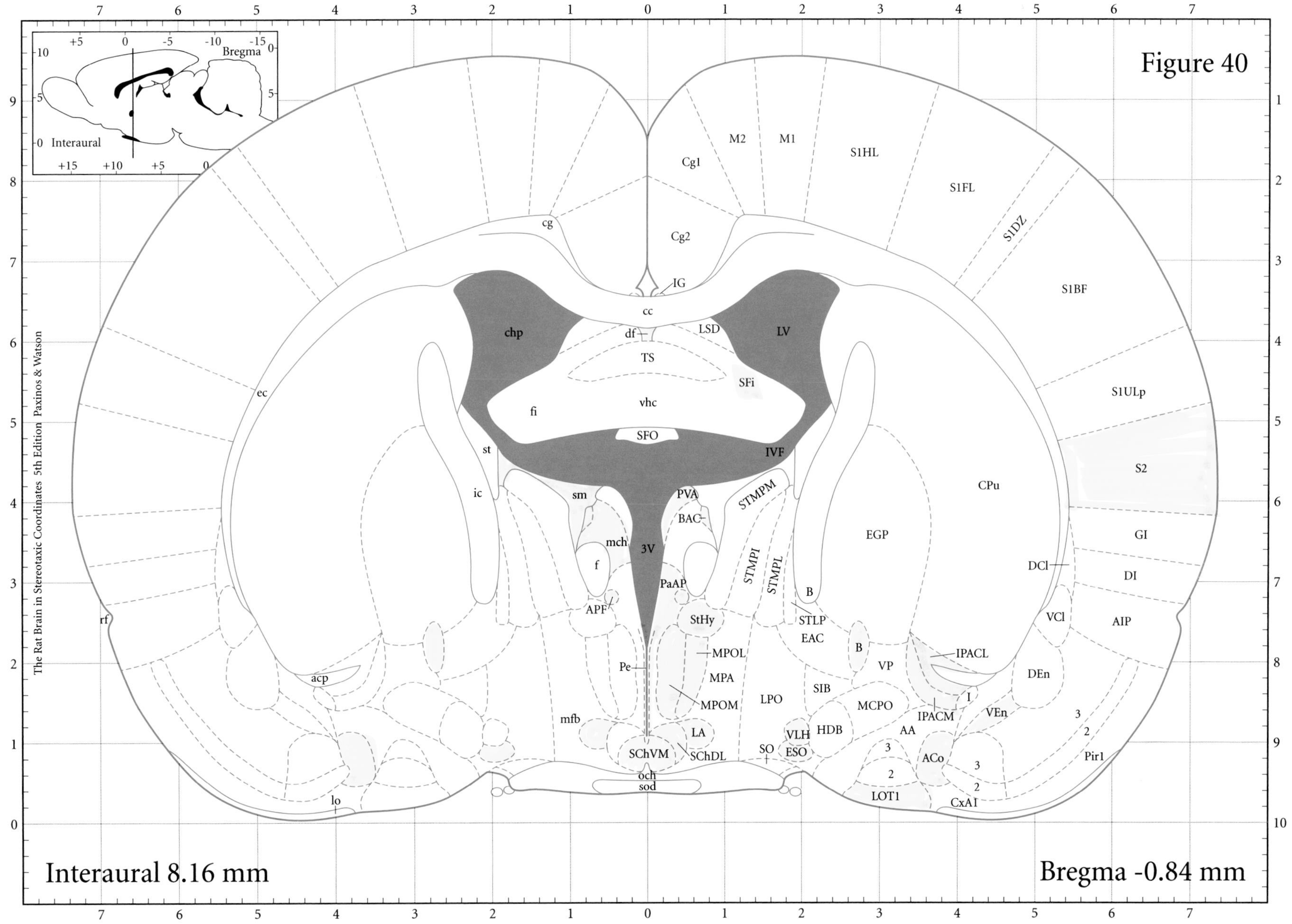

Interaural 8.16 mm

Bregma -0.84 mm

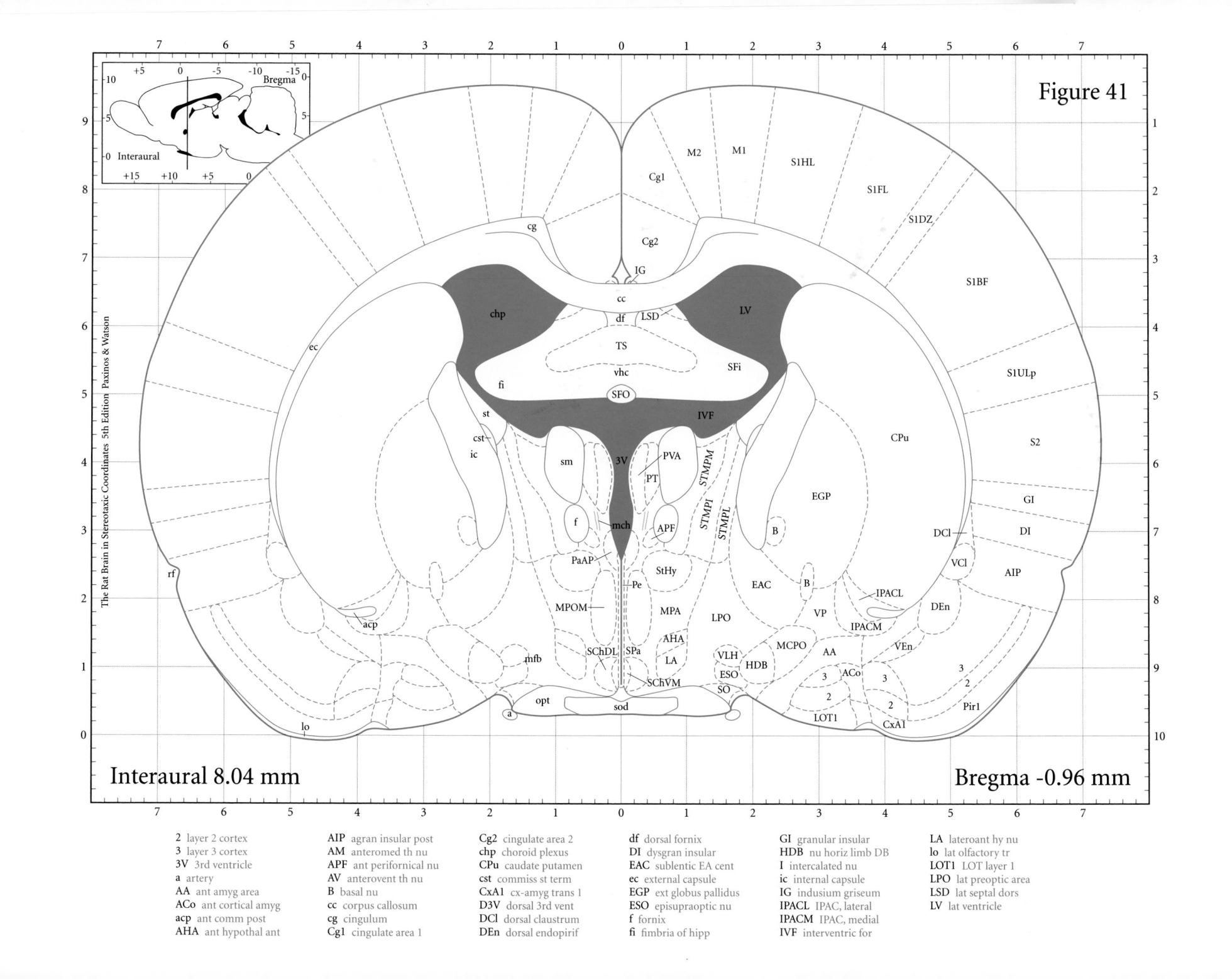

Figure 41

Interaural 8.04 mm

Bregma -0.96 mm

2 layer 2 cortex
3 layer 3 cortex
3V 3rd ventricle
a artery
AA ant amyg area
ACo ant cortical amyg
acp ant comm post
AHA ant hypothal ant

AIP agran insular post
AM anteromed th nu
APF ant perifornical nu
AV anterovent th nu
B basal nu
cc corpus callosum
cg cingulum
Cg1 cingulate area 1

Cg2 cingulate area 2
chp choroid plexus
CPu caudate putamen
cst commiss st term
CxA1 cx-amyg trans 1
D3V dorsal 3rd vent
DCl dorsal claustrum
DEn dorsal endopirif

df dorsal fornix
DI dysgran insular
EAC sublentic EA cent
ec external capsule
EGP ext globus pallidus
ESO episupraoptic nu
f fornix
fi fimbria of hipp

GI granular insular
HDB nu horiz limb DB
I intercalated nu
ic internal capsule
IG indusium griseum
IPACL IPAC, lateral
IPACM IPAC, medial
IVF interventric for

LA lateroant hy nu
lo lat olfactory tr
LOT1 LOT layer 1
LPO lat preoptic area
LSD lat septal dors
LV lat ventricle

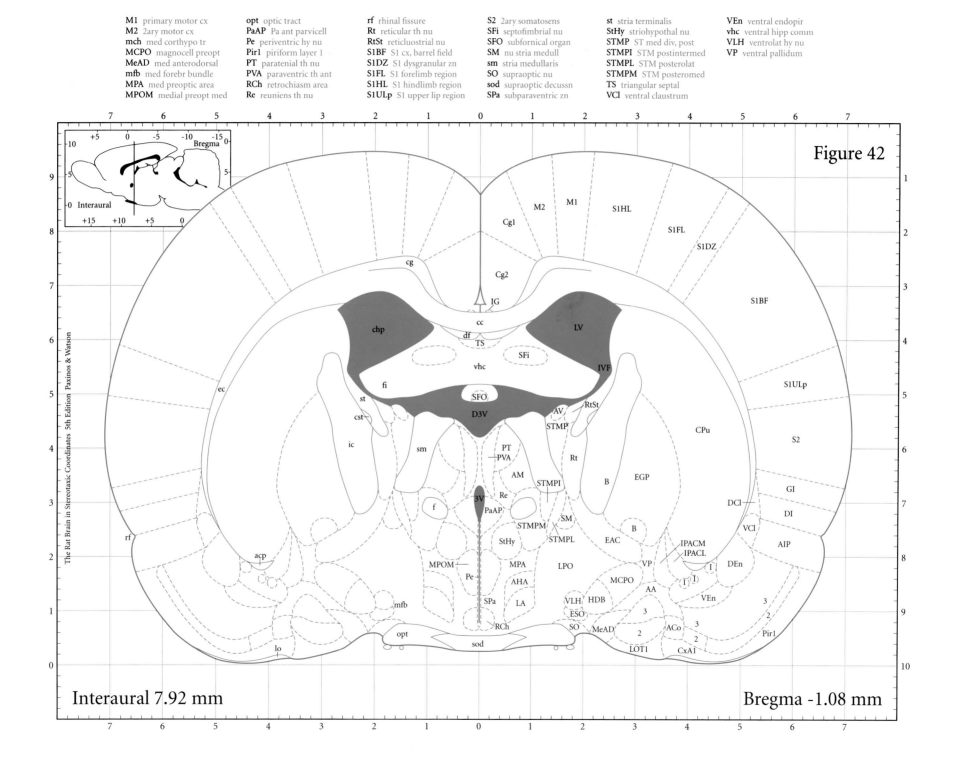

M1   primary motor cx
M2   2ary motor cx
mch   med corthypo tr
MCPO   magnocell preopt
MeAD   med anterodorsal
mfb   med forebr bundle
MPA   med preoptic area
MPOM   medial preopt med

opt   optic tract
PaAP   Pa ant parvicell
Pe   periventric hy nu
Pir1   piriform layer 1
PT   paratenial th nu
PVA   paraventric th ant
RCh   retrochiasm area
Re   reuniens th nu

rf   rhinal fissure
Rt   reticular th nu
RtSt   reticluostrial nu
S1BF   S1 cx, barrel field
S1DZ   S1 dysgranular zn
S1FL   S1 forelimb region
S1HL   S1 hindlimb region
S1ULp   S1 upper lip region

S2   2ary somatosens
SFi   septofimbrial nu
SFO   subfornical organ
SM   nu stria medull
sm   stria medullaris
SO   supraoptic nu
sod   supraoptic decussn
SPa   subparaventric zn

st   stria terminalis
StHy   striohypothal nu
STMP   ST med div, post
STMPI   STM postintermed
STMPL   STM posterolat
STMPM   STM posteromed
TS   triangular septal
VCl   ventral claustrum

VEn   ventral endopir
vhc   ventral hipp comm
VLH   ventrolat hy nu
VP   ventral pallidum

Figure 42

Interaural 7.92 mm

Bregma -1.08 mm

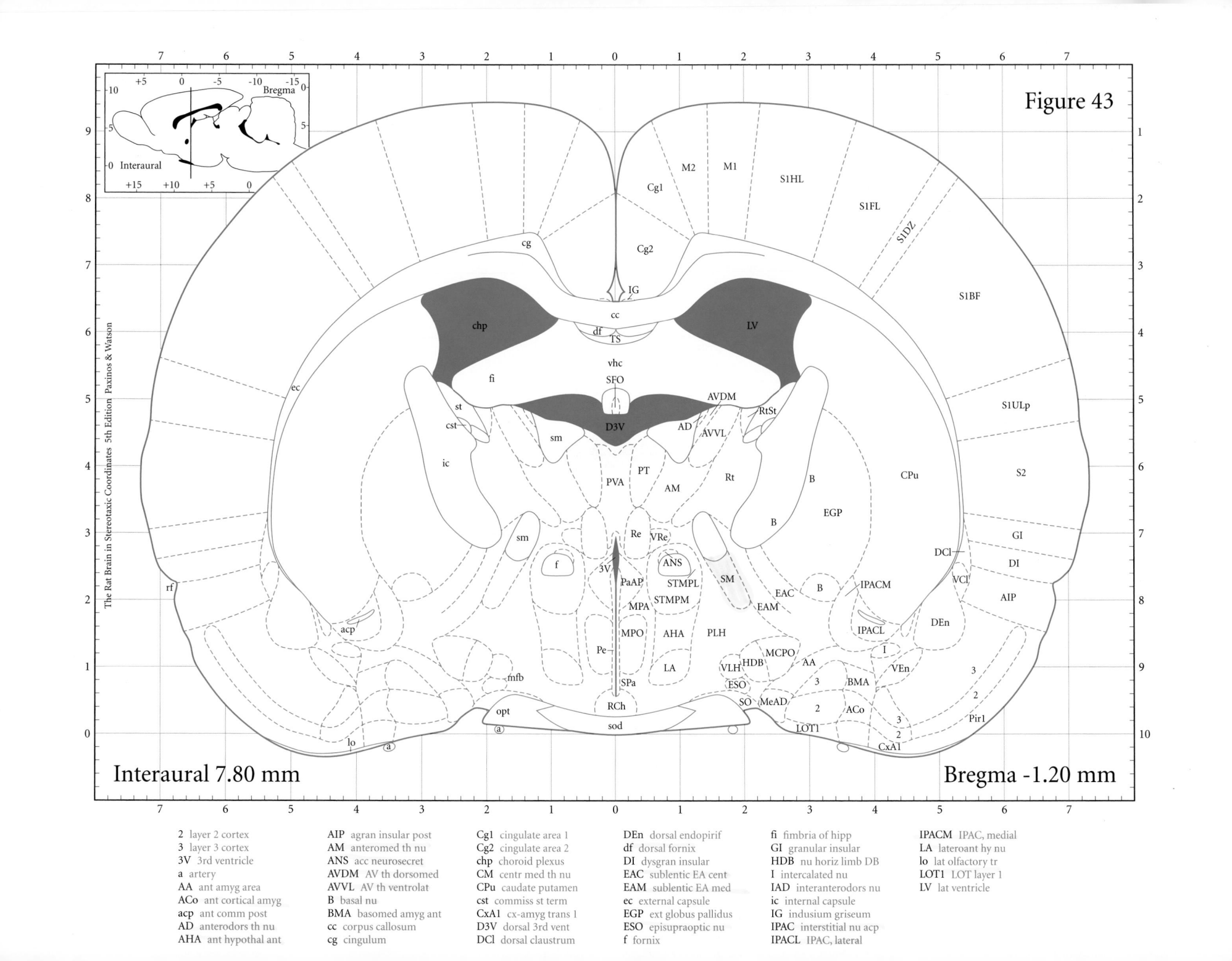

Figure 43

Interaural 7.80 mm

Bregma -1.20 mm

2   layer 2 cortex
3   layer 3 cortex
3V   3rd ventricle
a   artery
AA   ant amyg area
ACo ant cortical amyg
acp ant comm post
AD   anterodors th nu
AHA ant hypothal ant

AIP agran insular post
AM   anteromed th nu
ANS acc neurosecret
AVDM AV th dorsomed
AVVL AV th ventrolat
B   basal nu
BMA basomed amyg ant
cc   corpus callosum
cg   cingulum

Cg1 cingulate area 1
Cg2 cingulate area 2
chp choroid plexus
CM   centr med th nu
CPu caudate putamen
cst commiss st term
CxA1 cx-amyg trans 1
D3V dorsal 3rd vent
DCl dorsal claustrum

DEn dorsal endopirif
df   dorsal fornix
DI   dysgran insular
EAC sublentic EA cent
EAM sublentic EA med
ec   external capsule
EGP ext globus pallidus
ESO episupraoptic nu
f   fornix

fi   fimbria of hipp
GI   granular insular
HDB nu horiz limb DB
I   intercalated nu
IAD interanterodors nu
ic   internal capsule
IG   indusium griseum
IPAC interstitial nu acp
IPACL IPAC, lateral

IPACM IPAC, medial
LA   lateroant hy nu
lo   lat olfactory tr
LOT1 LOT layer 1
LV   lat ventricle

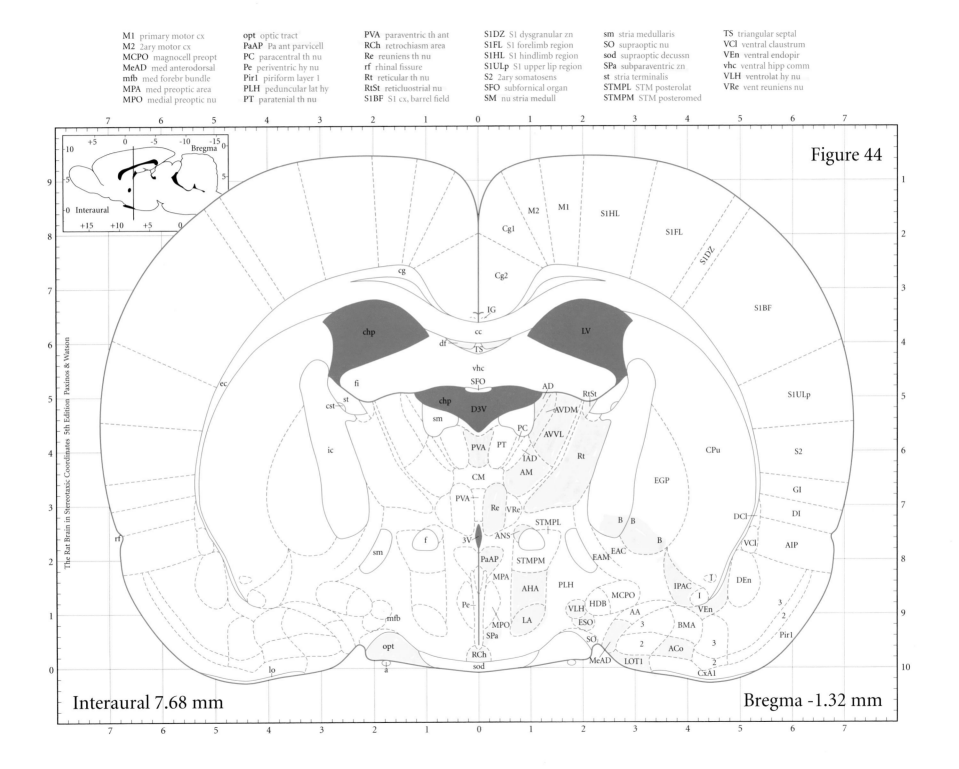

Figure 44

Interaural 7.68 mm

Bregma -1.32 mm

The Rat Brain in Stereotaxic Coordinates 5th Edition Paxinos & Watson

M1 primary motor cx
M2 2ary motor cx
MCPO magnocell preopt
MeAD med anterodorsal
mfb med forebr bundle
MPA med preoptic area
MPO medial preoptic nu

opt optic tract
PaAP Pa ant parvicell
PC paracentral th nu
Pe periventric hy nu
Pir1 piriform layer 1
PLH peduncular lat hy
PT paratenial th nu

PVA paraventric th ant
RCh retrochiasm area
Re reuniens th nu
rf rhinal fissure
Rt reticular th nu
RtSt reticluostrial nu
S1BF S1 cx, barrel field

S1DZ S1 dysgranular zn
S1FL S1 forelimb region
S1HL S1 hindlimb region
S1ULp S1 upper lip region
S2 2ary somatosens
SFO subfornical organ
SM nu stria medull

sm stria medullaris
SO supraoptic nu
sod supraoptic decussn
SPa subparaventric zn
st stria terminalis
STMPL STM posterolat
STMPM STM posteromed

TS triangular septal
VCl ventral claustrum
VEn ventral endopir
vhc ventral hipp comm
VLH ventrolat hy nu
VRe vent reuniens nu

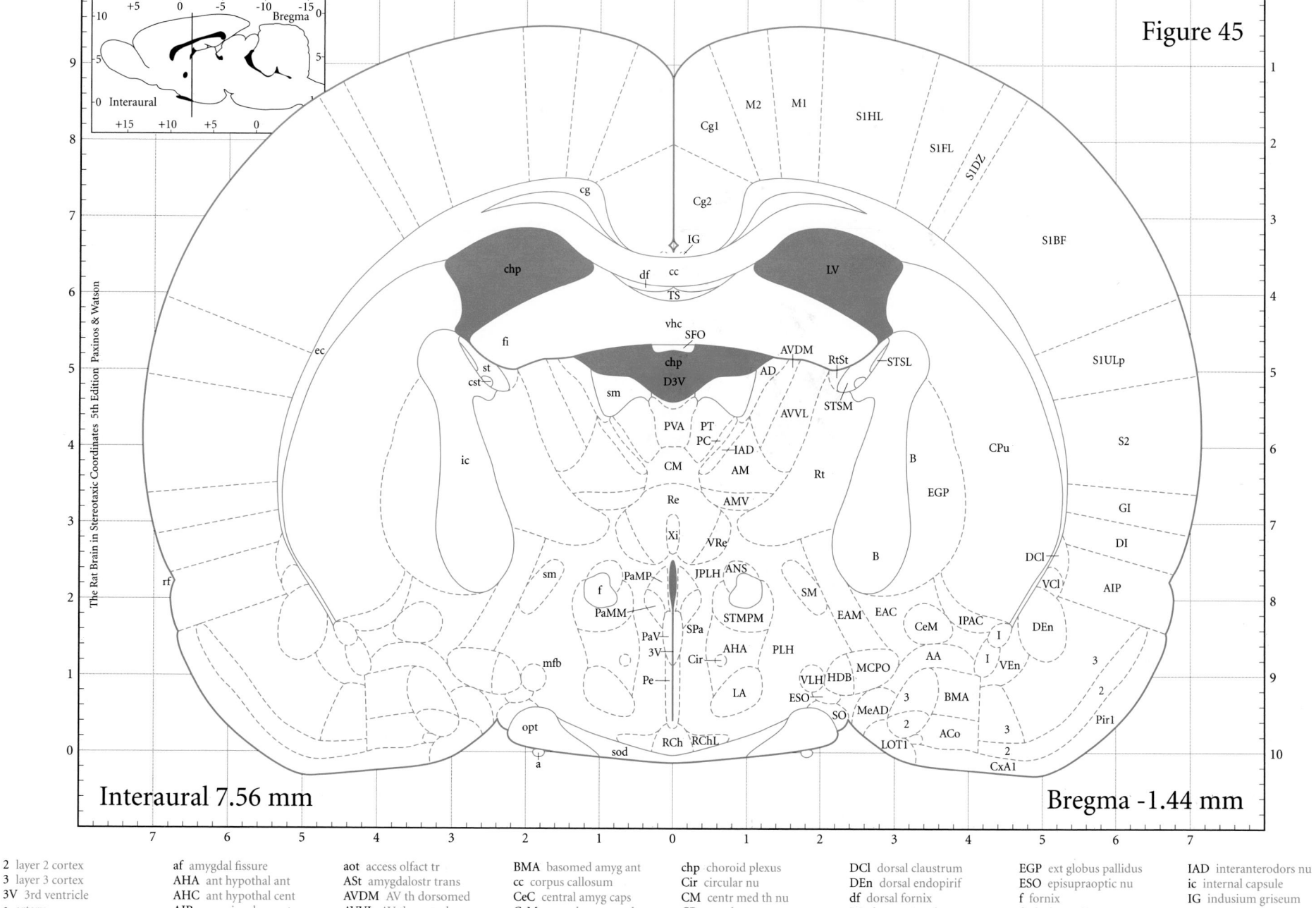

Figure 45

Interaural 7.56 mm

Bregma -1.44 mm

The Rat Brain in Stereotaxic Coordinates 5th Edition Paxinos & Watson

2 layer 2 cortex
3 layer 3 cortex
3V 3rd ventricle
a artery
AA ant amyg area
ACo ant cortical amyg
AD anterodors th nu

af amygdal fissure
AHA ant hypothal ant
AHC ant hypothal cent
AIP agran insular post
AM anteromed th nu
AMV anteromed th vent
ANS acc neurosecret

aot access olfact tr
ASt amygdalostr trans
AVDM AV th dorsomed
AVVL AV th ventrolat
B basal nu
BAOT bed nu acc olf tr
BLA basolat amyg ant

BMA basomed amyg ant
cc corpus callosum
CeC central amyg caps
CeM central amyg med
cg cingulum
Cg1 cingulate area 1
Cg2 cingulate area 2

chp choroid plexus
Cir circular nu
CM centr med th nu
CPu caudate putamen
cst commiss st term
CxA1 cx-amyg trans 1
D3V dorsal 3rd vent

DCl dorsal claustrum
DEn dorsal endopirif
df dorsal fornix
DI dysgran insular
EAC sublentic EA cent
EAM sublentic EA med
ec external capsule

EGP ext globus pallidus
ESO episupraoptic nu
f fornix
fi fimbria of hipp
GI granular insular
HDB nu horiz limb DB
I intercalated nu

IAD interanterodors nu
ic internal capsule
IG indusium griseum
IPAC interstitial nu acp
JPLH juxtaparav lat hy
LA lateroant hy nu
LOT1 LOT layer 1
LV lat ventricle

Figure 46

M1 primary motor cx
M2 2ary motor cx
MCPO magnocell preopt
MeAD med anterodorsal
mfb med forebr bundle
MoDG molecular dent gy
opt optic tract
PaMM Pa med magno
PaMP Pa med parvicell

PaV Pa, ventral part
PaXi paraxiphoid nu
PC paracentral th nu
Pe periventric hy nu
Pir1 piriform layer 1
PLH peduncular lat hy
PT paratenial th nu
PV paraventric th nu
PVA paraventric th ant

RCh retrochiasm area
RChL retrochiasm lat
Re reuniens th nu
rf rhinal fissure
Rt reticular th nu
RtSt reticluostrial nu
S1BF S1 cx, barrel field
S1DZ S1 dysgranular zn
S1FL S1 forelimb region

S1HL S1 hindlimb region
S1Sh S1 shoulder region
S1ULp S1 upper lip region
S2 2ary somatosens
SFO subfornical organ
SM nu stria medull
sm stria medullaris
SO supraoptic nu
sod supraoptic decussn

SPa subparaventric zn
st stria terminalis
STMPM STM posteromed
STSL ST supracaps lat
STSM ST supracaps med
TS triangular septal
TuLH tuberal lat hy
VA ventral ant th nu
VCl ventral claustrum

VEn ventral endopir
vhc ventral hipp comm
VLH ventrolat hy nu
VRe vent reuniens nu
Xi xiphoid th nu

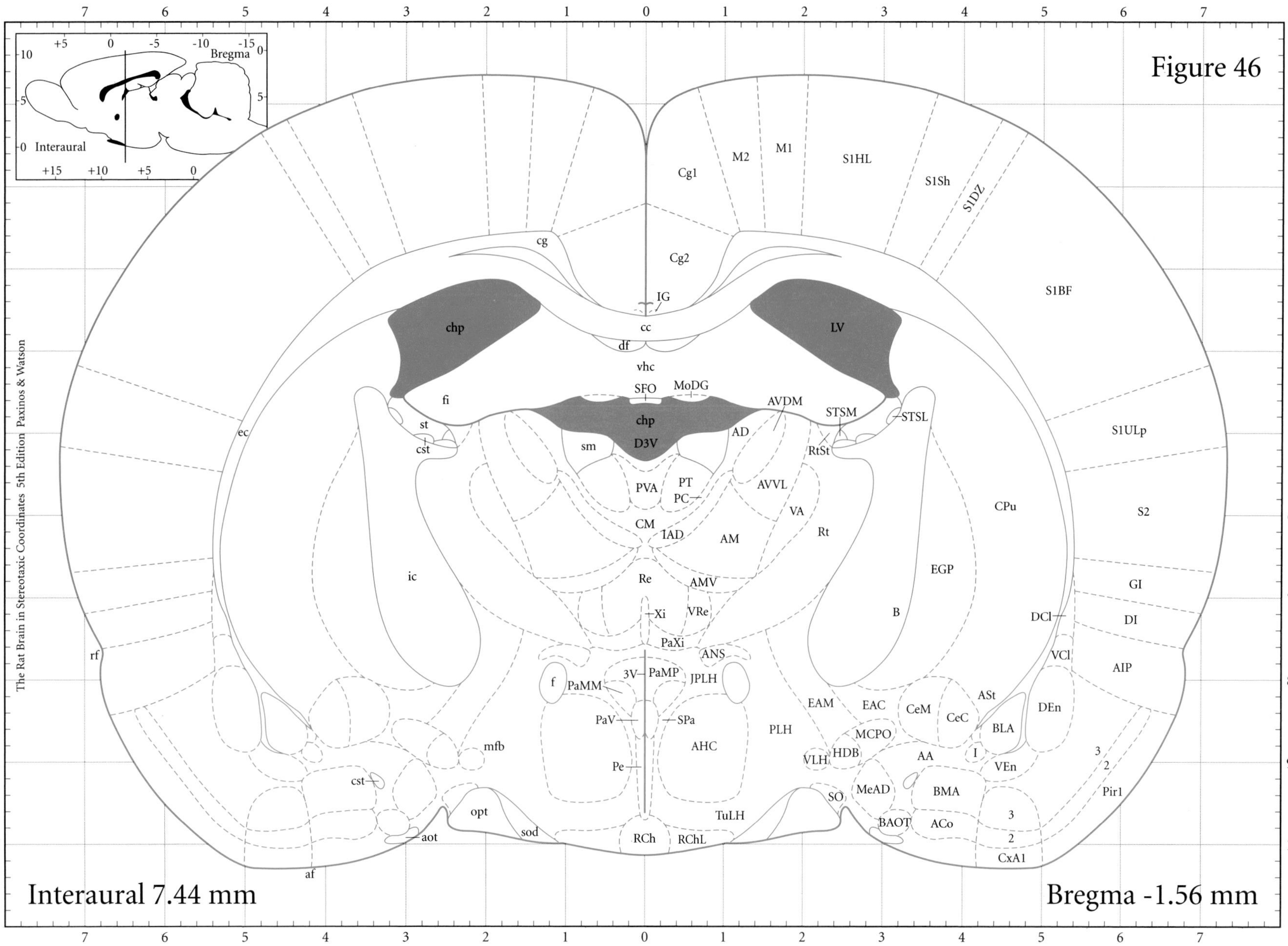

Interaural 7.44 mm

Bregma -1.56 mm

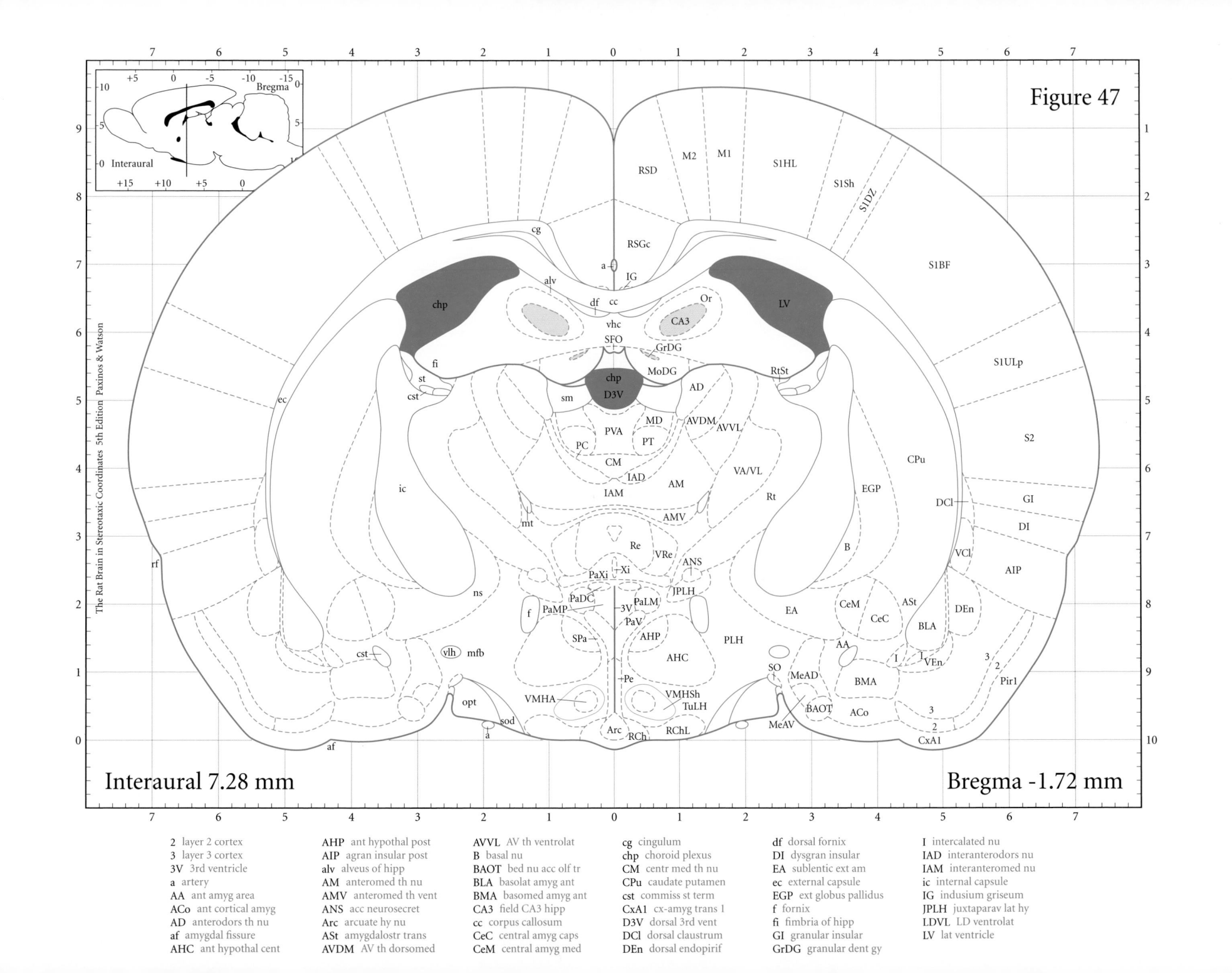

Figure 47

The Rat Brain in Stereotaxic Coordinates 5th Edition Paxinos & Watson

Interaural 7.28 mm

Bregma -1.72 mm

2 layer 2 cortex
3 layer 3 cortex
3V 3rd ventricle
a artery
AA ant amyg area
ACo ant cortical amyg
AD anterodors th nu
af amygdal fissure
AHC ant hypothal cent

AHP ant hypothal post
AIP agran insular post
alv alveus of hipp
AM anteromed th nu
AMV anteromed th vent
ANS acc neurosecret
Arc arcuate hy nu
ASt amygdalostr trans
AVDM AV th dorsomed

AVVL AV th ventrolat
B basal nu
BAOT bed nu acc olf tr
BLA basolat amyg ant
BMA basomed amyg ant
CA3 field CA3 hipp
cc corpus callosum
CeC central amyg caps
CeM central amyg med

cg cingulum
chp choroid plexus
CM centr med th nu
CPu caudate putamen
cst commiss st term
CxA1 cx-amyg trans 1
D3V dorsal 3rd vent
DCl dorsal claustrum
DEn dorsal endopirif

df dorsal fornix
DI dysgran insular
EA sublentic ext am
ec external capsule
EGP ext globus pallidus
f fornix
fi fimbria of hipp
GI granular insular
GrDG granular dent gy

I intercalated nu
IAD interanterodors nu
IAM interanteromed nu
ic internal capsule
IG indusium griseum
JPLH juxtaparav lat hy
LDVL LD ventrolat
LV lat ventricle

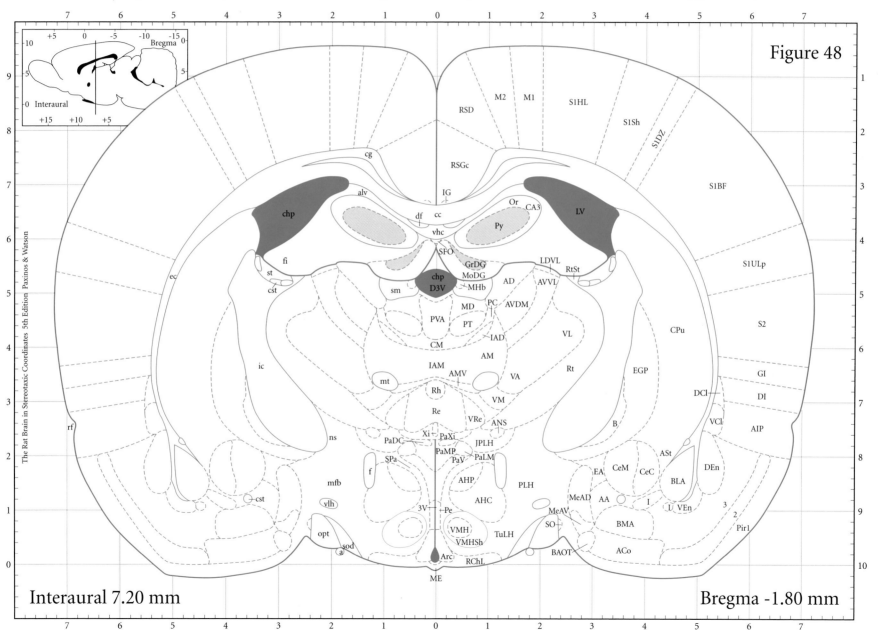

Figure 48

Interaural 7.20 mm

Bregma -1.80 mm

The Rat Brain in Stereotaxic Coordinates 5th Edition Paxinos & Watson

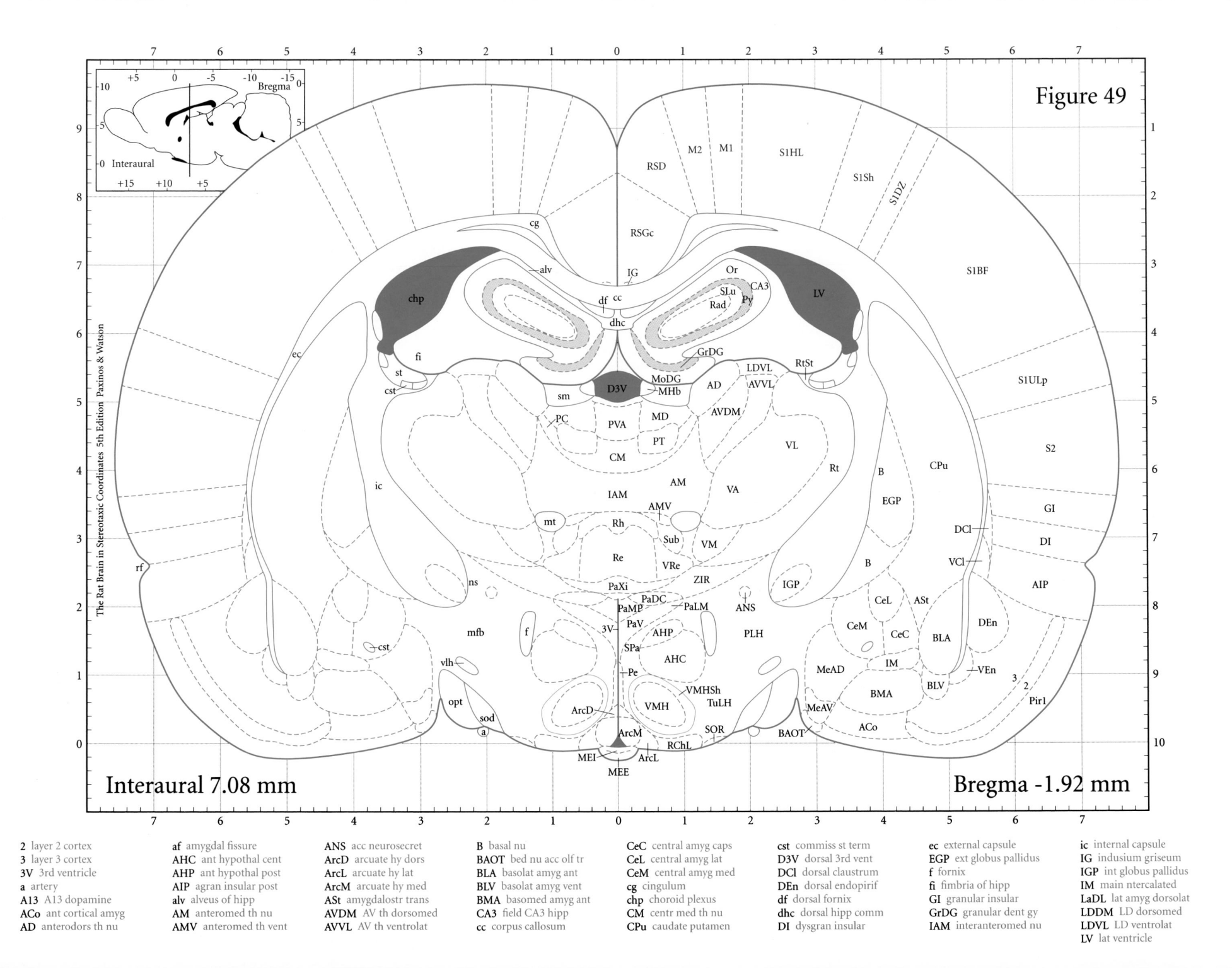

Figure 49

Interaural 7.08 mm

Bregma -1.92 mm

2 layer 2 cortex	**af** amygdal fissure	**ANS** acc neurosecret	**B** basal nu	**CeC** central amyg caps	**cst** commiss st term	**ec** external capsule	**ic** internal capsule
3 layer 3 cortex	**AHC** ant hypothal cent	**ArcD** arcuate hy dors	**BAOT** bed nu acc olf tr	**CeL** central amyg lat	**D3V** dorsal 3rd vent	**EGP** ext globus pallidus	**IG** indusium griseum
3V 3rd ventricle	**AHP** ant hypothal post	**ArcL** arcuate hy lat	**BLA** basolat amyg ant	**CeM** central amyg med	**DCl** dorsal claustrum	**f** fornix	**IGP** int globus pallidus
a artery	**AIP** agran insular post	**ArcM** arcuate hy med	**BLV** basolat amyg vent	**cg** cingulum	**DEn** dorsal endopirif	**fi** fimbria of hipp	**IM** main ntercalated
A13 A13 dopamine	**alv** alveus of hipp	**ASt** amygdalostr trans	**BMA** basomed amyg ant	**chp** choroid plexus	**df** dorsal fornix	**GI** granular insular	**LaDL** lat amyg dorsolat
ACo ant cortical amyg	**AM** anteromed th nu	**AVDM** AV th dorsomed	**CA3** field CA3 hipp	**CM** centr med th nu	**dhc** dorsal hipp comm	**GrDG** granular dent gy	**LDDM** LD dorsomed
AD anterodors th nu	**AMV** anteromed th vent	**AVVL** AV th ventrolat	**cc** corpus callosum	**CPu** caudate putamen	**DI** dysgran insular	**IAM** interanteromed nu	**LDVL** LD ventrolat
							LV lat ventricle

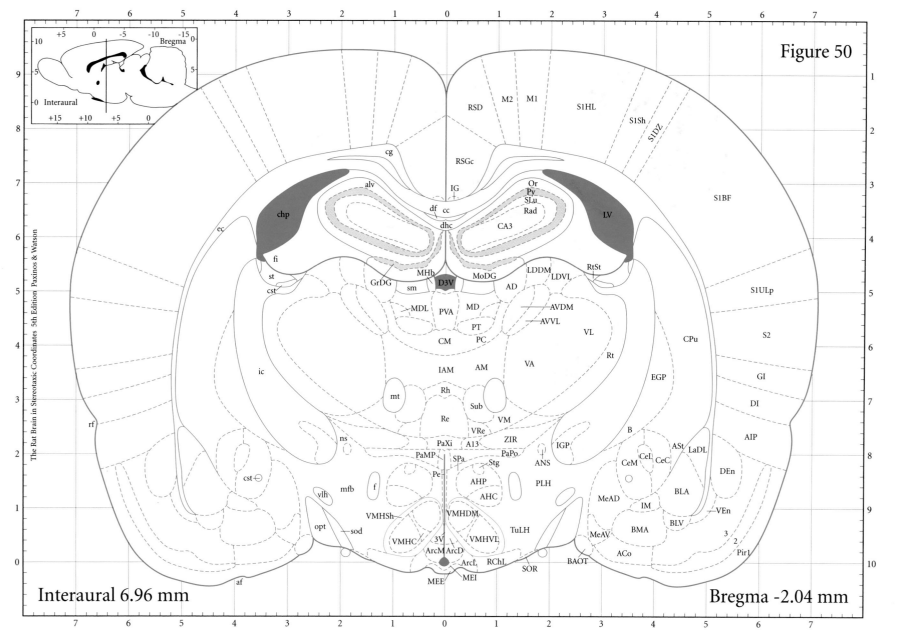

Figure 50

Interaural 6.96 mm

Bregma -2.04 mm

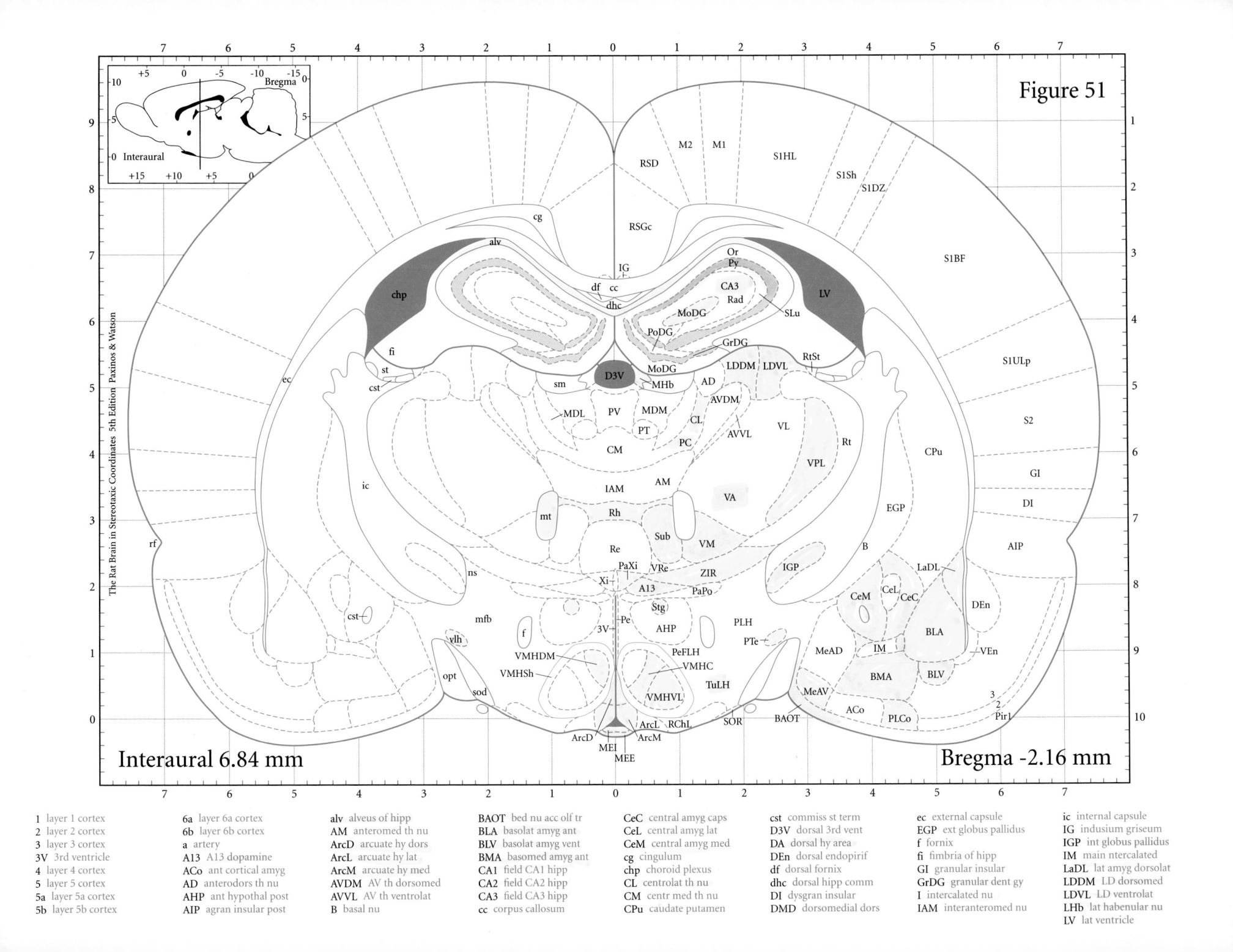

Figure 51

Interaural 6.84 mm

Bregma -2.16 mm

1 layer 1 cortex
2 layer 2 cortex
3 layer 3 cortex
3V 3rd ventricle
4 layer 4 cortex
5 layer 5 cortex
5a layer 5a cortex
5b layer 5b cortex

6a layer 6a cortex
6b layer 6b cortex
a artery
A13 A13 dopamine
ACo ant cortical amyg
AD anterodors th nu
AHP ant hypothal post
AIP agran insular post

alv alveus of hipp
AM anteromed th nu
ArcD arcuate hy dors
ArcL arcuate hy lat
ArcM arcuate hy med
AVDM AV th dorsomed
AVVL AV th ventrolat
B basal nu

BAOT bed nu acc olf tr
BLA basolat amyg ant
BLV basolat amyg vent
BMA basomed amyg ant
CA1 field CA1 hipp
CA2 field CA2 hipp
CA3 field CA3 hipp
cc corpus callosum

CeC central amyg caps
CeL central amyg lat
CeM central amyg med
cg cingulum
chp choroid plexus
CL centrolat th nu
CM centr med th nu
CPu caudate putamen

cst commiss st term
D3V dorsal 3rd vent
DA dorsal hy area
DEn dorsal endopirif
df dorsal fornix
dhc dorsal hipp comm
DI dysgran insular
DMD dorsomedial dors

ec external capsule
EGP ext globus pallidus
f fornix
fi fimbria of hipp
GI granular insular
GrDG granular dent gy
I intercalated nu
IAM interanteromed nu

ic internal capsule
IG indusium griseum
IGP int globus pallidus
IM main ntercalated
LaDL lat amyg dorsolat
LDDM LD dorsomed
LDVL LD ventrolat
LHb lat habenular nu
LV lat ventricle

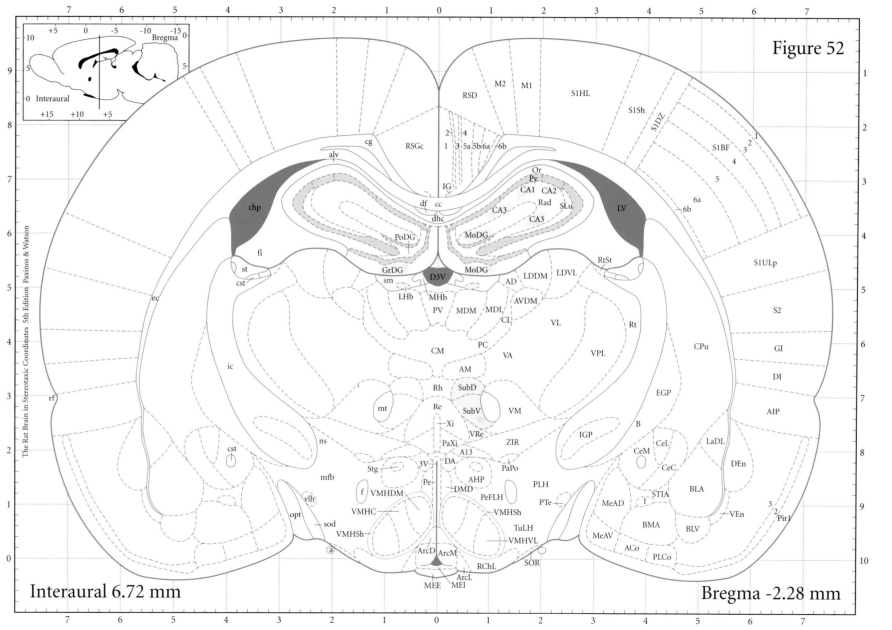

Figure 52

Interaural 6.72 mm

Bregma -2.28 mm

The Rat Brain in Stereotaxic Coordinates 5th Edition Paxinos & Watson

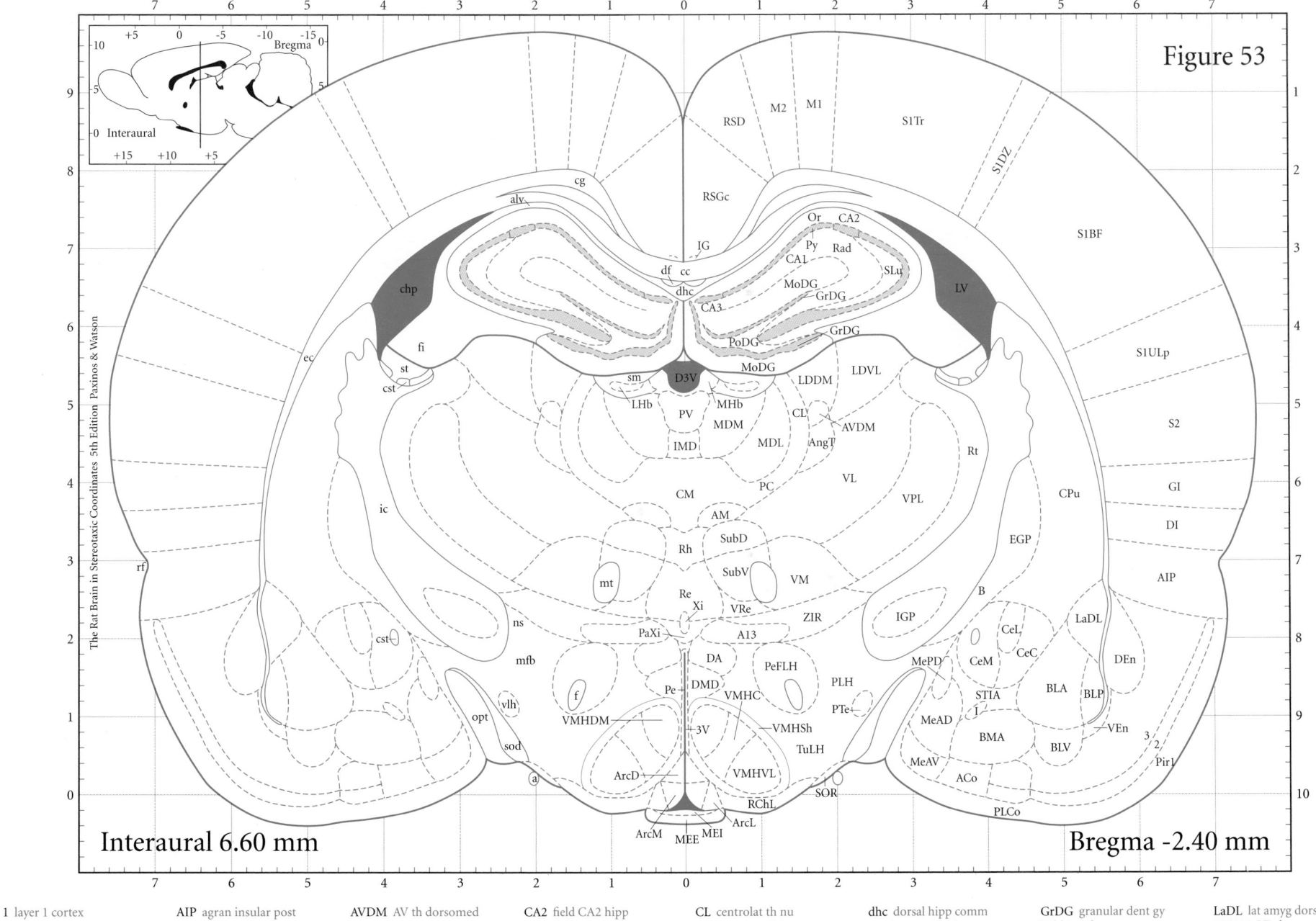

Figure 53

Interaural 6.60 mm

Bregma -2.40 mm

1 layer 1 cortex
2 layer 2 cortex
3 layer 3 cortex
3V 3rd ventricle
A13 A13 dopamine
ACo ant cortical amyg
af amygdal fissure

AIP agran insular post
alv alveus of hipp
AM anteromed th nu
AngT angular th nu
ArcD arcuate hy dors
ArcL arcuate hy lat
ArcM arcuate hy med
ASt amygdalostr trans

AVDM AV th dorsomed
B basal nu
BLA basolat amyg ant
BLP basolat amyg post
BLV basolat amyg vent
BMA basomed amyg ant
BMP basomed am post
CA1 field CA1 hipp

CA2 field CA2 hipp
CA3 field CA3 hipp
cc corpus callosum
CeC central amyg caps
CeL central amyg lat
CeM central amyg med
cg cingulum
chp choroid plexus

CL centrolat th nu
CM centr med th nu
CPu caudate putamen
cst commiss st term
D3V dorsal 3rd vent
DA dorsal hy area
DEn dorsal endopirif
df dorsal fornix

dhc dorsal hipp comm
DI dysgran insular
DMD dorsomedial dors
ec external capsule
EGP ext globus pallidus
f fornix
fi fimbria of hipp
GI granular insular

GrDG granular dent gy
hif hipp fissure
I intercalated nu
ic internal capsule
IG indusium griseum
IGP int globus pallidus
IMD intermediodors nu
IMG intramedull gray

LaDL lat amyg dorsolat
LDDM LD dorsomed
LDVL LD ventrolat
LHb lat habenular nu
LMol lacunosum molec
LV lat ventricle

Figure 54

Interaural 6.48 mm

Bregma -2.52 mm

The Rat Brain in Stereotaxic Coordinates 5th Edition Paxinos & Watson

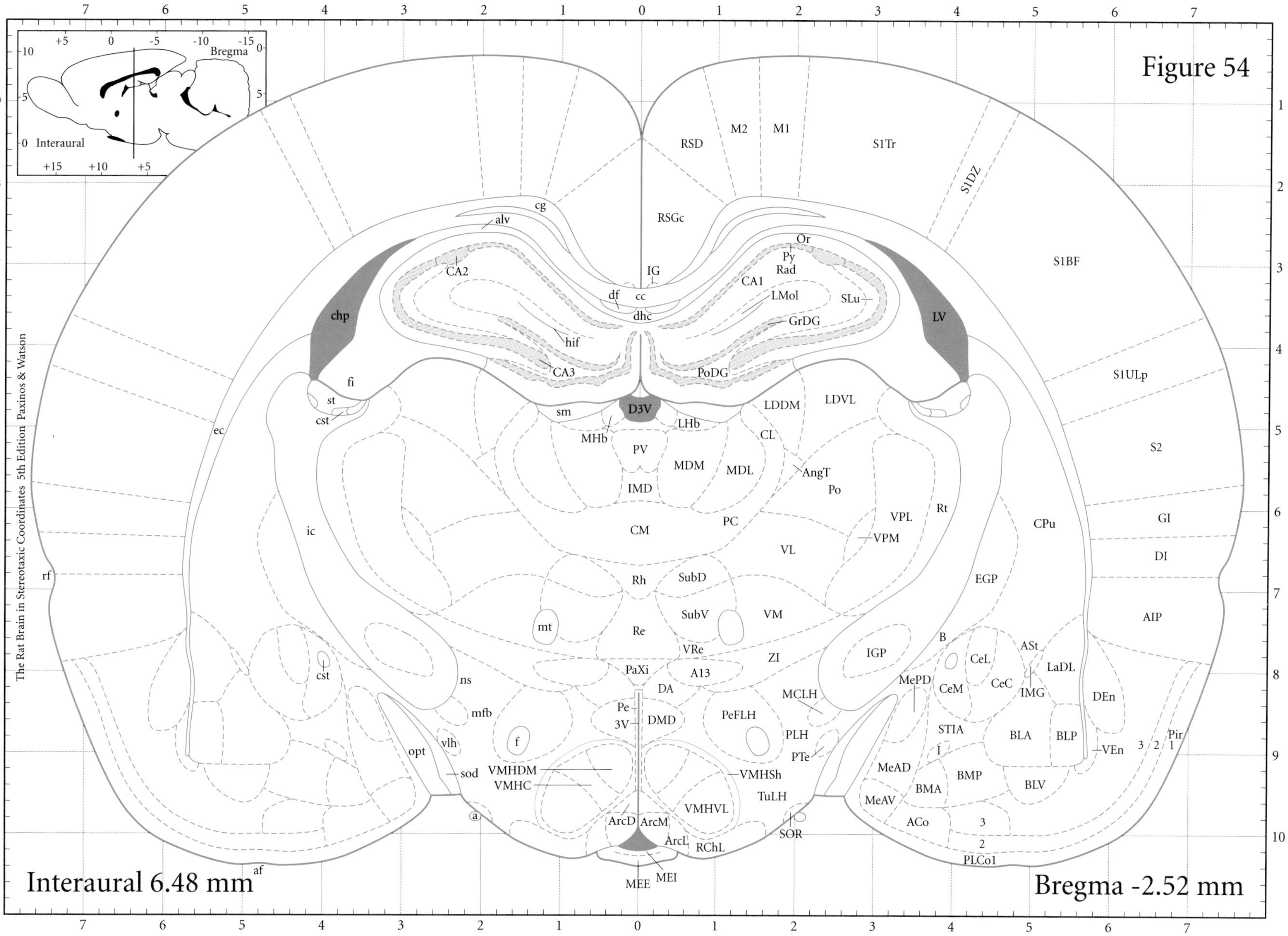

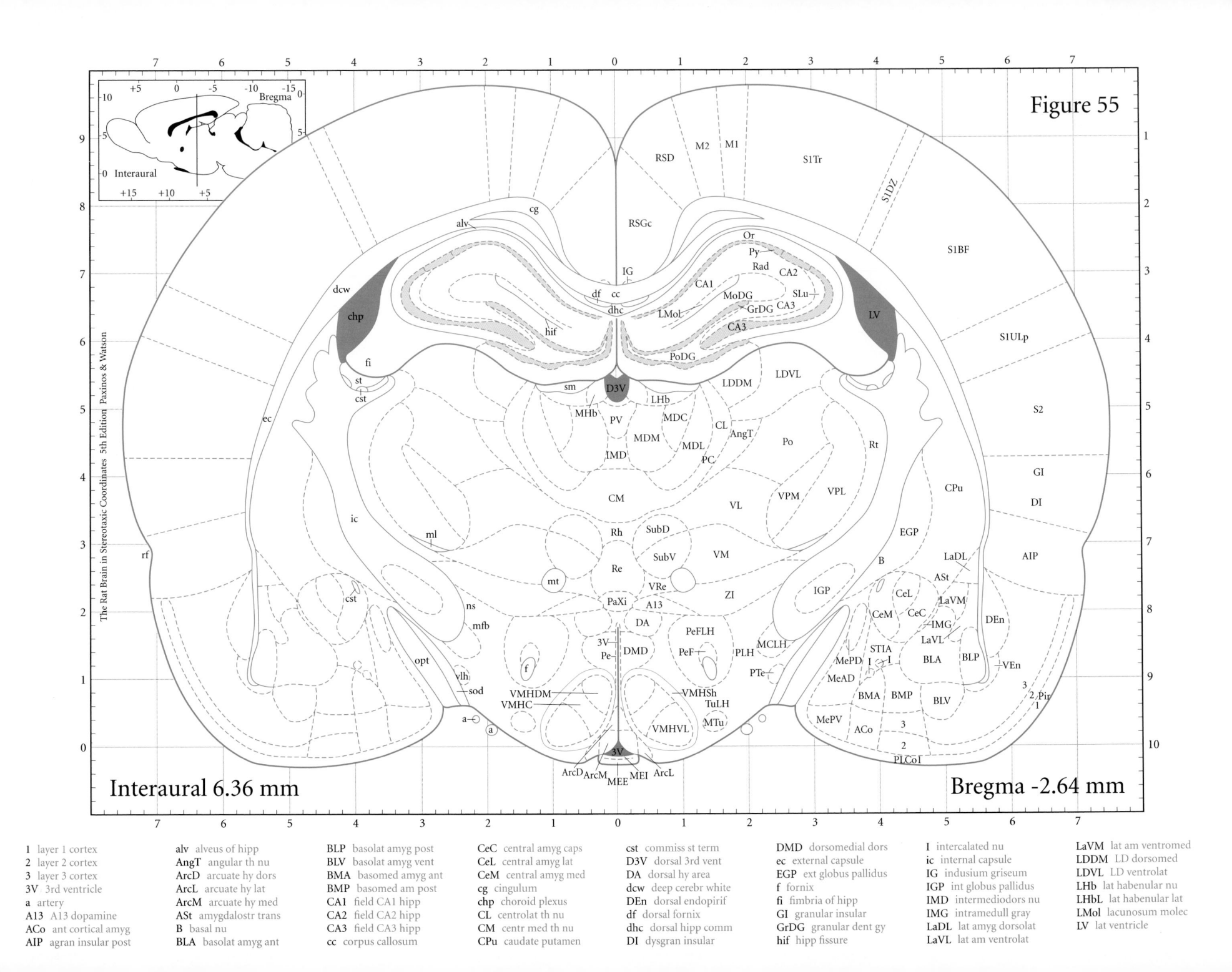

Figure 55

Interaural 6.36 mm

Bregma -2.64 mm

1	layer 1 cortex	alv	alveus of hipp	BLP	basolat amyg post	CeC	central amyg caps	cst	commiss st term	DMD	dorsomedial dors	I	intercalated nu	LaVM	lat am ventromed
2	layer 2 cortex	AngT	angular th nu	BLV	basolat amyg vent	CeL	central amyg lat	D3V	dorsal 3rd vent	ec	external capsule	ic	internal capsule	LDDM	LD dorsomed
3	layer 3 cortex	ArcD	arcuate hy dors	BMA	basomed amyg ant	CeM	central amyg med	DA	dorsal hy area	EGP	ext globus pallidus	IG	indusium griseum	LDVL	LD ventrolat
3V	3rd ventricle	ArcL	arcuate hy lat	BMP	basomed am post	cg	cingulum	dcw	deep cerebr white	f	fornix	IGP	int globus pallidus	LHb	lat habenular nu
a	artery	ArcM	arcuate hy med	CA1	field CA1 hipp	chp	choroid plexus	DEn	dorsal endopirif	fi	fimbria of hipp	IMD	intermediodors nu	LHbL	lat habenular lat
A13	A13 dopamine	ASt	amygdalostr trans	CA2	field CA2 hipp	CL	centrolat th nu	df	dorsal fornix	GI	granular insular	IMG	intramedull gray	LMol	lacunosum molec
ACo	ant cortical amyg	B	basal nu	CA3	field CA3 hipp	CM	centr med th nu	dhc	dorsal hipp comm	GrDG	granular dent gy	LaDL	lat amyg dorsolat	LV	lat ventricle
AIP	agran insular post	BLA	basolat amyg ant	cc	corpus callosum	CPu	caudate putamen	DI	dysgran insular	hif	hipp fissure	LaVL	lat am ventrolat		

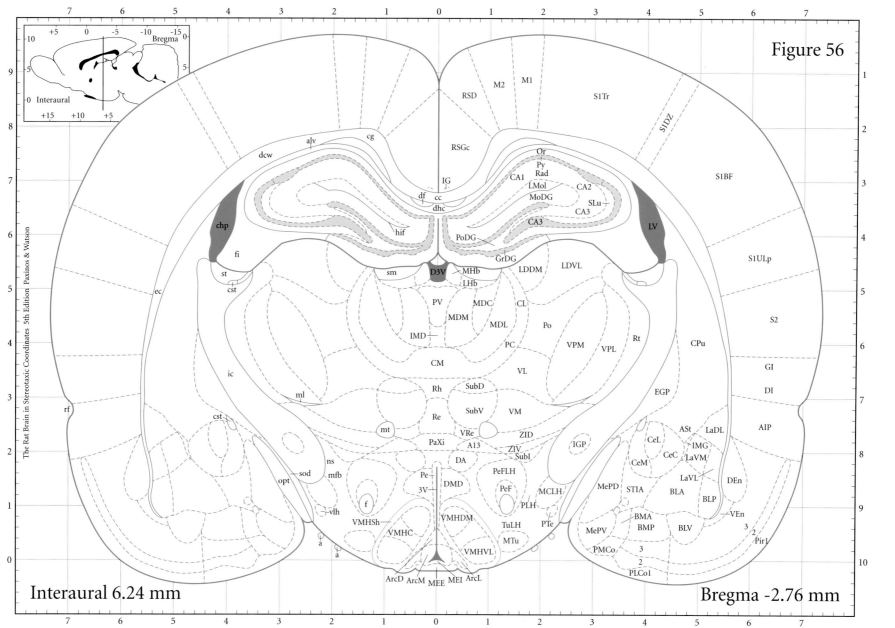

Figure 56

Interaural 6.24 mm

Bregma -2.76 mm

M1	primary motor cx
M2	2ary motor cx
MCLH	magnocell lat hy
MDC	mediodorsal cent
MDL	mediodorsal lat
MDM	mediodorsal med
MeAD	med anterodorsal
MEE	med eminence ext
MEI	med eminence int
MePD	med posterodors
MePV	med posterovent
mfb	med forebr bundle
MHb	med habenular nu
ml	medial lemniscus
MoDG	molecular dent gy
mt	mammillothal tr
MTu	medial tuberal nu
ns	nigrostriat bundle
opt	optic tract
Or	oriens layer hipp
PaXi	paraxiphoid nu
PC	paracentral th nu
Pe	periventric hy nu
PeF	periformical nu
PeFLH	periformical lat hy
Pir	piriform cx
Pir1	piriform layer 1
PLCo1	postlat cort am 1
PLH	peduncular lat hy
PMCo	postmed cort am
Po	post thalamic nu
PoDG	polymorph dent gy
PTe	paraterete nu
PV	paraventric th nu
Py	pyramidal cells
Rad	radiatum layer
Re	reuniens th nu
rf	rhinal fissure
Rh	rhomboid thal nu
RSD	retrosple dysgran
RSGc	RSG, c region
Rt	reticular th nu
S1BF	S1 cx, barrel field
S1DZ	S1 dysgranular zn
S1Tr	S1 trunk region
S1ULp	S1 upper lip region
S2	2ary somatosens
SLu	stratum lucidum
sm	stria medullaris
sod	supraoptic decussn
st	stria terminalis
STIA	ST intraamygdal
SubD	submedius dors
SubI	subincertal nu
SubV	submedius vent
TuLH	tuberal lat hy
VEn	ventral endopir
VL	ventrolat th nu
vlh	vent lat hy tr
VM	ventromed nu
VMHC	VMH, central
VMHDM	VMH, dorsomed
VMHSh	VMH shell
VMHVL	VMH, ventrolat
VPL	vent posterolat nu
VPM	vent posteromed
VRe	vent reuniens nu
ZI	zona incerta
ZID	zona incerta does
ZIV	zona incerta vent

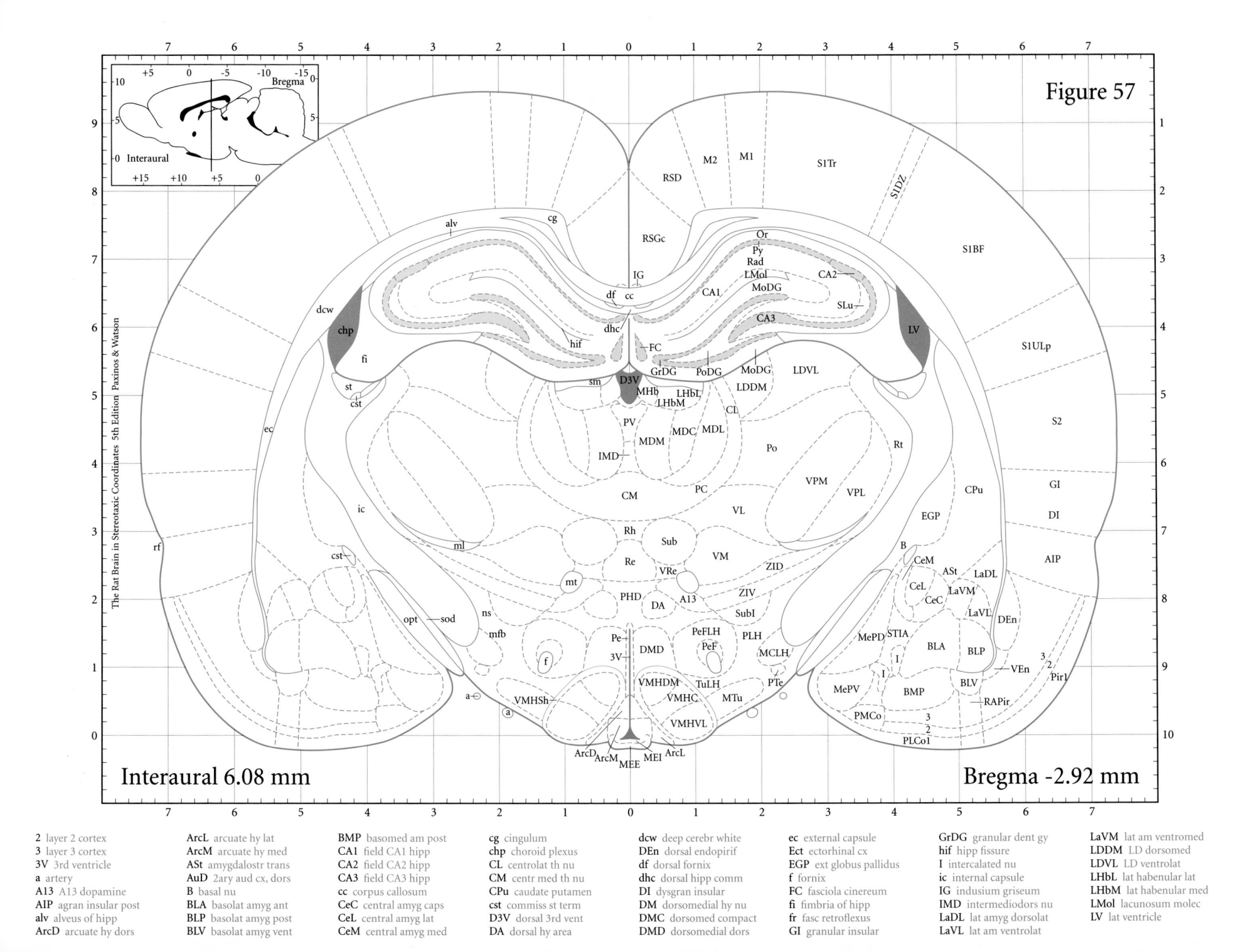

Figure 57

Interaural 6.08 mm

Bregma -2.92 mm

The Rat Brain in Stereotaxic Coordinates 5th Edition Paxinos & Watson

2 layer 2 cortex
3 layer 3 cortex
3V 3rd ventricle
a artery
A13 A13 dopamine
AIP agran insular post
alv alveus of hipp
ArcD arcuate hy dors

ArcL arcuate hy lat
ArcM arcuate hy med
ASt amygdalostr trans
AuD 2ary aud cx, dors
B basal nu
BLA basolat amyg ant
BLP basolat amyg post
BLV basolat amyg vent

BMP basomed am post
CA1 field CA1 hipp
CA2 field CA2 hipp
CA3 field CA3 hipp
cc corpus callosum
CeC central amyg caps
CeL central amyg lat
CeM central amyg med

cg cingulum
chp choroid plexus
CL centrolat th nu
CM centr med th nu
CPu caudate putamen
cst commiss st term
D3V dorsal 3rd vent
DA dorsal hy area

dcw deep cerebr white
DEn dorsal endopirif
df dorsal fornix
dhc dorsal hipp comm
DI dysgran insular
DM dorsomedial hy nu
DMC dorsomed compact
DMD dorsomedial dors

ec external capsule
Ect ectorhinal cx
EGP ext globus pallidus
f fornix
FC fasciola cinereum
fi fimbria of hipp
fr fasc retroflexus
GI granular insular

GrDG granular dent gy
hif hipp fissure
I intercalated nu
ic internal capsule
IG indusium griseum
IMD intermediodors nu
LaDL lat amyg dorsolat
LaVL lat am ventrolat

LaVM lat am ventromed
LDDM LD dorsomed
LDVL LD ventrolat
LHbL lat habenular lat
LHbM lat habenular med
LMol lacunosum molec
LV lat ventricle

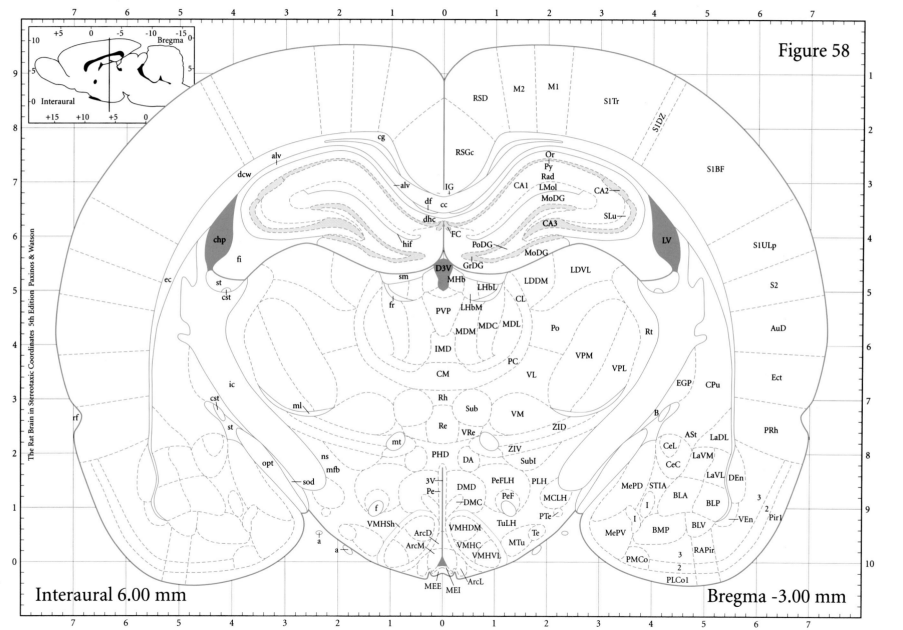

Figure 58

The Rat Brain in Stereotaxic Coordinates 5th Edition Paxinos & Watson

Interaural 6.00 mm

Bregma -3.00 mm

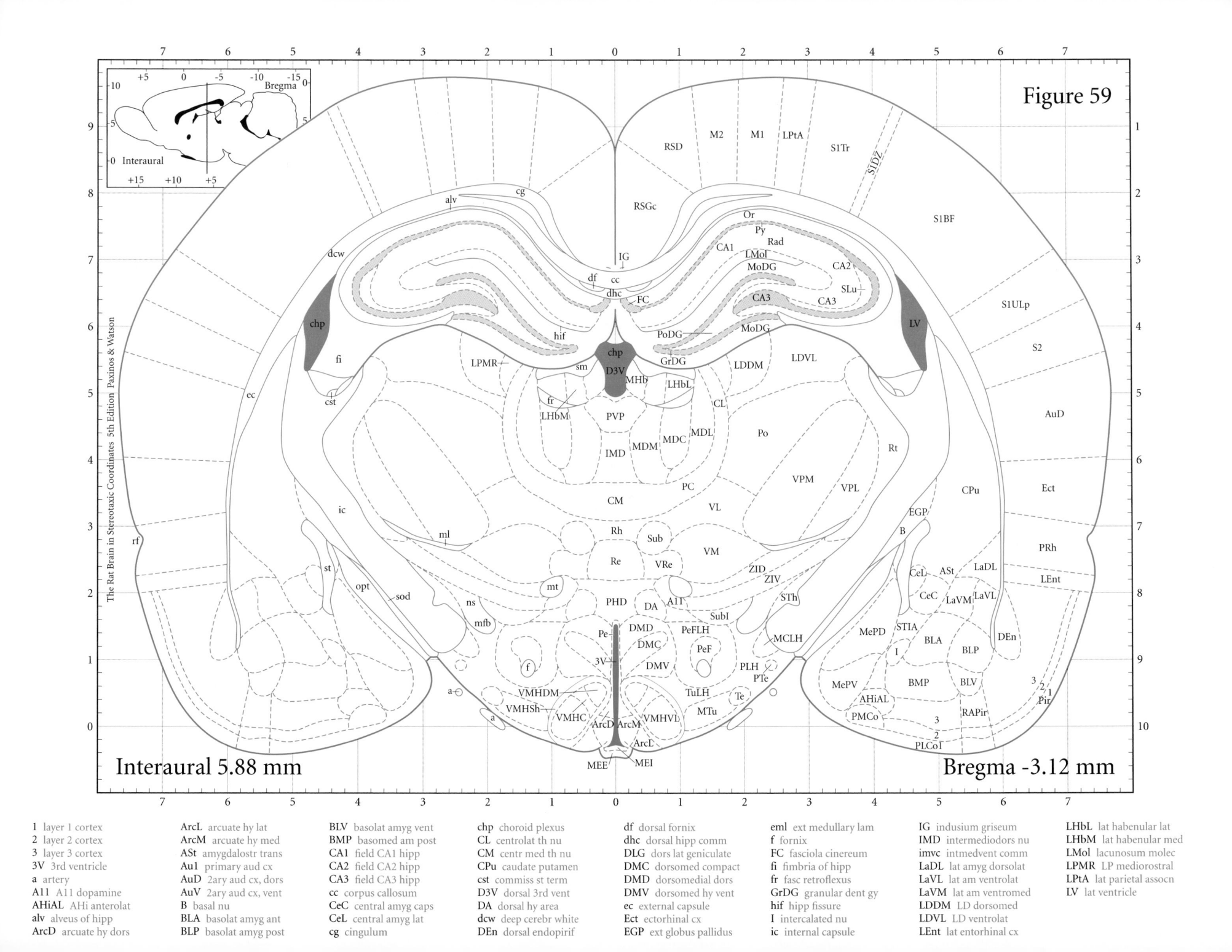

Figure 59

Interaural 5.88 mm

Bregma -3.12 mm

1 layer 1 cortex
2 layer 2 cortex
3 layer 3 cortex
3V 3rd ventricle
a artery
A11 A11 dopamine
AHiAL AHi anterolat
alv alveus of hipp
ArcD arcuate hy dors

ArcL arcuate hy lat
ArcM arcuate hy med
ASt amygdalostr trans
Au1 primary aud cx
AuD 2ary aud cx, dors
AuV 2ary aud cx, vent
B basal nu
BLA basolat amyg ant
BLP basolat amyg post

BLV basolat amyg vent
BMP basomed am post
CL centrolat th nu
CM centr med th nu
CPu caudate putamen
CA1 field CA1 hipp
CA2 field CA2 hipp
CA3 field CA3 hipp
cc corpus callosum
CeC central amyg caps
CeL central amyg lat
cg cingulum

chp choroid plexus
CL centrolat th nu
CM centr med th nu
CPu caudate putamen
cst commiss st term
D3V dorsal 3rd vent
DA dorsal hy area
dcw deep cerebr white
DEn dorsal endopirif

df dorsal fornix
dhc dorsal hipp comm
DLG dors lat geniculate
DMC dorsomed compact
DMD dorsomedial dors
DMV dorsomed hy vent
ec external capsule
Ect ectorhinal cx
EGP ext globus pallidus

eml ext medullary lam
f fornix
FC fasciola cinereum
fi fimbria of hipp
fr fasc retroflexus
GrDG granular dent gy
hif hipp fissure
I intercalated nu
ic internal capsule

IG indusium griseum
IMD intermediodors nu
imvc intmedvent comm
LaDL lat amyg dorsolat
LaVL lat am ventrolat
LaVM lat am ventromed
LDDM LD dorsomed
LDVL LD ventrolat
LEnt lat entorhinal cx

LHbL lat habenular lat
LHbM lat habenular med
LMol lacunosum molec
LPMR LP mediorostral
LPtA lat parietal assocn
LV lat ventricle

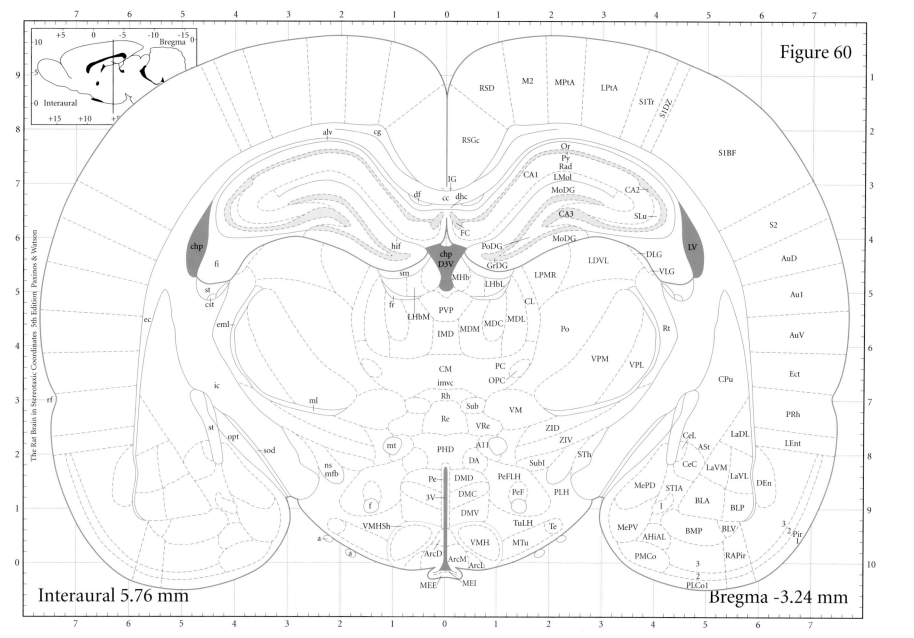

M1 primary motor cx
M2 2ary motor cx
MCLH magnocell lat hy
MDC mediodorsal cent
MDL mediodorsal lat
MDM mediodorsal med
MEE med eminence ext
MEI med eminence int
MePD med posterodors

MePV med posterovent
mfb med forebr bundle
MHb med habenular nu
ml medial lemniscus
MoDG molecular dent gy
MPtA med parietal assn
mt mammillothal tr
MTu medial tuberal nu
ns nigrostriat bundle

OPC oval paracent nu
opt optic tract
Or oriens layer hipp
PC paracentric th nu
Pe periventric hy nu
PeF periformical nu
PeFLH periformical lat hy
PHD post hy area dors
Pir piriform cx

PLCo1 postlat cort am 1
PLH peduncular lat hy
PMCo postmed cort am
Po post thalamic nu
PoDG polymorph dent gy
PRh perirhinal cx
PTe paraterete nu
PVP paraventr th post
Py pyramidal cells

Rad radiatum layer
RAPir rost amygdalopir
Re reuniens th nu
rf rhinal fissure
Rh rhomboid thal nu
RSD retrosple dysgran
RSGc RSG, c region
Rt reticular th nu
S1BF S1 cx, barrel field

S1DZ S1 dysgranular zn
S1Tr S1 trunk region
S1ULp S1 upper lip region
S2 2ary somatosens
SLu stratum lucidum
sm stria medullaris
sod supraoptic decussn
st stria terminalis
STh subthalamic nu

STIA ST intraamygdal
Sub submedius th nu
SubI subincertal nu
Te terete hypothal nu
TuLH tuberal lat hy
VL ventrolat th nu
VLG vent lat genic nu
VM ventromed nu
VMH ventromed hy nu

VMHC VMH, central
VMHDM VMH, dorsomed
VMHSh VMH shell
VMHVL VMH, ventrolat
VPL vent posterolat nu
VPM vent posteromed
VRe vent reuniens nu
ZID zona incerta does
ZIV zona incerta vent

Figure 60

The Rat Brain in Stereotaxic Coordinates 5th Edition Paxinos & Watson

Interaural 5.76 mm

Bregma -3.24 mm

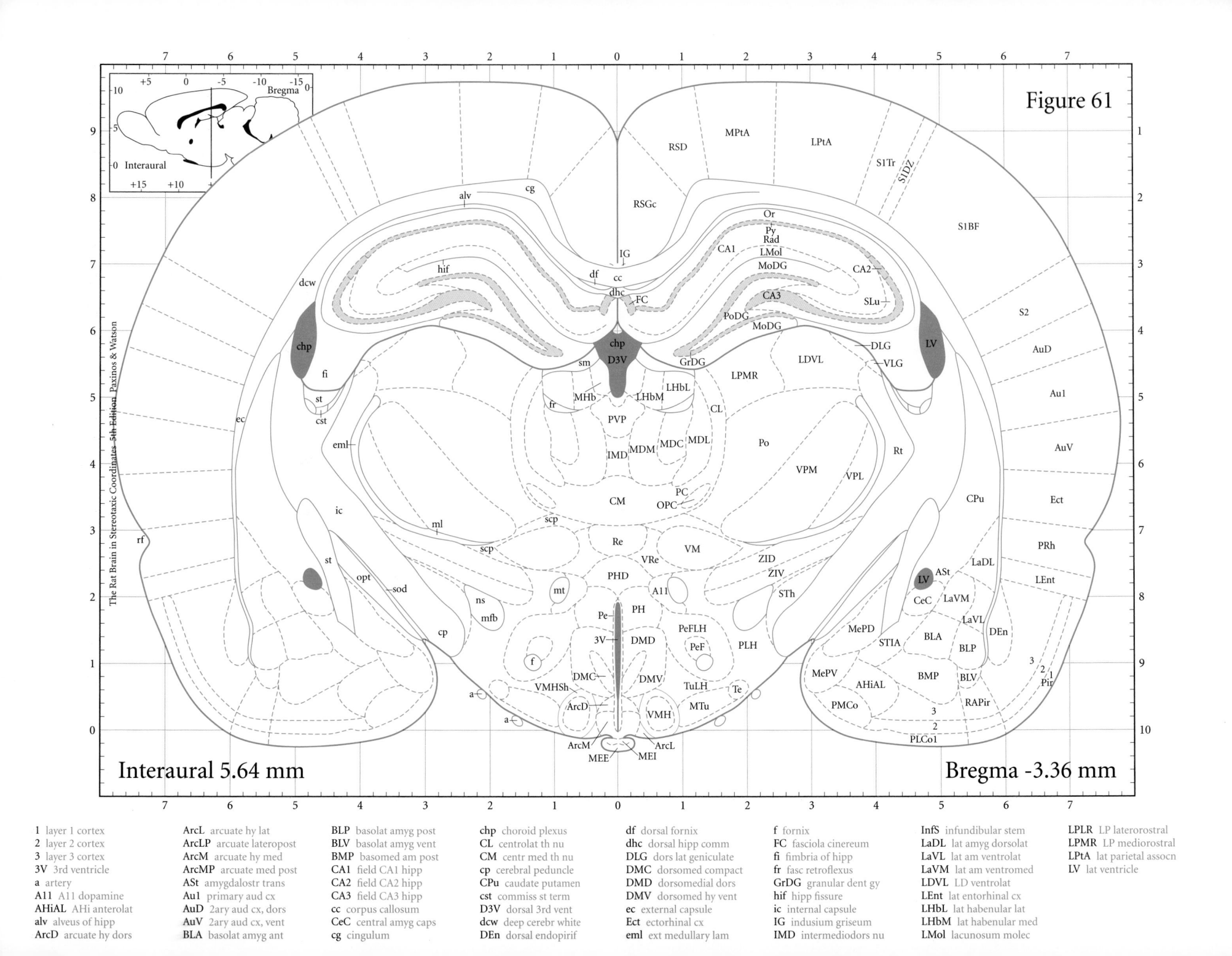

Figure 61

Interaural 5.64 mm

Bregma -3.36 mm

The Rat Brain in Stereotaxic Coordinates, 5th Edition, Paxinos & Watson

1 layer 1 cortex
2 layer 2 cortex
3 layer 3 cortex
3V 3rd ventricle
a artery
A11 A11 dopamine
AHiAL AHi anterolat
alv alveus of hipp
ArcD arcuate hy dors

ArcL arcuate hy lat
ArcLP arcuate lateropost
ArcM arcuate hy med
ArcMP arcuate med post
ASt amygdalostr trans
Au1 primary aud cx
AuD 2ary aud cx, dors
AuV 2ary aud cx, vent
BLA basolat amyg ant

BLP basolat amyg post
BLV basolat amyg vent
BMP basomed am post
CA1 field CA1 hipp
CA2 field CA2 hipp
CA3 field CA3 hipp
cc corpus callosum
CeC central amyg caps
cg cingulum

chp choroid plexus
CL centrolat th nu
CM centr med th nu
cp cerebral peduncle
CPu caudate putamen
cst commiss st term
D3V dorsal 3rd vent
dcw deep cerebr white
DEn dorsal endopirif

df dorsal fornix
dhc dorsal hipp comm
DLG dors lat geniculate
DMC dorsomed compact
DMD dorsomedial dors
DMV dorsomed hy vent
ec external capsule
Ect ectorhinal cx
eml ext medullary lam

f fornix
FC fasciola cinereum
fi fimbria of hipp
fr fasc retroflexus
GrDG granular dent gy
hif hipp fissure
ic internal capsule
IG indusium griseum
IMD intermediodors nu

InfS infundibular stem
LaDL lat amyg dorsolat
LaVL lat am ventrolat
LaVM lat am ventromed
LDVL LD ventrolat
LEnt lat entorhinal cx
LHbL lat habenular lat
LHbM lat habenular med
LMol lacunosum molec

LPLR LP laterorostral
LPMR LP mediorostral
LPtA lat parietal assocn
LV lat ventricle

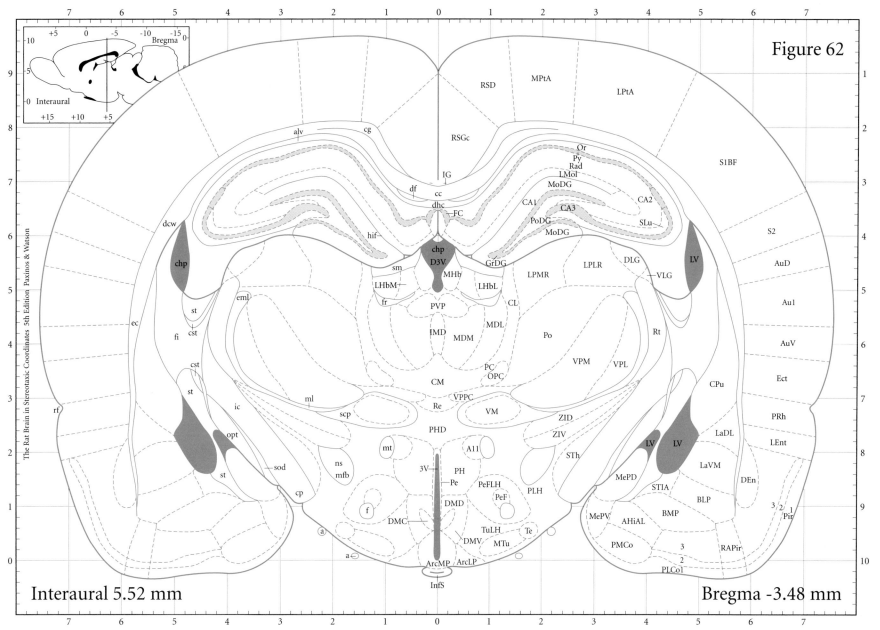

Figure 62

The Rat Brain in Stereotaxic Coordinates 5th Edition Paxinos & Watson

Interaural 5.52 mm

Bregma -3.48 mm

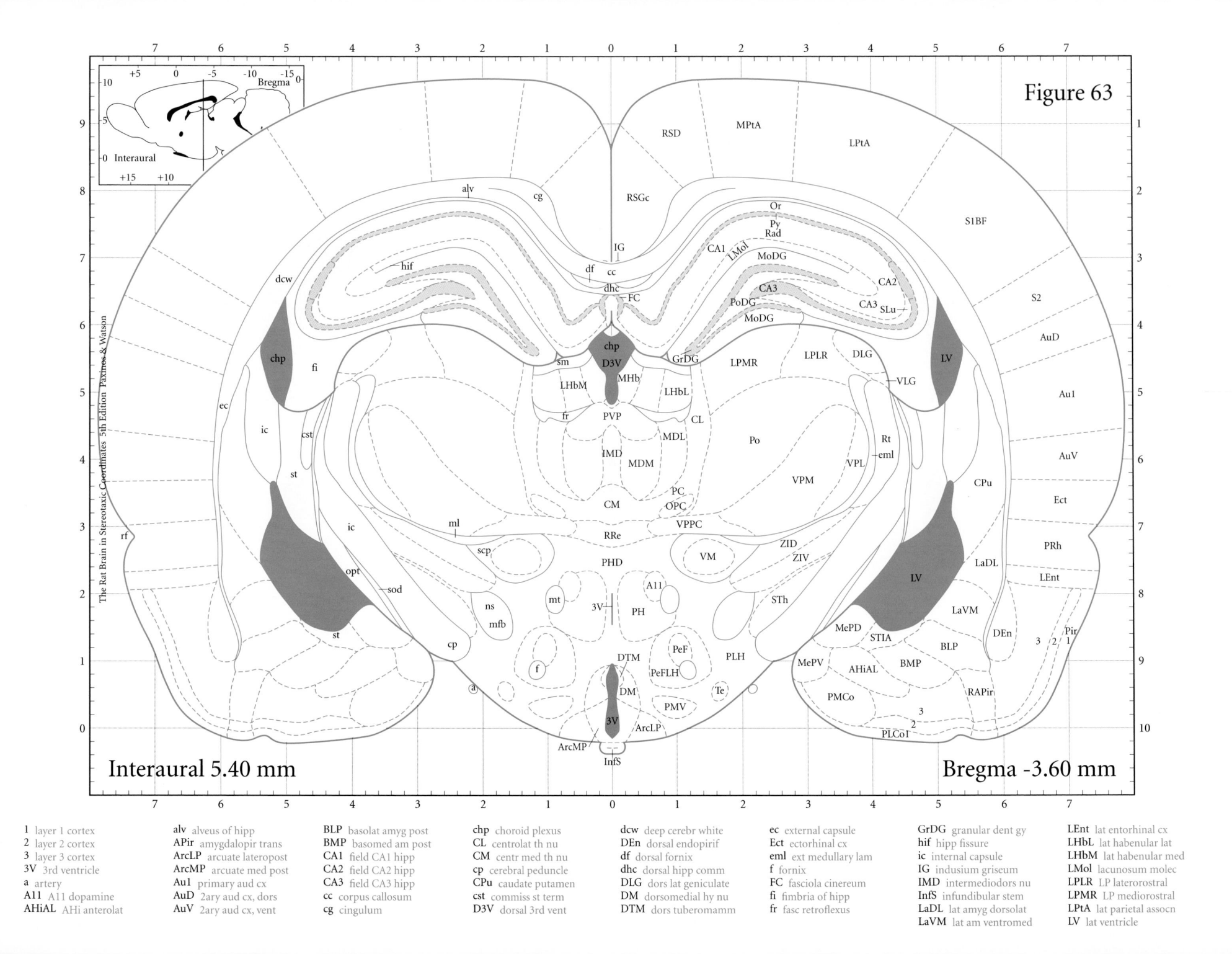

Figure 63

Interaural 5.40 mm

Bregma -3.60 mm

The Rat Brain in Stereotaxic Coordinates 5th Edition Paxinos & Watson

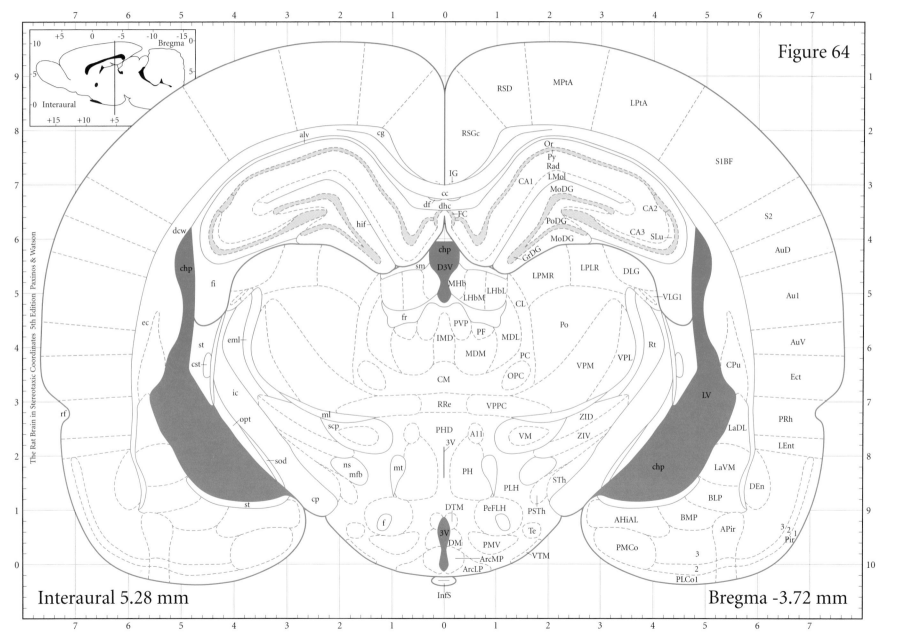

Figure 64

Interaural 5.28 mm

Bregma -3.72 mm

The Rat Brain in Stereotaxic Coordinates 5th Edition Paxinos & Watson

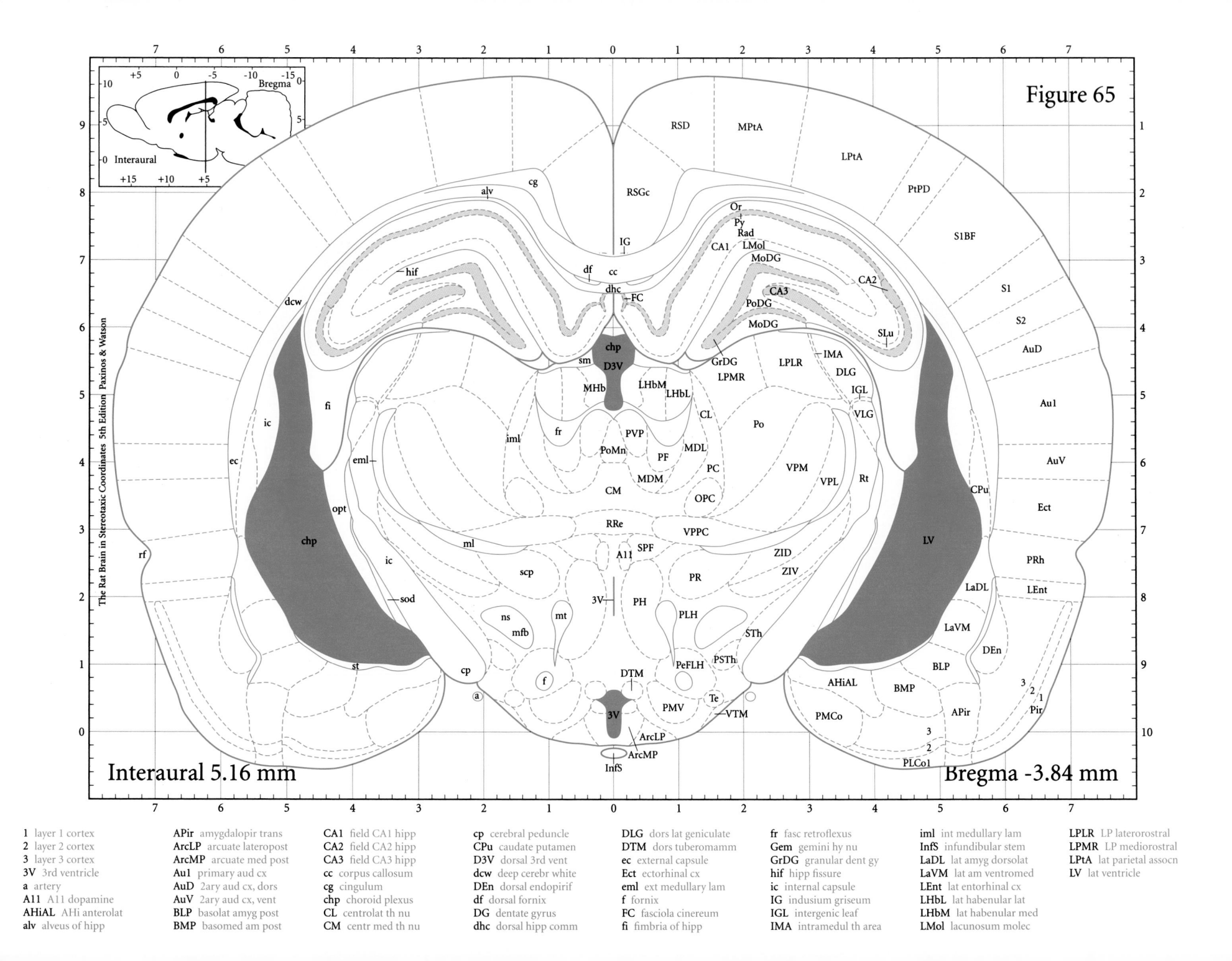

Figure 65

Interaural 5.16 mm

Bregma -3.84 mm

1 layer 1 cortex
2 layer 2 cortex
3 layer 3 cortex
3V 3rd ventricle
a artery
A11 A11 dopamine
AHiAL AHi anterolat
alv alveus of hipp

APir amygdalopir trans
ArcLP arcuate lateropost
ArcMP arcuate med post
Au1 primary aud cx
AuD 2ary aud cx, dors
AuV 2ary aud cx, vent
BLP basolat amyg post
BMP basomed am post

CA1 field CA1 hipp
CA2 field CA2 hipp
CA3 field CA3 hipp
cc corpus callosum
cg cingulum
chp choroid plexus
CL centrolat th nu
CM centr med th nu

cp cerebral peduncle
CPu caudate putamen
D3V dorsal 3rd vent
dcw deep cerebr white
df dorsal fornix
DEn dorsal endopirif
DG dentate gyrus
dhc dorsal hipp comm

DLG dors lat geniculate
DTM dors tuberomamm
ec external capsule
Ect ectorhinal cx
eml ext medullary lam
f fornix
FC fasciola cinereum
fi fimbria of hipp

fr fasc retroflexus
Gem gemini hy nu
GrDG granular dent gy
hif hipp fissure
ic internal capsule
IG indusium griseum
IGL intergenic leaf
IMA intramedul th area

iml int medullary lam
InfS infundibular stem
LaDL lat amyg dorsolat
LaVM lat am ventromed
LEnt lat entorhinal cx
LHbL lat habenular lat
LHbM lat habenular med
LMol lacunosum molec

LPLR LP laterorostral
LPMR LP mediorostral
LPtA lat parietal assocn
LV lat ventricle

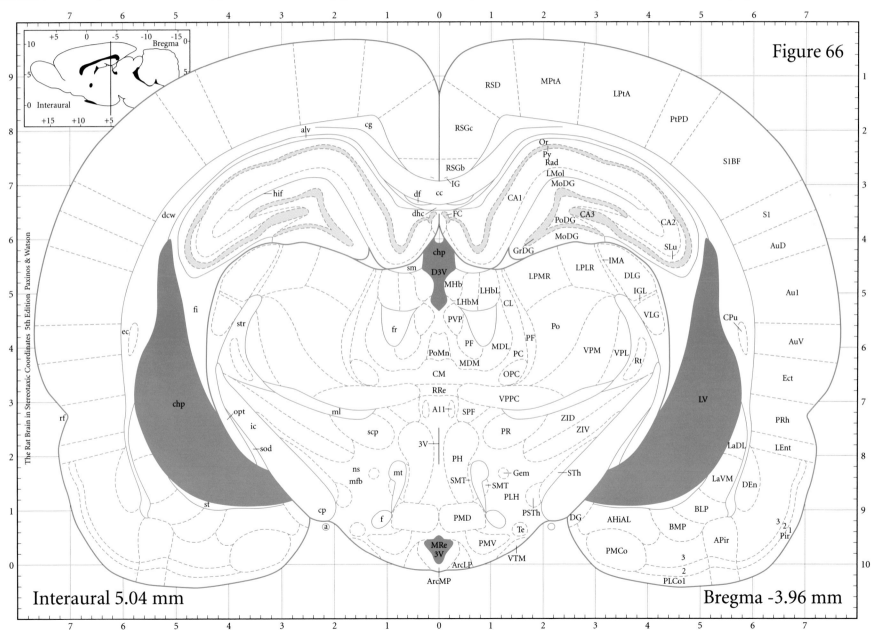

Figure 66

Interaural 5.04 mm

Bregma -3.96 mm

The Rat Brain in Stereotaxic Coordinates 5th Edition Paxinos & Watson

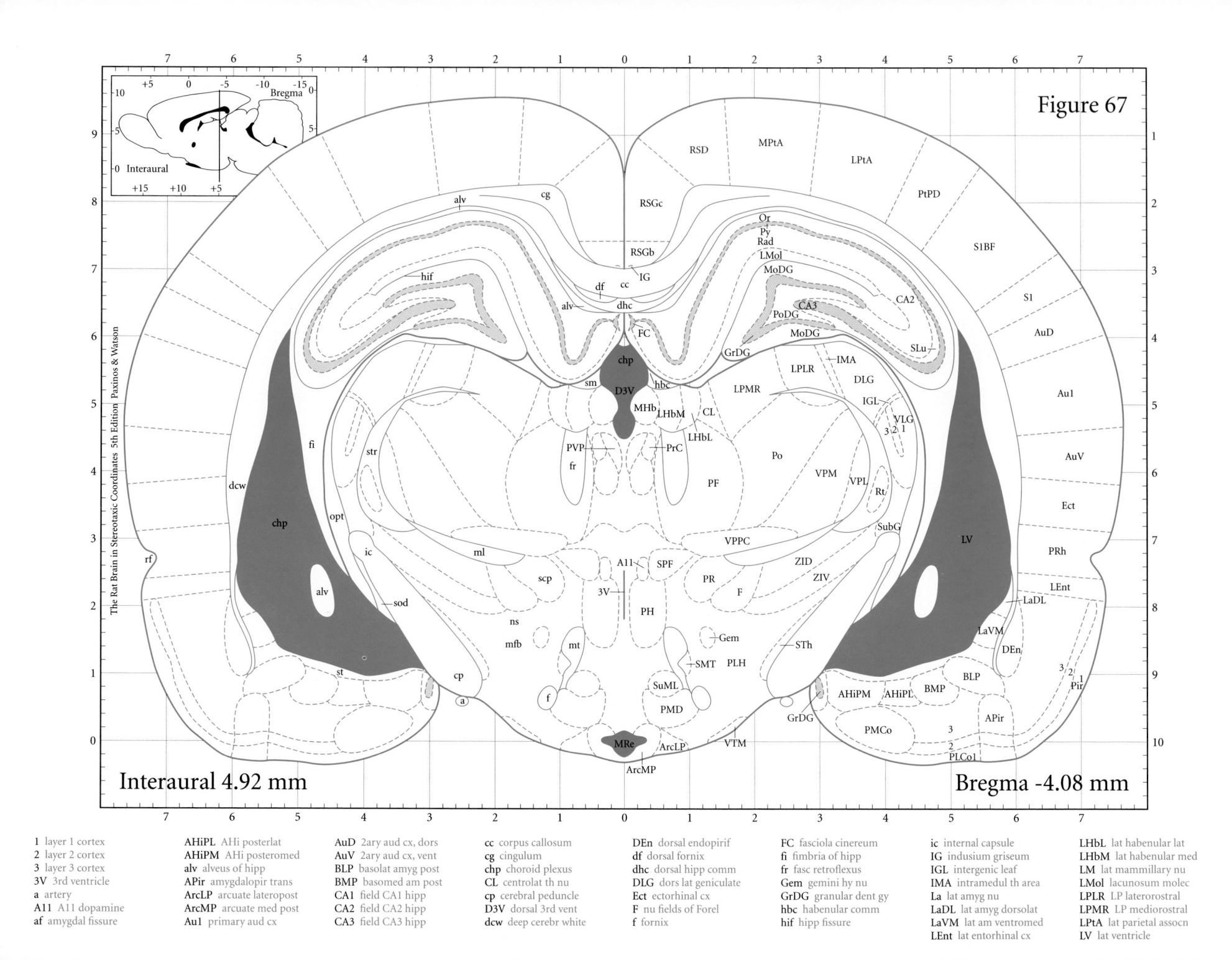

Figure 67

Interaural 4.92 mm

Bregma -4.08 mm

The Rat Brain in Stereotaxic Coordinates 5th Edition Paxinos & Watson

1 layer 1 cortex
2 layer 2 cortex
3 layer 3 cortex
3V 3rd ventricle
a artery
A11 A11 dopamine
af amygdal fissure

AHiPL AHi posterlat
AHiPM AHi posteromed
alv alveus of hipp
APir amygdalopir trans
ArcLP arcuate lateropost
ArcMP arcuate med post
Au1 primary aud cx

AuD 2ary aud cx, dors
AuV 2ary aud cx, vent
BLP basolat amyg post
BMP basomed am post
CA1 field CA1 hipp
CA2 field CA2 hipp
CA3 field CA3 hipp

cc corpus callosum
cg cingulum
chp choroid plexus
CL centrolat th nu
cp cerebral peduncle
D3V dorsal 3rd vent
dcw deep cerebr white

DEn dorsal endopirif
df dorsal fornix
dhc dorsal hipp comm
DLG dors lat geniculate
Ect ectorhinal cx
F nu fields of Forel
f fornix

FC fasciola cinereum
fi fimbria of hipp
fr fasc retroflexus
Gem gemini hy nu
GrDG granular dent gy
hbc habenular comm
hif hipp fissure

ic internal capsule
IG indusium griseum
IGL intergenic leaf
IMA intramedul th area
La lat amyg nu
LaDL lat amyg dorsolat
LaVM lat am ventromed
LEnt lat entorhinal cx

LHbL lat habenular lat
LHbM lat habenular med
LM lat mammillary nu
LMol lacunosum molec
LPLR LP laterorostral
LPMR LP mediorostral
LPtA lat parietal assocn
LV lat ventricle

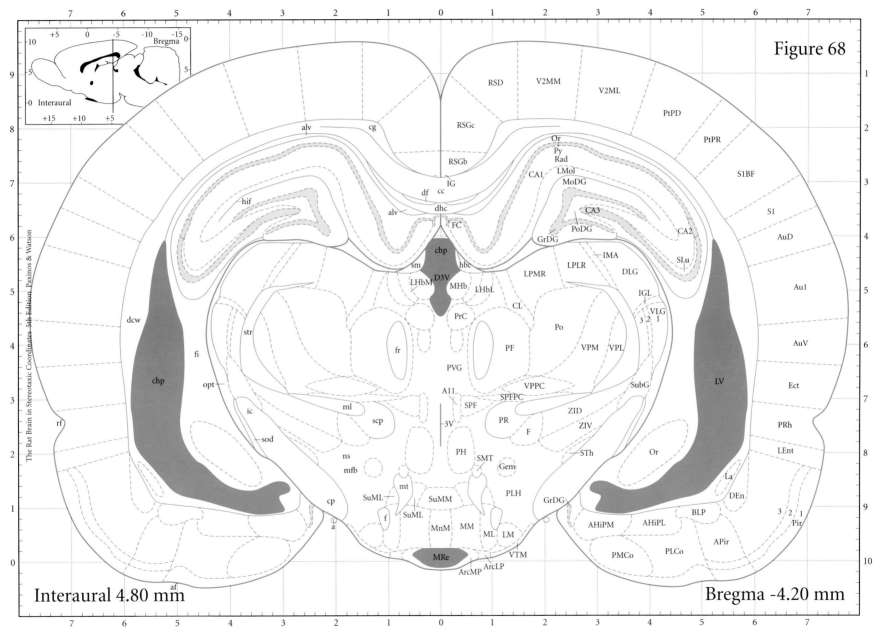

Figure 68

Interaural 4.80 mm

Bregma -4.20 mm

The Rat Brain in Stereotaxic Coordinates 5th Edition Paxinos & Watson

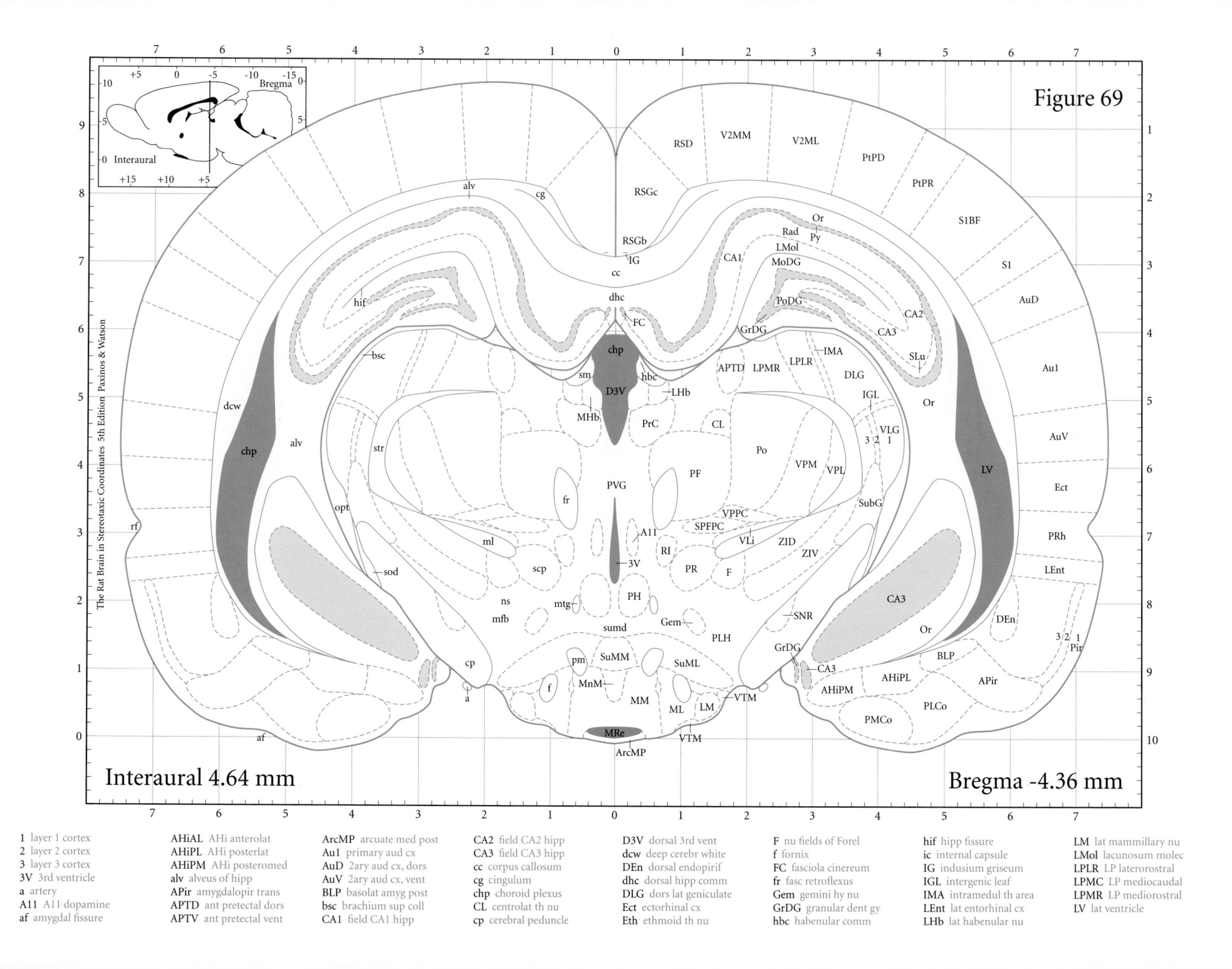

Figure 69

Interaural 4.64 mm

Bregma -4.36 mm

The Rat Brain in Stereotaxic Coordinates 5th Edition Paxinos & Watson

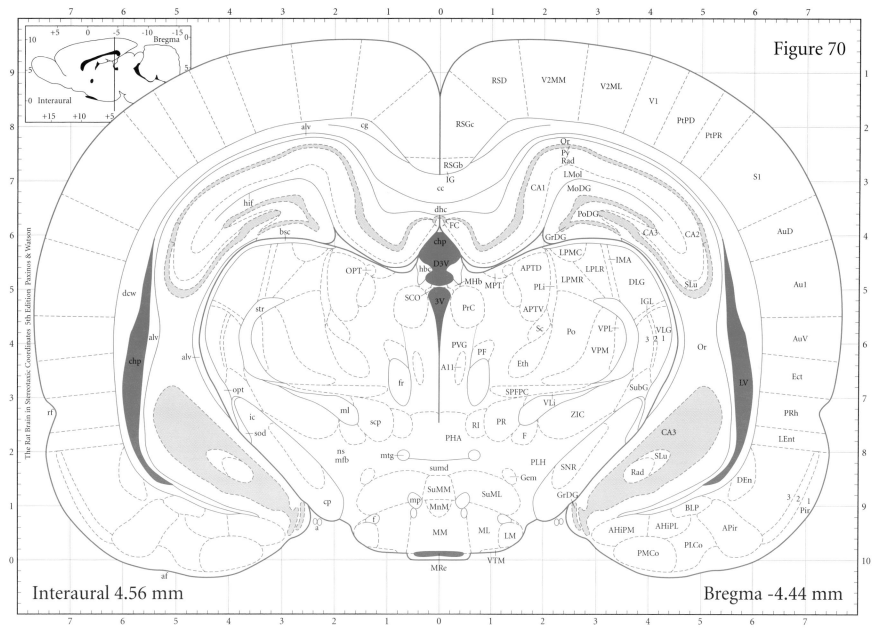

Figure 70

Interaural 4.56 mm

Bregma -4.44 mm

The Rat Brain in Stereotaxic Coordinates 5th Edition Paxinos & Watson

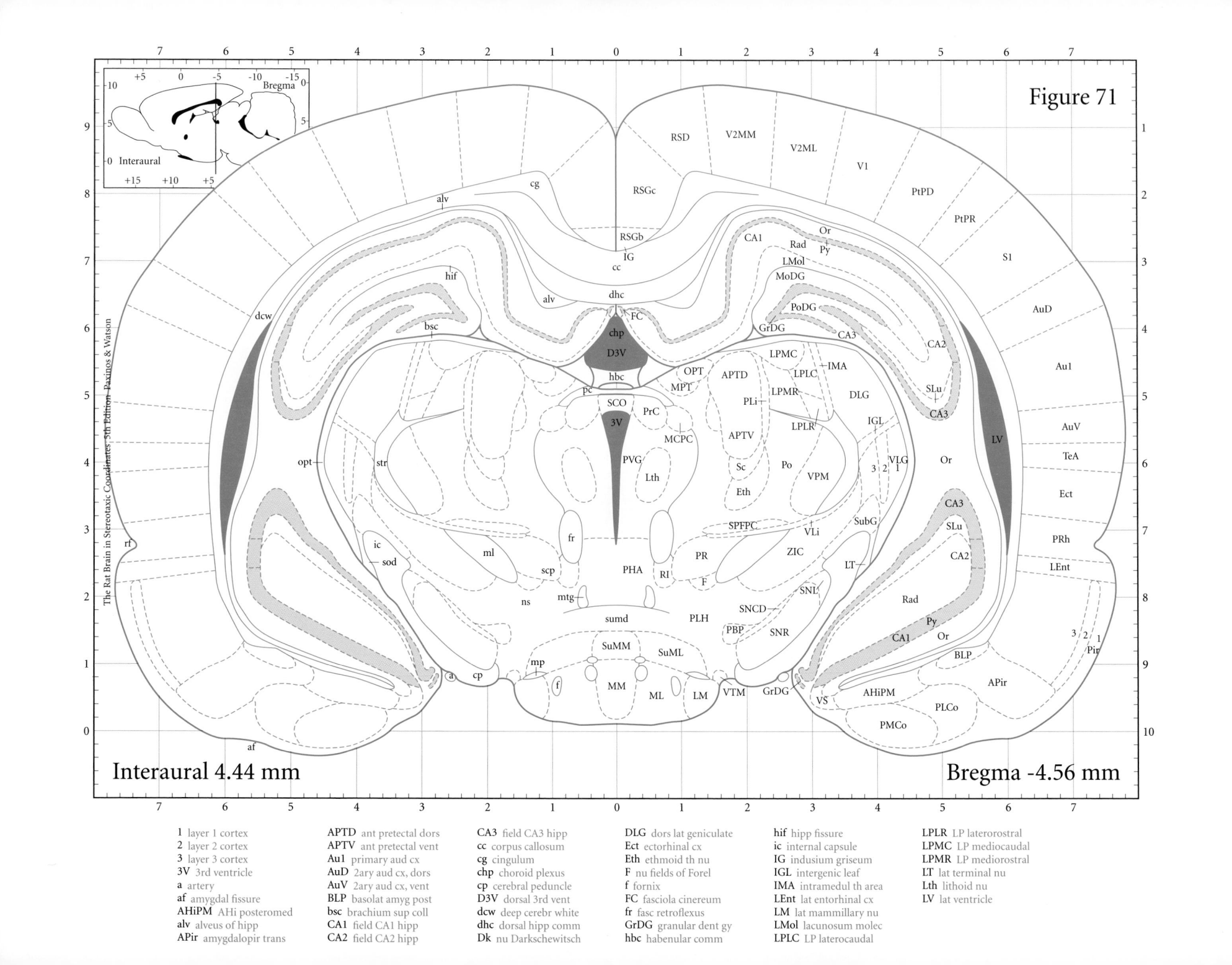

Figure 71

Interaural 4.44 mm

Bregma -4.56 mm

1 layer 1 cortex
2 layer 2 cortex
3 layer 3 cortex
3V 3rd ventricle
a artery
af amygdal fissure
AHiPM AHi posteromed
alv alveus of hipp
APir amygdalopir trans

APTD ant pretectal dors
APTV ant pretectal vent
Au1 primary aud cx
AuD 2ary aud cx, dors
AuV 2ary aud cx, vent
BLP basolat amyg post
bsc brachium sup coll
CA1 field CA1 hipp
CA2 field CA2 hipp

CA3 field CA3 hipp
cc corpus callosum
cg cingulum
chp choroid plexus
cp cerebral peduncle
D3V dorsal 3rd vent
dcw deep cerebr white
dhc dorsal hipp comm
Dk nu Darkschewitsch

DLG dors lat geniculate
Ect ectorhinal cx
Eth ethmoid th nu
F nu fields of Forel
f fornix
FC fasciola cinereum
fr fasc retroflexus
GrDG granular dent gy
hbc habenular comm

hif hipp fissure
ic internal capsule
IG indusium griseum
IGL intergenic leaf
IMA intramedul th area
LEnt lat entorhinal cx
LM lat mammillary nu
LMol lacunosum molec
LPLC LP laterocaudal

LPLR LP laterorostral
LPMC LP mediocaudal
LPMR LP mediorostral
LT lat terminal nu
Lth lithoid nu
LV lat ventricle

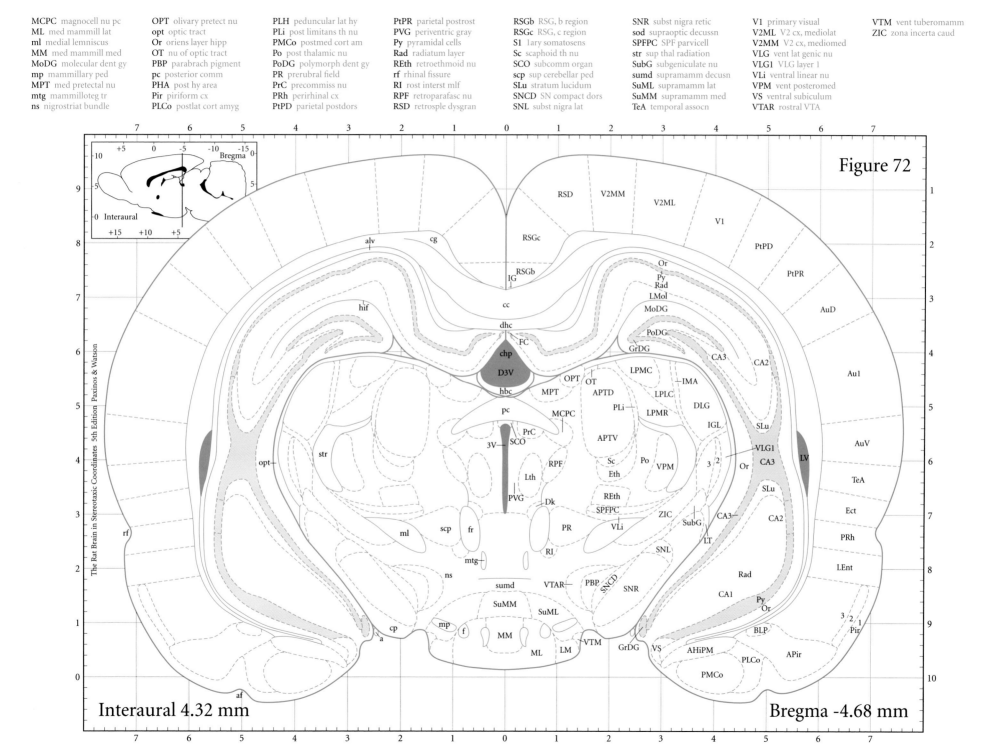

Figure 72

The Rat Brain in Stereotaxic Coordinates 5th Edition Paxinos & Watson

Interaural 4.32 mm

Bregma -4.68 mm

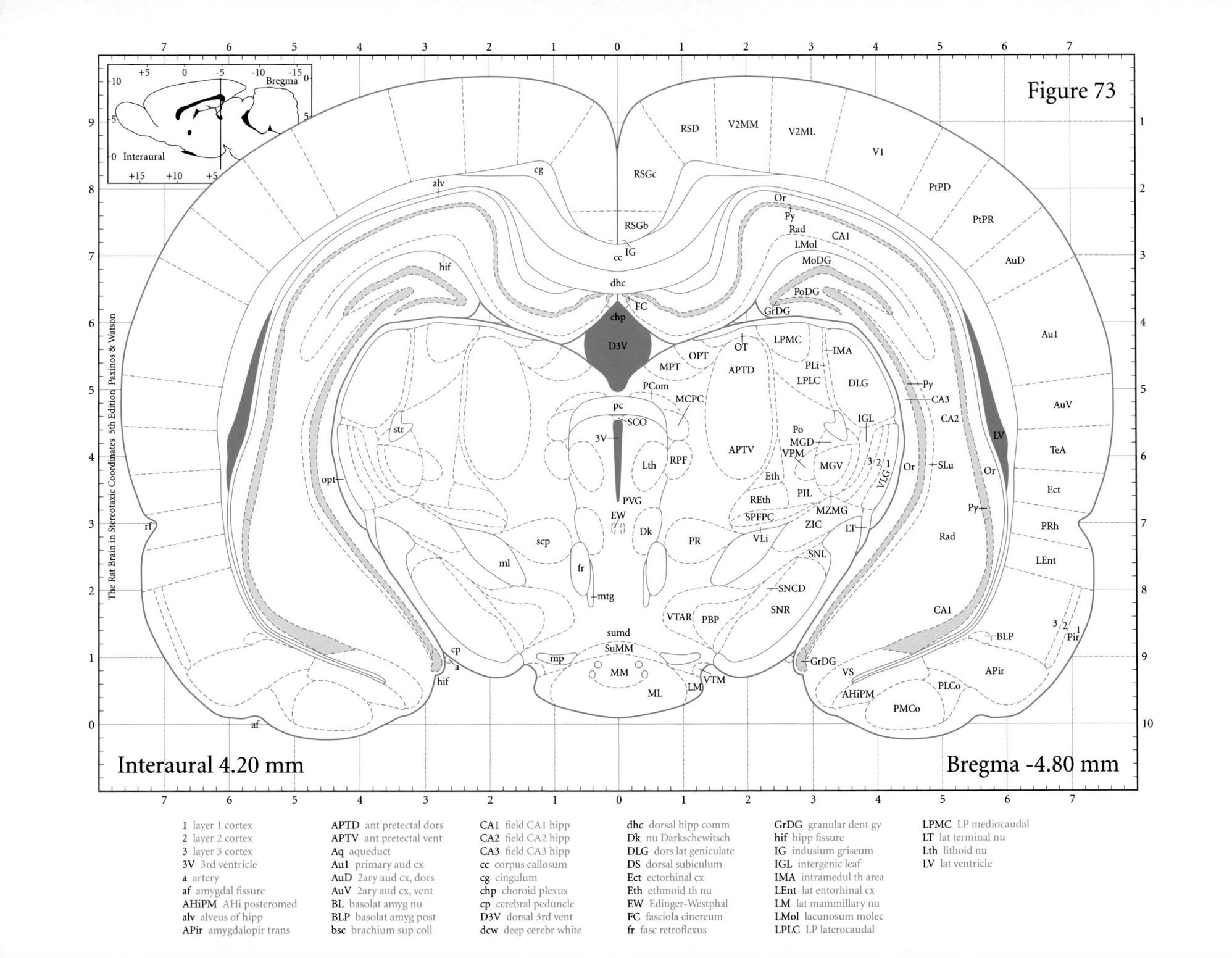

Figure 73

Interaural 4.20 mm

Bregma -4.80 mm

1 layer 1 cortex
2 layer 2 cortex
3 layer 3 cortex
3V 3rd ventricle
a artery
af amygdal fissure
AHiPM AHi posteromed
alv alveus of hipp
APir amygdalopir trans

APTD ant pretectal dors
APTV ant pretectal vent
Aq aqueduct
Au1 primary aud cx
AuD 2ary aud cx, dors
AuV 2ary aud cx, vent
BL basolat amyg nu
BLP basolat amyg post
bsc brachium sup coll

CA1 field CA1 hipp
CA2 field CA2 hipp
CA3 field CA3 hipp
cc corpus callosum
cg cingulum
chp choroid plexus
cp cerebral peduncle
D3V dorsal 3rd vent
dcw deep cerebr white

dhc dorsal hipp comm
Dk nu Darkschewitsch
DLG dors lat geniculate
DS dorsal subiculum
Ect ectorhinal cx
Eth ethmoid th nu
EW Edinger-Westphal
FC fasciola cinereum
fr fasc retroflexus

GrDG granular dent gy
hif hipp fissure
IG indusium griseum
IGL intergenic leaf
IMA intramedul th area
LEnt lat entorhinal cx
LM lat mammillary nu
LMol lacunosum molec
LPLC LP laterocaudal

LPMC LP mediocaudal
LT lat terminal nu
Lth lithoid nu
LV lat ventricle

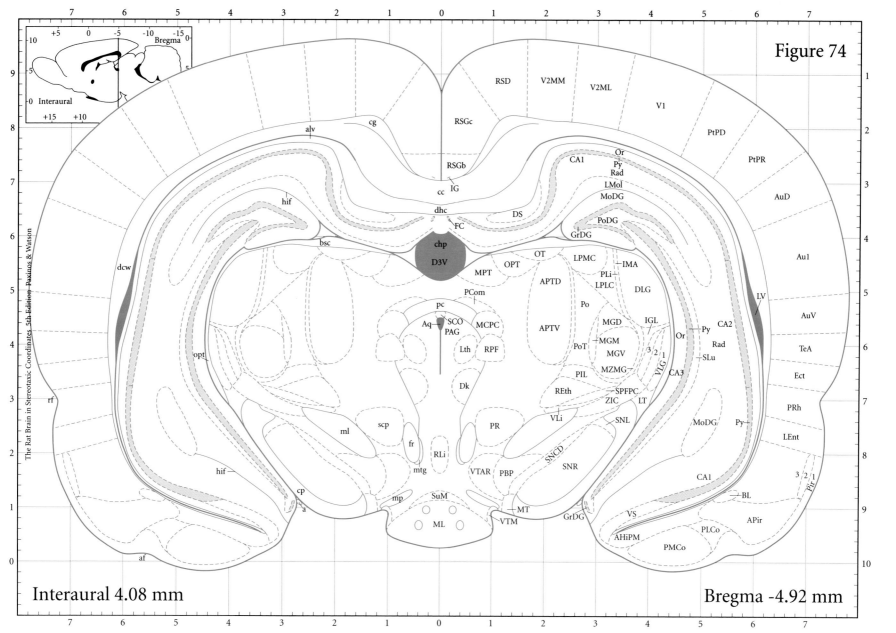

Figure 74

Interaural 4.08 mm

Bregma -4.92 mm

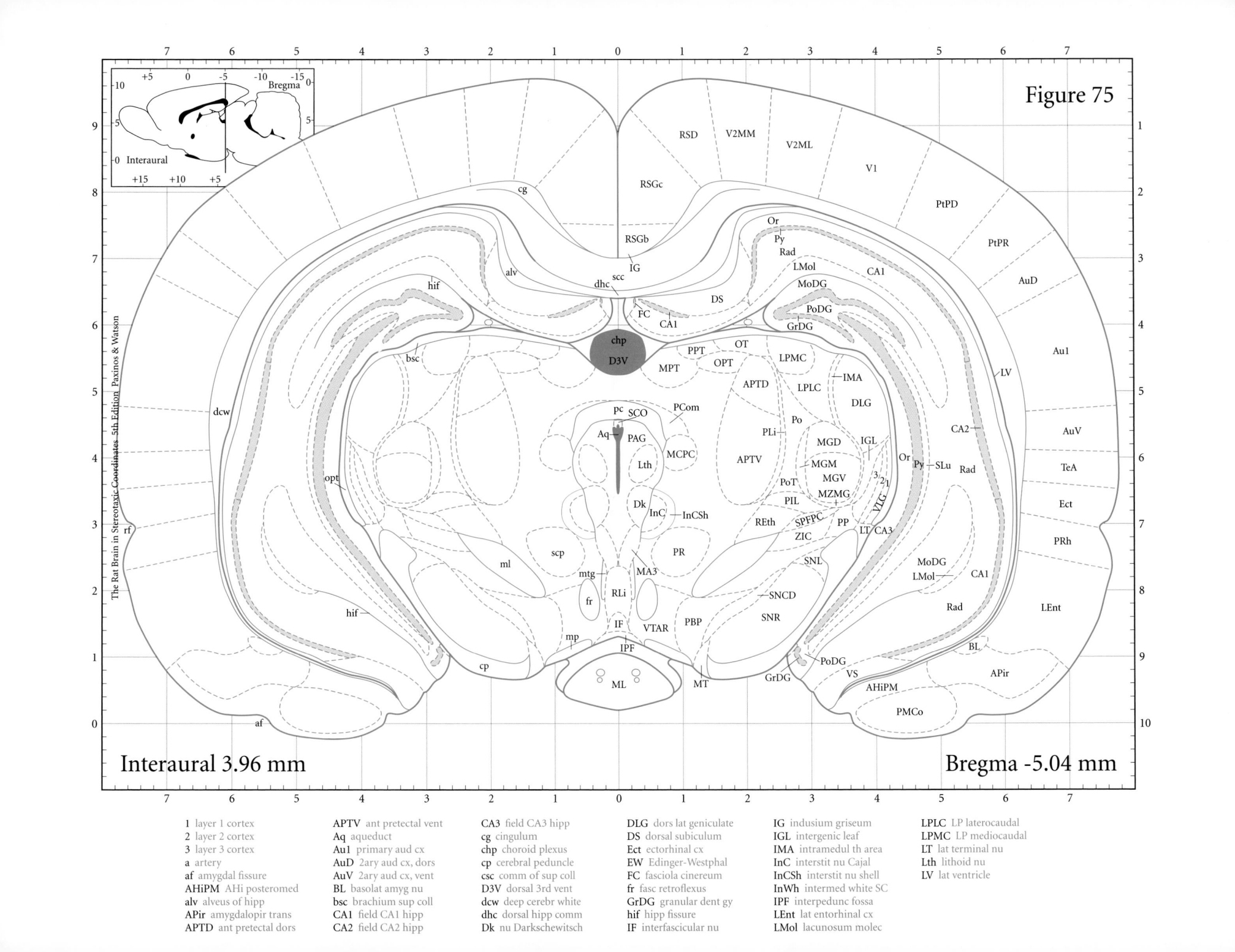

Figure 75

Interaural 3.96 mm

Bregma -5.04 mm

1   layer 1 cortex
2   layer 2 cortex
3   layer 3 cortex
a   artery
af   amygdal fissure
AHiPM   AHi posteromed
alv   alveus of hipp
APir   amygdalopir trans
APTD   ant pretectal dors

APTV   ant pretectal vent
Aq   aqueduct
Au1   primary aud cx
AuD   2ary aud cx, dors
AuV   2ary aud cx, vent
BL   basolat amyg nu
bsc   brachium sup coll
CA1   field CA1 hipp
CA2   field CA2 hipp

CA3   field CA3 hipp
cg   cingulum
chp   choroid plexus
cp   cerebral peduncle
csc   comm of sup coll
D3V   dorsal 3rd vent
dcw   deep cerebr white
dhc   dorsal hipp comm
Dk   nu Darkschewitsch

DLG   dors lat geniculate
DS   dorsal subiculum
Ect   ectorhinal cx
EW   Edinger-Westphal
FC   fasciola cinereum
fr   fasc retroflexus
GrDG   granular dent gy
hif   hipp fissure
IF   interfascicular nu

IG   indusium griseum
IGL   intergenic leaf
IMA   intramedul th area
InC   interstit nu Cajal
InCSh   interstit nu shell
InWh   intermed white SC
IPF   interpedunc fossa
LEnt   lat entorhinal cx
LMol   lacunosum molec

LPLC   LP laterocaudal
LPMC   LP mediocaudal
LT   lat terminal nu
Lth   lithoid nu
LV   lat ventricle

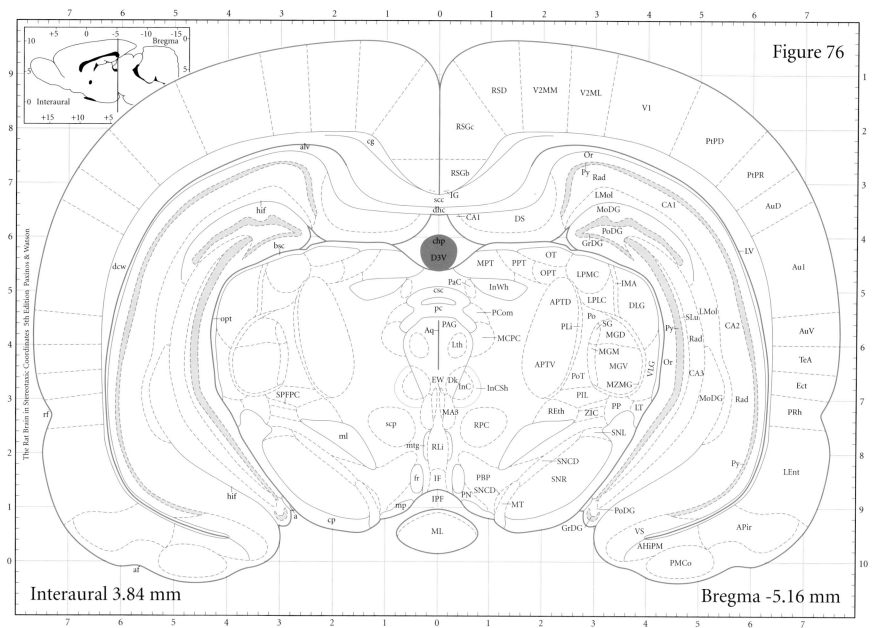

Figure 76

Interaural 3.84 mm

Bregma -5.16 mm

The Rat Brain in Stereotaxic Coordinates 5th Edition Paxinos & Watson

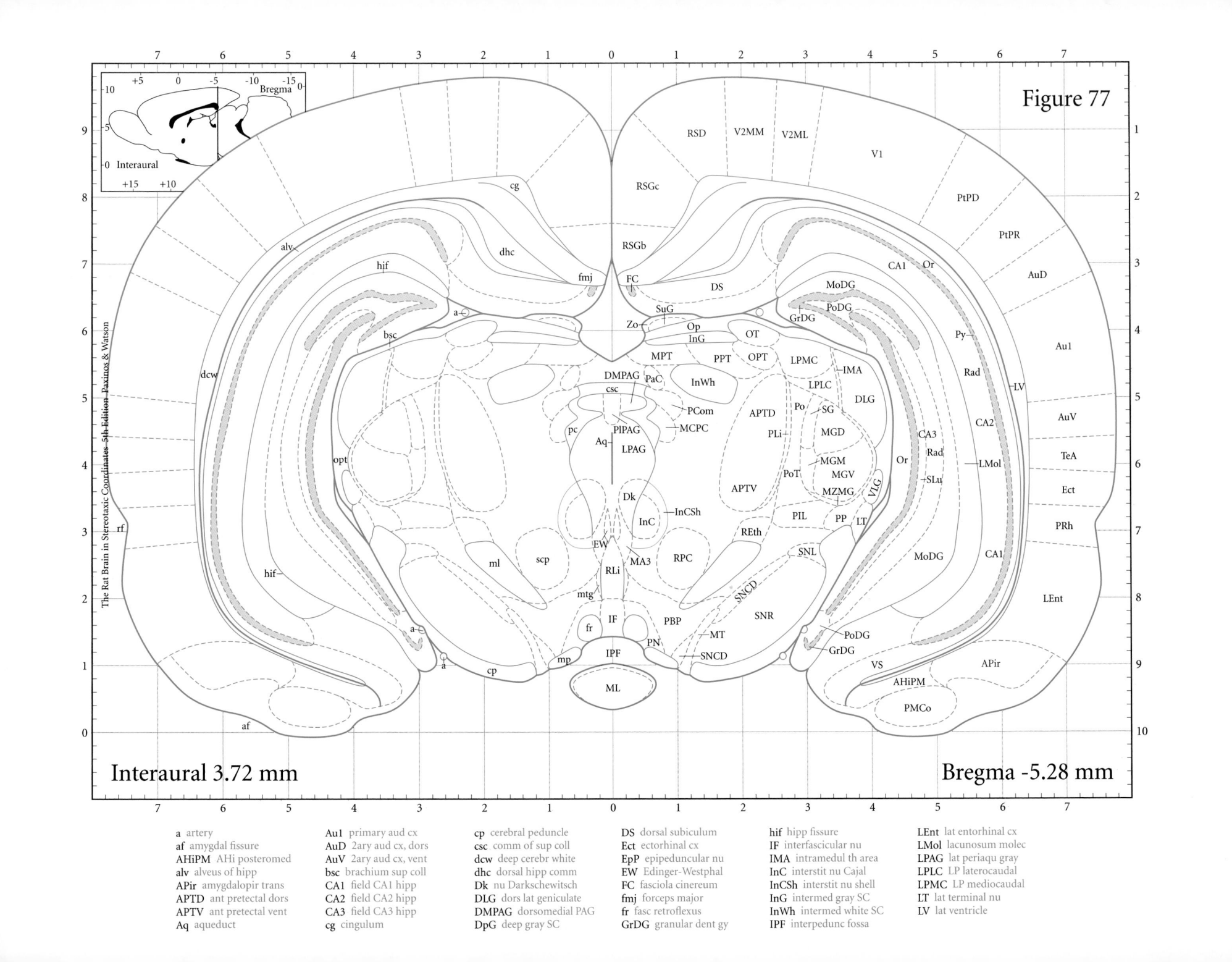

Figure 77

Interaural 3.72 mm

Bregma -5.28 mm

a artery
af amygdal fissure
AHiPM AHi posteromed
alv alveus of hipp
APir amygdalopir trans
APTD ant pretectal dors
APTV ant pretectal vent
Aq aqueduct

Au1 primary aud cx
AuD 2ary aud cx, dors
AuV 2ary aud cx, vent
bsc brachium sup coll
CA1 field CA1 hipp
CA2 field CA2 hipp
CA3 field CA3 hipp
cg cingulum

cp cerebral peduncle
csc comm of sup coll
dcw deep cerebr white
dhc dorsal hipp comm
Dk nu Darkschewitsch
DLG dors lat geniculate
DMPAG dorsomedial PAG
DpG deep gray SC

DS dorsal subiculum
Ect ectorhinal cx
EpP epipeduncular nu
EW Edinger-Westphal
FC fasciola cinereum
fmj forceps major
fr fasc retroflexus
GrDG granular dent gy

hif hipp fissure
IF interfascicular nu
IMA intramedul th area
InC interstit nu Cajal
InCSh interstit nu shell
InG intermed gray SC
InWh intermed white SC
IPF interpedunc fossa

LEnt lat entorhinal cx
LMol lacunosum molec
LPAG lat periaqu gray
LPLC LP laterocaudal
LPMC LP mediocaudal
LT lat terminal nu
LV lat ventricle

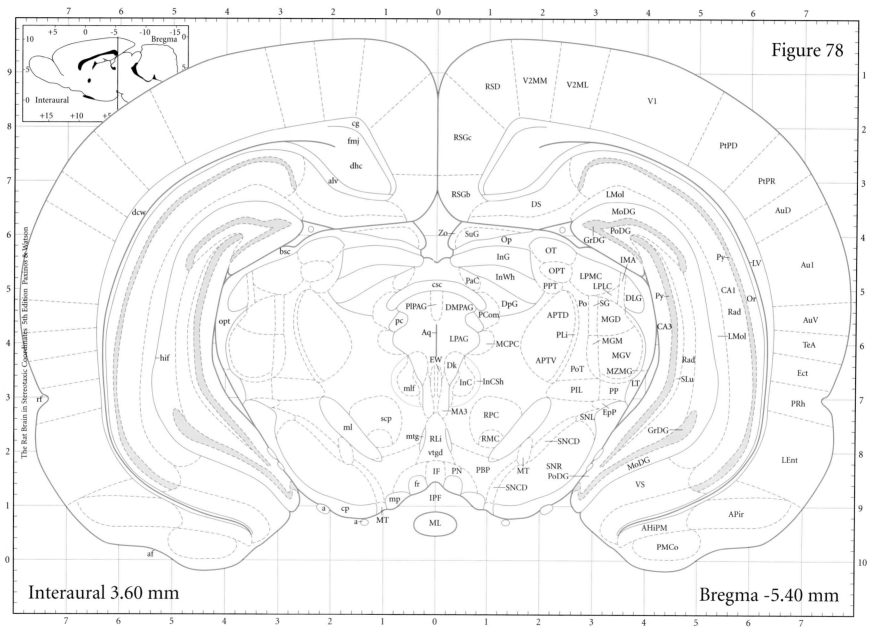

Figure 78

Interaural 3.60 mm

Bregma -5.40 mm

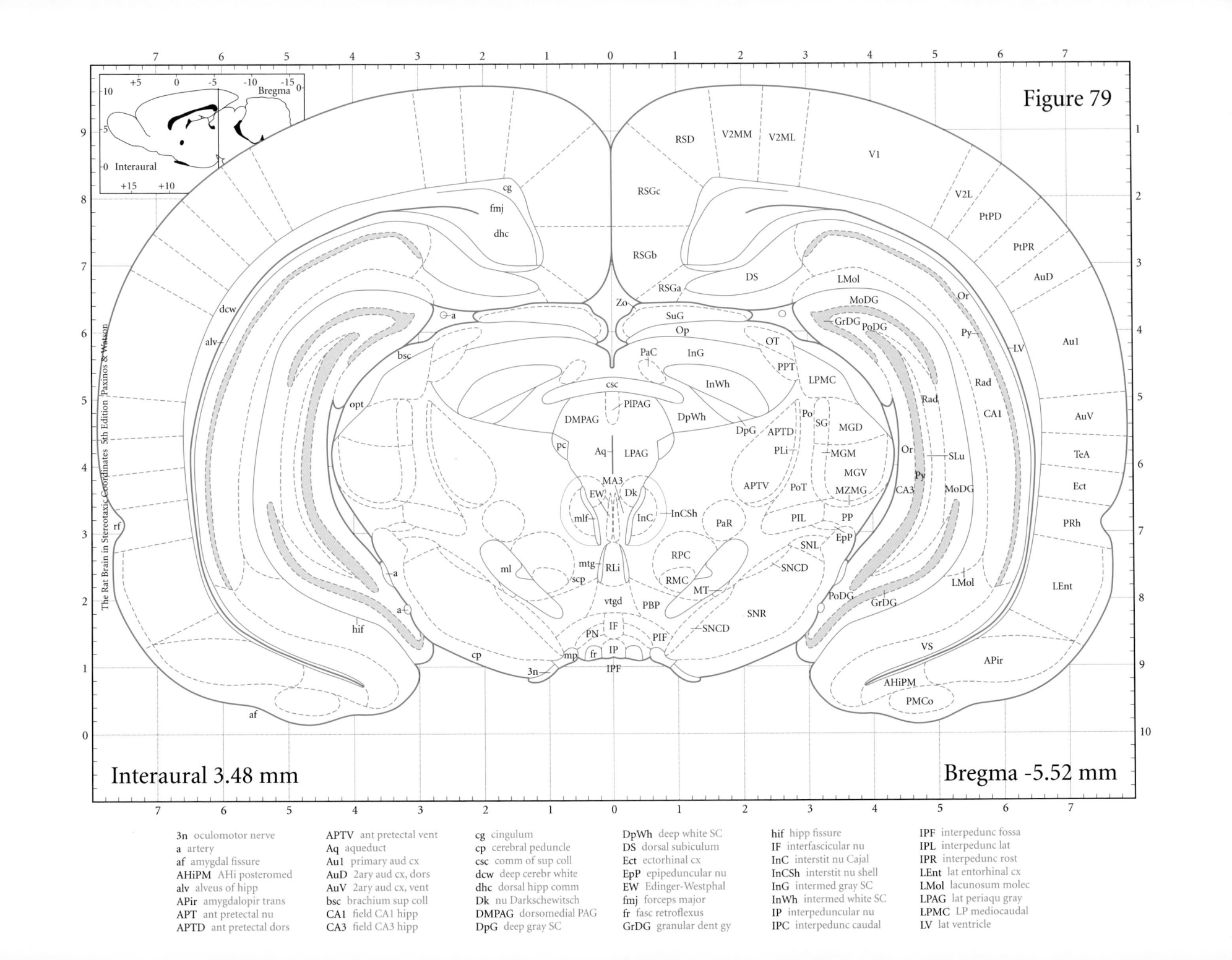

Figure 79

Interaural 3.48 mm

Bregma -5.52 mm

The Rat Brain in Stereotaxic Coordinates 5th Edition 'Paxinos & Watson

3n   oculomotor nerve
a   artery
af   amygdal fissure
AHiPM   AHi posteromed
alv   alveus of hipp
APir   amygdalopir trans
APT   ant pretectal nu
APTD   ant pretectal dors

APTV   ant pretectal vent
Aq   aqueduct
Au1   primary aud cx
AuD   2ary aud cx, dors
AuV   2ary aud cx, vent
bsc   brachium sup coll
CA1   field CA1 hipp
CA3   field CA3 hipp

cg   cingulum
cp   cerebral peduncle
csc   comm of sup coll
dcw   deep cerebr white
dhc   dorsal hipp comm
Dk   nu Darkschewitsch
DMPAG   dorsomedial PAG
DpG   deep gray SC

DpWh   deep white SC
DS   dorsal subiculum
Ect   ectorhinal cx
EpP   epipeduncular nu
EW   Edinger-Westphal
fmj   forceps major
fr   fasc retroflexus
GrDG   granular dent gy

hif   hipp fissure
IF   interfascicular nu
InC   interstit nu Cajal
InCSh   interstit nu shell
InG   intermed gray SC
InWh   intermed white SC
IP   interpeduncular nu
IPC   interpedunc caudal

IPF   interpedunc fossa
IPL   interpedunc lat
IPR   interpedunc rost
LEnt   lat entorhinal cx
LMol   lacunosum molec
LPAG   lat periaqu gray
LPMC   LP mediocaudal
LV   lat ventricle

MA3 medial acc oculom
MGD med genic dorsal
MGM med genic medial
MGV med genic ventral
ml medial lemniscus
mlf med long fasc
MoDG molecular dent gy
mp mammillary ped

MT med terminal nu
mtg mammilloteg tr
MZMG marg zn med genic
Op optic n layer SC
opt optic tract
Or oriens layer hipp
OT nu of optic tract
PaC paracommissural

PaR pararubral nu
PBP parabrach pigment
pc posterior comm
PIF parainterfascic nu
PIL post intralam th nu
PlPAG plioglial PAG
PMCo postmed cort am

PN paranigral nu
Po post thalamic nu
PoDG polymorph dent gy
PoT post th nu triang
PP peripeduncular nu
PPT post pretectal nu
PRh perirhinal cx
PtPD parietal postdors

PtPR parietal postrost
Py pyramidal cells
Rad radiatum layer
rf rhinal fissure
RLi rost linear raphe
RMC red nu magnocell
RPC red nu parvicell
RSD retrosple dysgran

RSGa RSG, a region
RSGb RSG, b region
RSGc RSG, c region
scp sup cerebellar ped
SG supragenic th nu
SLu stratum lucidum
SNCD SN compact dors
SNL subst nigra lat

SNR subst nigra retic
SuG superfic gray SC
TeA temporal assocn
V1 primary visual
V2L V2, lateral
V2ML V2 cx, mediolat
V2MM V2 cx, mediomed
VS ventral subiculum

vtgd vent teg decussn
Zo zonal layer SC

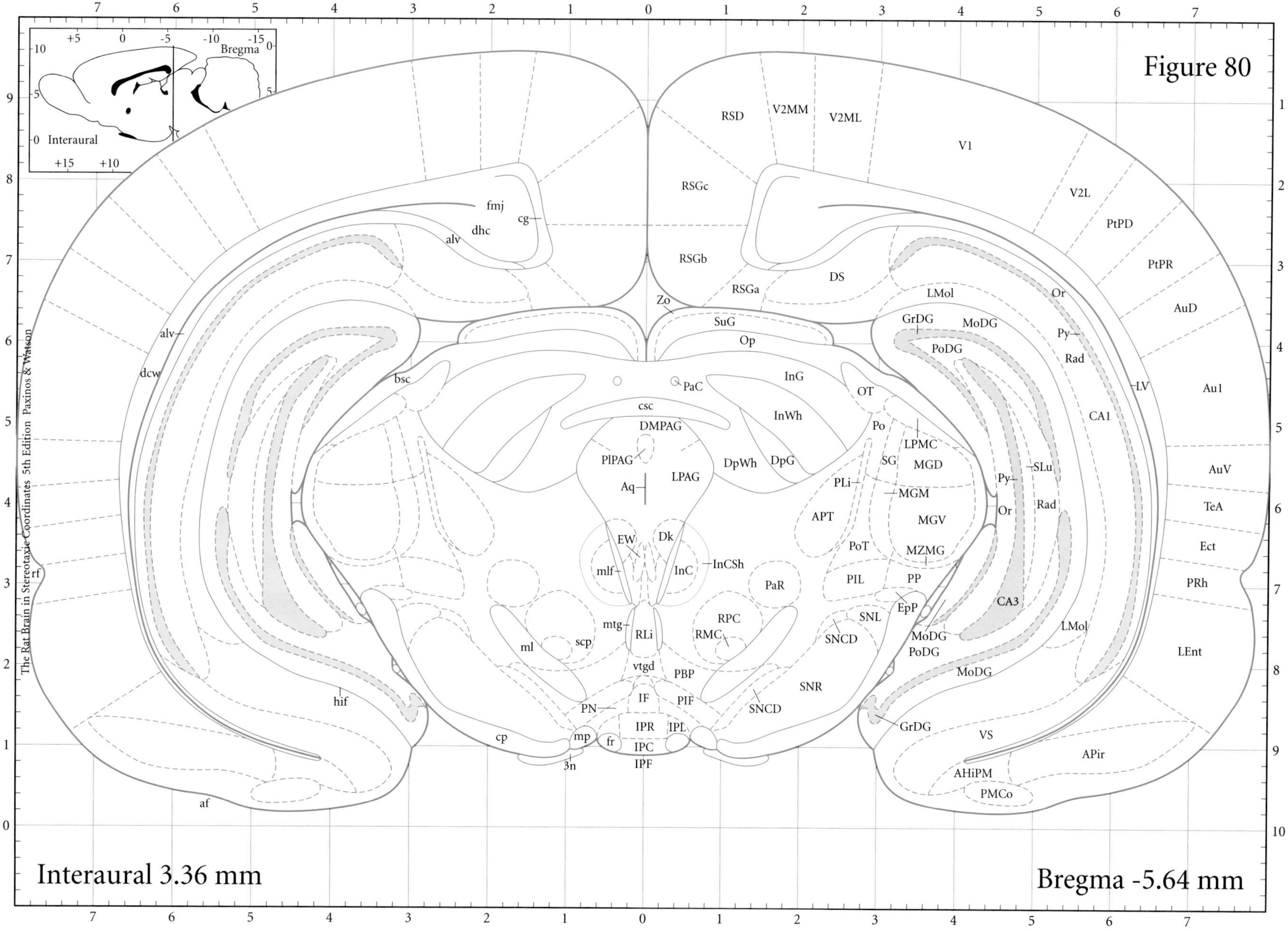

Figure 80

Interaural 3.36 mm

Bregma -5.64 mm

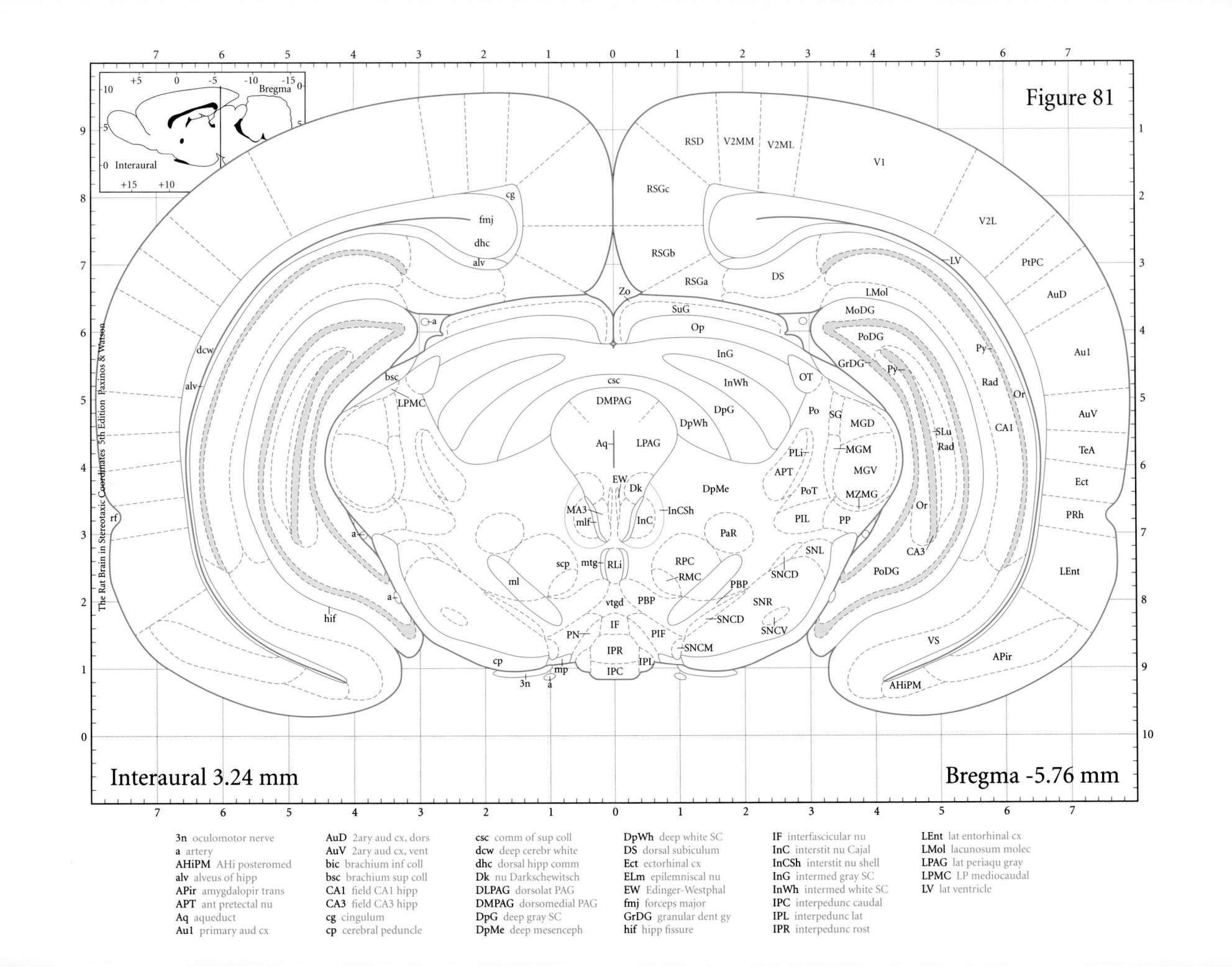

Figure 81

The Rat Brain in Stereotaxic Coordinates 5th Edition Paxinos & Watson

3n oculomotor nerve	AuD 2ary aud cx, dors	csc comm of sup coll	DpWh deep white SC	IF interfascicular nu	LEnt lat entorhinal cx
a artery	AuV 2ary aud cx, vent	dcw deep cerebr white	DS dorsal subiculum	InC interstit nu Cajal	LMol lacunosum molec
AHiPM AHi posteromed	bic brachium inf coll	dhc dorsal hipp comm	Ect ectorhinal cx	InCSh interstit nu shell	LPAG lat periaqu gray
alv alveus of hipp	bsc brachium sup coll	Dk nu Darkschewitsch	ELm epilemniscal nu	InG intermed gray SC	LPMC LP mediocaudal
APir amygdalopir trans	CA1 field CA1 hipp	DLPAG dorsolat PAG	EW Edinger-Westphal	InWh intermed white SC	LV lat ventricle
APT ant pretectal nu	CA3 field CA3 hipp	DMPAG dorsomedial PAG	fmj forceps major	IPC interpedunc caudal	
Aq aqueduct	cg cingulum	DpG deep gray SC	GrDG granular dent gy	IPL interpedunc lat	
Au1 primary aud cx	cp cerebral peduncle	DpMe deep mesenceph	hif hipp fissure	IPR interpedunc rost	

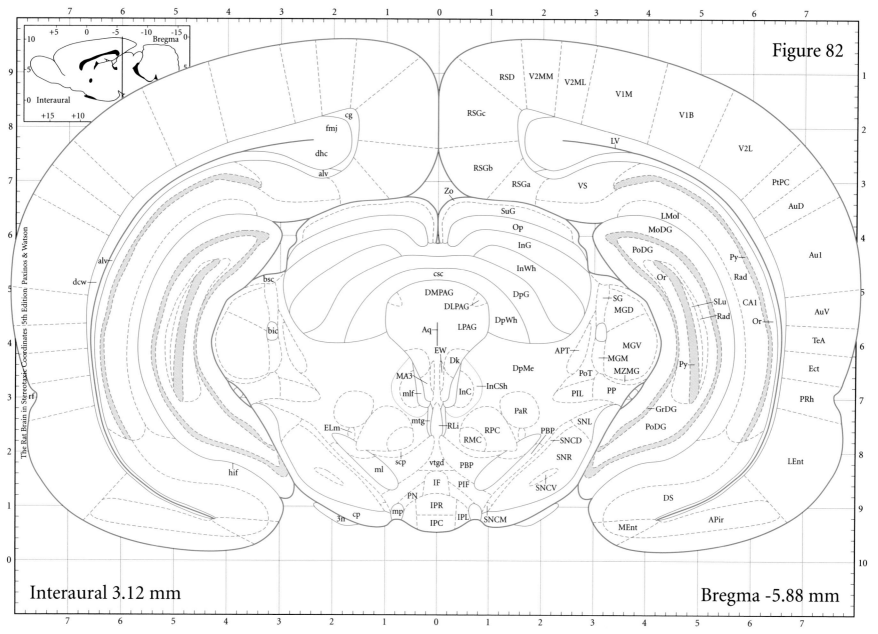

Figure 82

The Rat Brain in Stereotaxic Coordinates 5th Edition Paxinos & Watson

Interaural 3.12 mm

Bregma -5.88 mm

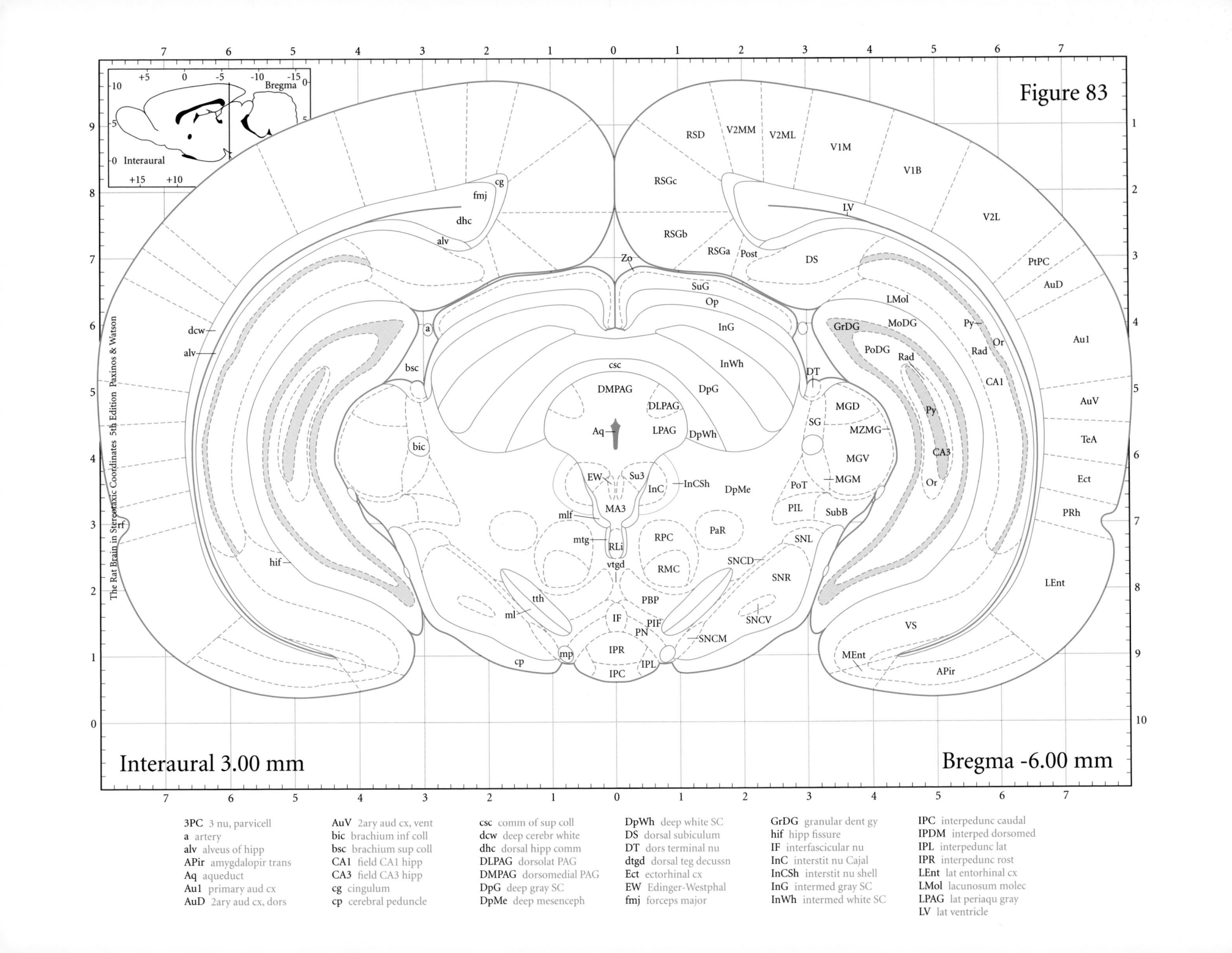

Figure 83

Interaural 3.00 mm

Bregma -6.00 mm

3PC 3 nu, parvicell
a artery
alv alveus of hipp
APir amygdalopir trans
Aq aqueduct
Au1 primary aud cx
AuD 2ary aud cx, dors

AuV 2ary aud cx, vent
bic brachium inf coll
bsc brachium sup coll
CA1 field CA1 hipp
CA3 field CA3 hipp
cg cingulum
cp cerebral peduncle

csc comm of sup coll
dcw deep cerebr white
dhc dorsal hipp comm
DLPAG dorsolat PAG
DMPAG dorsomedial PAG
DpG deep gray SC
DpMe deep mesenceph

DpWh deep white SC
DS dorsal subiculum
DT dors terminal nu
dtgd dorsal teg decussn
Ect ectorhinal cx
EW Edinger-Westphal
fmj forceps major

GrDG granular dent gy
hif hipp fissure
IF interfascicular nu
InC interstit nu Cajal
InCSh interstit nu shell
InG intermed gray SC
InWh intermed white SC

IPC interpedunc caudal
IPDM interped dorsomed
IPL interpedunc lat
IPR interpedunc rost
LEnt lat entorhinal cx
LMol lacunosum molec
LPAG lat periaqu gray
LV lat ventricle

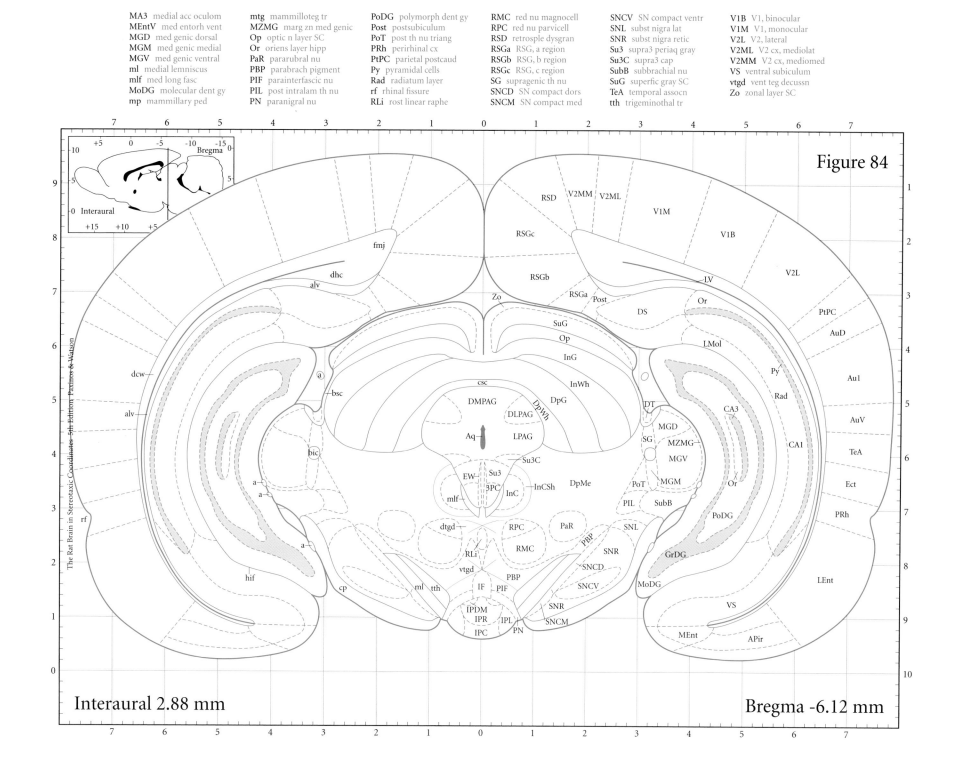

MA3 medial acc oculom
MEntV med entorh vent
MGD med genic dorsal
MGM med genic medial
MGV med genic ventral
ml medial lemniscus
mlf med long fasc
MoDG molecular dent gy
mp mammillary ped

mtg mammilloteg tr
MZMG marg zn med genic
Op optic n layer SC
Or oriens layer hipp
PaR pararubral nu
PBP parabrach pigment
PIF parainterfascic nu
PIL post intralam th nu
PN paranigral nu

PoDG polymorph dent gy
Post postsubiculum
PoT post th nu triang
PRh perirhinal cx
PtPC parietal postcaud
Py pyramidal cells
Rad radiatum layer
rf rhinal fissure
RLi rost linear raphe

RMC red nu magnocell
RPC red nu parvicell
RSD retrosple dysgran
RSGa RSG, a region
RSGb RSG, b region
RSGc RSG, c region
SG supragenic th nu
SNCD SN compact dors
SNCM SN compact med

SNCV SN compact ventr
SNL subst nigra lat
SNR subst nigra retic
Su3 supra3 periaq gray
Su3C supra3 cap
SubB subbrachial nu
SuG superfic gray SC
TeA temporal assocn
tth trigeminothal tr

V1B V1, binocular
V1M V1, monocular
V2L V2, lateral
V2ML V2 cx, mediolat
V2MM V2 cx, mediomed
VS ventral subiculum
vtgd vent teg decussn
Zo zonal layer SC

Figure 84

Interaural 2.88 mm

Bregma -6.12 mm

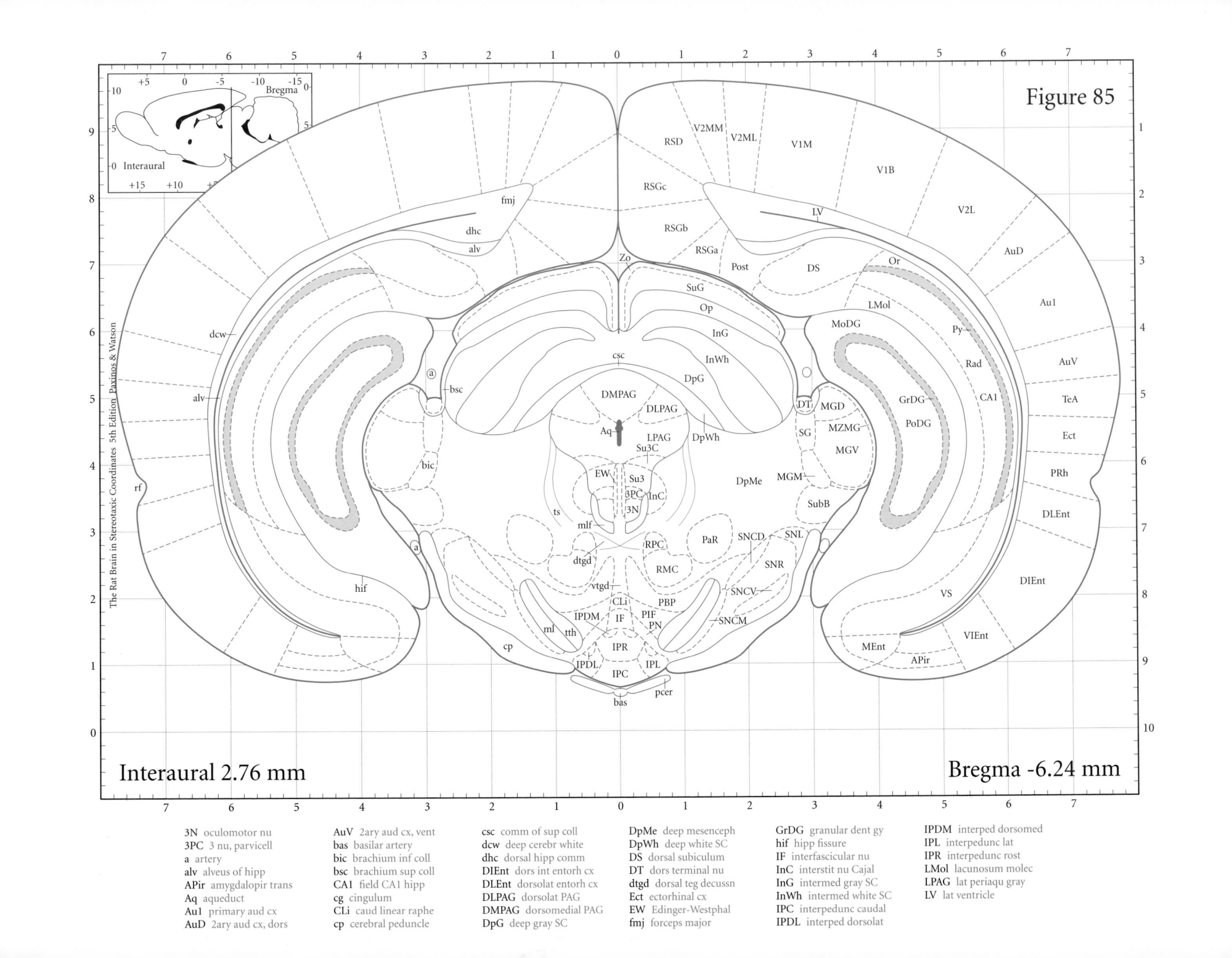

Figure 85

Interaural 2.76 mm

Bregma -6.24 mm

3N oculomotor nu
3PC 3 nu, parvicell
a artery
alv alveus of hipp
APir amygdalopir trans
Aq aqueduct
Au1 primary aud cx
AuD 2ary aud cx, dors

AuV 2ary aud cx, vent
bas basilar artery
bic brachium inf coll
bsc brachium sup coll
CA1 field CA1 hipp
cg cingulum
CLi caud linear raphe
cp cerebral peduncle

csc comm of sup coll
dcw deep cerebr white
dhc dorsal hipp comm
DIEnt dors int entorh cx
DLEnt dorsolat entorh cx
DLPAG dorsolat PAG
DMPAG dorsomedial PAG
DpG deep gray SC

DpMe deep mesenceph
DpWh deep white SC
DS dorsal subiculum
DT dors terminal nu
dtgd dorsal teg decussn
Ect ectorhinal cx
EW Edinger-Westphal
fmj forceps major

GrDG granular dent gy
hif hipp fissure
IF interfascicular nu
InC interstit nu Cajal
InG intermed gray SC
InWh intermed white SC
IPC interpedunc caudal
IPDL interped dorsolat

IPDM interped dorsomed
IPL interpedunc lat
IPR interpedunc rost
LMol lacunosum molec
LPAG lat periaqu gray
LV lat ventricle

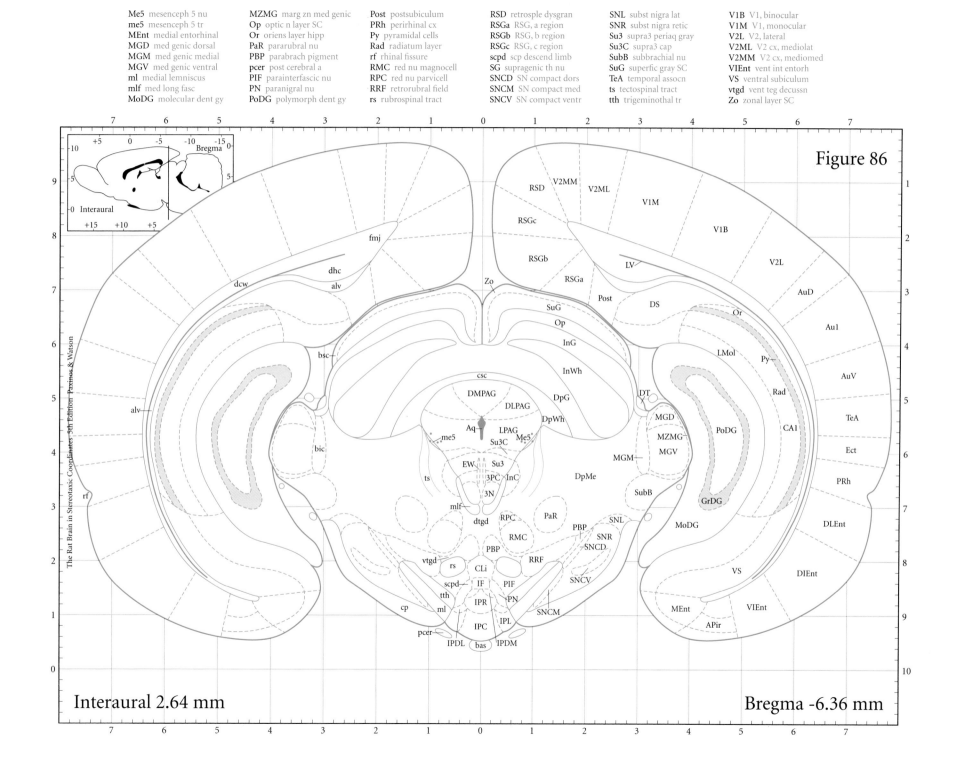

Figure 86

Interaural 2.64 mm

Bregma -6.36 mm

Me5 mesenceph 5 nu
me5 mesenceph 5 tr
MEnt medial entorhinal
MGD med genic dorsal
MGM med genic medial
MGV med genic ventral
ml medial lemniscus
mlf med long fasc
MoDG molecular dent gy

MZMG marg zn med genic
Op optic n layer SC
Or oriens layer hipp
PaR pararubral nu
PBP parabrach pigment
pcer post cerebral a
PIF parainterfascic nu
PN paranigral nu
PoDG polymorph dent gy

Post postsubiculum
PRh perirhinal cx
Py pyramidal cells
Rad radiatum layer
rf rhinal fissure
RMC red nu magnocell
RPC red nu parvicell
RRF retrorubral field
rs rubrospinal tract

RSD retrosple dysgran
RSGa RSG, a region
RSGb RSG, b region
RSGc RSG, c region
scpd scp descend limb
SG supragenic th nu
SNCD SN compact dors
SNCM SN compact med
SNCV SN compact ventr

SNL subst nigra lat
SNR subst nigra retic
Su3 supra3 periaq gray
Su3C supra3 cap
SubB subbrachial nu
SuG superfic gray SC
TeA temporal assocn
ts tectospinal tract
tth trigeminothal tr

V1B V1, binocular
V1M V1, monocular
V2L V2, lateral
V2ML V2 cx, mediolat
V2MM V2 cx, mediomed
VIEnt vent int entorh
VS ventral subiculum
vtgd vent teg decussn
Zo zonal layer SC

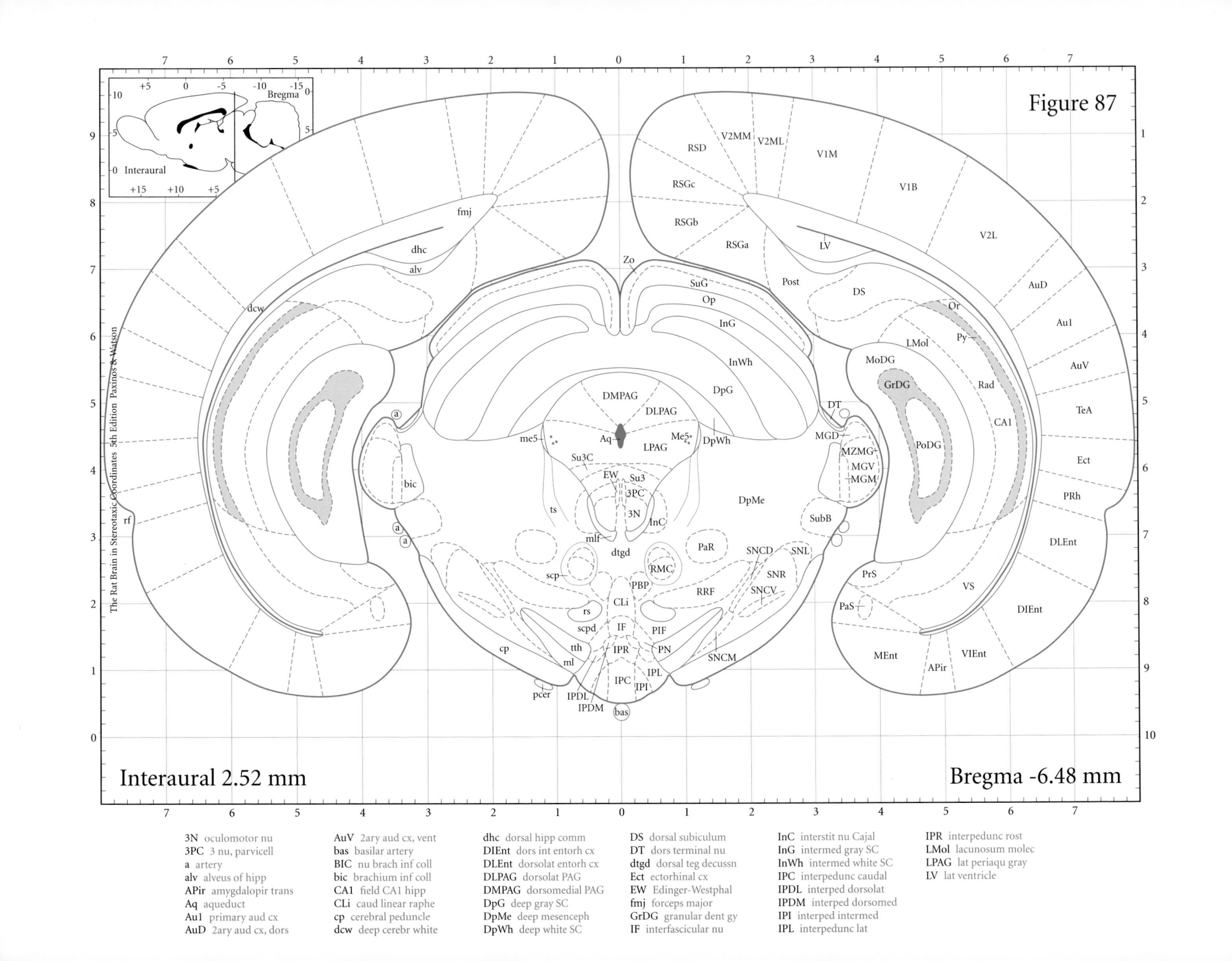

Figure 87

Interaural 2.52 mm

Bregma -6.48 mm

The Rat Brain in Stereotaxic Coordinates 5th Edition Paxinos & Watson

3N oculomotor nu
3PC 3 nu, parvicell
a artery
alv alveus of hipp
APir amygdalopir trans
Aq aqueduct
Au1 primary aud cx
AuD 2ary aud cx, dors

AuV 2ary aud cx, vent
bas basilar artery
BIC nu brach inf coll
bic brachium inf coll
CA1 field CA1 hipp
CLi caud linear raphe
cp cerebral peduncle
dcw deep cerebr white

dhc dorsal hipp comm
DIEnt dors int entorh cx
DLEnt dorsolat entorh cx
DLPAG dorsolat PAG
DMPAG dorsomedial PAG
DpG deep gray SC
DpMe deep mesenceph
DpWh deep white SC

DS dorsal subiculum
DT dors terminal nu
dtgd dorsal teg decussn
Ect ectorhinal cx
EW Edinger-Westphal
fmj forceps major
GrDG granular dent gy
IF interfascicular nu

InC interstit nu Cajal
InG intermed gray SC
InWh intermed white SC
IPC interpedunc caudal
IPDL interped dorsolat
IPDM interped dorsomed
IPI interped intermed
IPL interpedunc lat

IPR interpedunc rost
LMol lacunosum molec
LPAG lat periaqu gray
LV lat ventricle

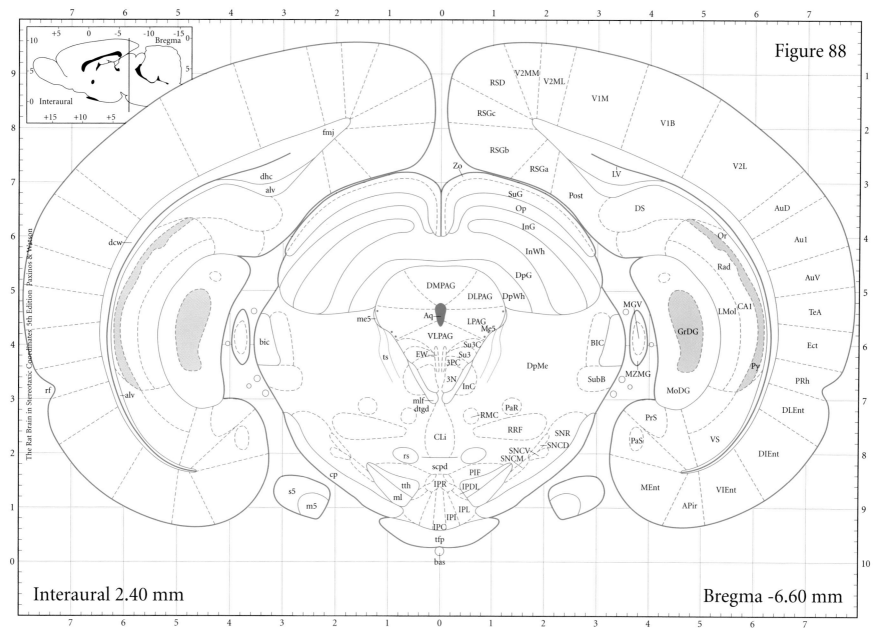

Figure 88

Interaural 2.40 mm

Bregma -6.60 mm

m5 motor root 5n
Me5 mesenceph 5 nu
me5 mesenceph 5 tr
MEnt medial entorhinal
MGD med genic dorsal
MGM med genic medial
MGV med genic ventral
ml medial lemniscus

mlf med long fasc
MoDG molecular dent gy
MZMG marg zn med genic
Op optic n layer SC
Or oriens layer hipp
PaR pararubral nu
PaS parasubiculum
PBP parabrach pigment

pcer post cerebral a
PIF parainterfascic nu
PN paranigral nu
PoDG polymorph dent gy
Post postsubiculum
PRh perirhinal cx
PrS presubiculum
Py pyramidal cells

Rad radiatum layer
rf rhinal fissure
RMC red nu magnocell
RRF retrorubral field
rs rubrospinal tract
RSD retrosple dysgran
RSGa RSG, a region
RSGb RSG, b region

RSGc RSG, c region
s5 sensory root of 5n
scp sup cerebellar ped
scpd scp descend limb
SNCD SN compact dors
SNCM SN compact med
SNCV SN compact ventr
SNL subst nigra lat

SNR subst nigra retic
Su3 supra3 periaq gray
Su3C supra3 cap
SubB subbrachial nu
SuG superfic gray SC
TeA temporal assocn
tfp trans fibers pons
ts tectospinal tract

tth trigeminothal tr
V1B V1, binocular
V1M V1, monocular
V2L V2, lateral
V2ML V2 cx, mediolat
V2MM V2 cx, mediomed
VIEnt vent int entorh
VLPAG ventrolat PAG

VS ventral subiculum
Zo zonal layer SC

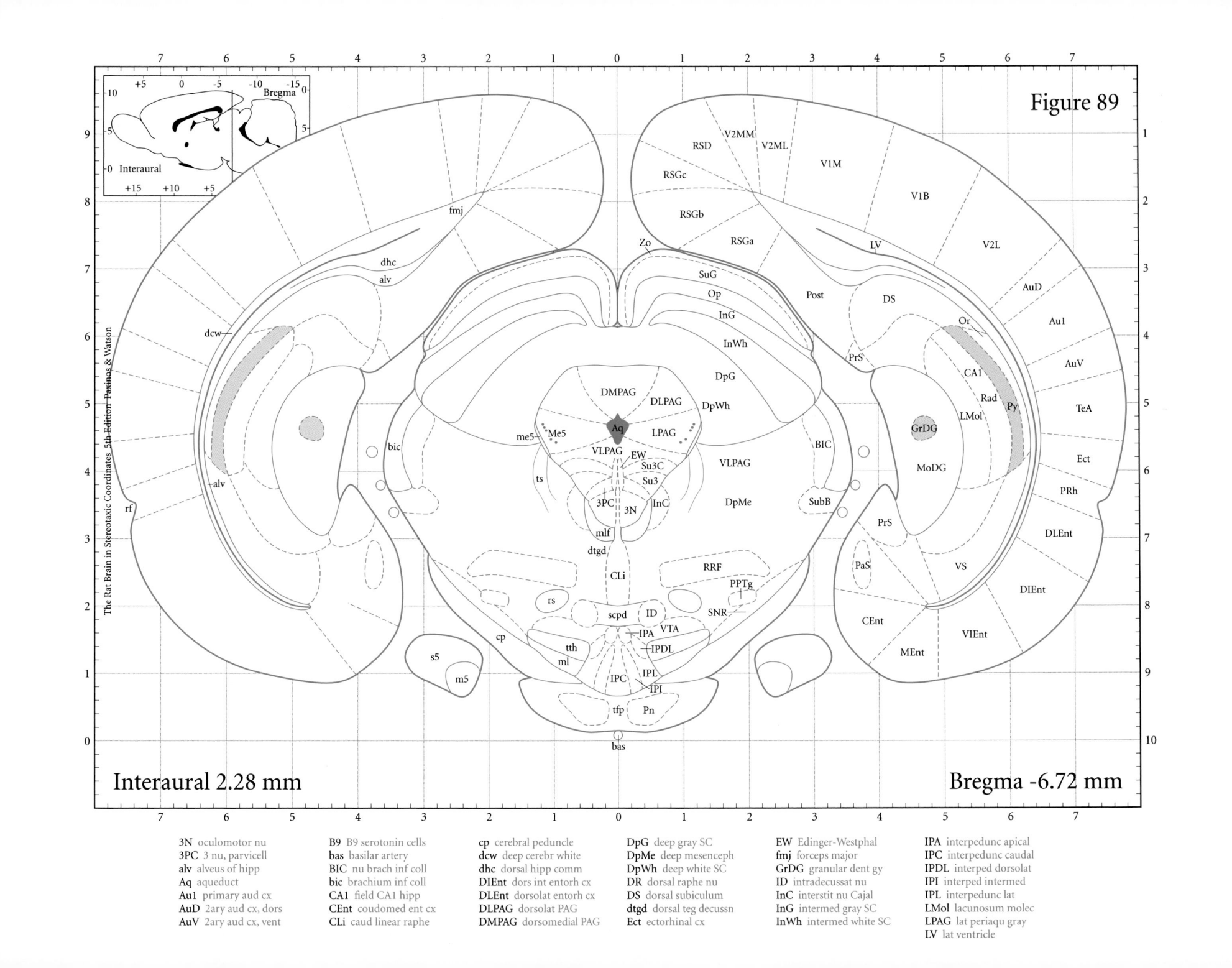

Figure 89

Interaural 2.28 mm

Bregma -6.72 mm

3N oculomotor nu
3PC 3 nu, parvicell
alv alveus of hipp
Aq aqueduct
Au1 primary aud cx
AuD 2ary aud cx, dors
AuV 2ary aud cx, vent

B9 B9 serotonin cells
bas basilar artery
BIC nu brach inf coll
bic brachium inf coll
CA1 field CA1 hipp
CEnt coudomed ent cx
CLi caud linear raphe

cp cerebral peduncle
dcw deep cerebr white
dhc dorsal hipp comm
DIEnt dors int entorh cx
DLEnt dorsolat entorh cx
DLPAG dorsolat PAG
DMPAG dorsomedial PAG

DpG deep gray SC
DpMe deep mesenceph
DpWh deep white SC
DR dorsal raphe nu
DS dorsal subiculum
dtgd dorsal teg decussn
Ect ectorhinal cx

EW Edinger-Westphal
fmj forceps major
GrDG granular dent gy
ID intradecussat nu
InC interstit nu Cajal
InG intermed gray SC
InWh intermed white SC

IPA interpedunc apical
IPC interpedunc caudal
IPDL interped dorsolat
IPI interped intermed
IPL interpedunc lat
LMol lacunosum molec
LPAG lat periaqu gray
LV lat ventricle

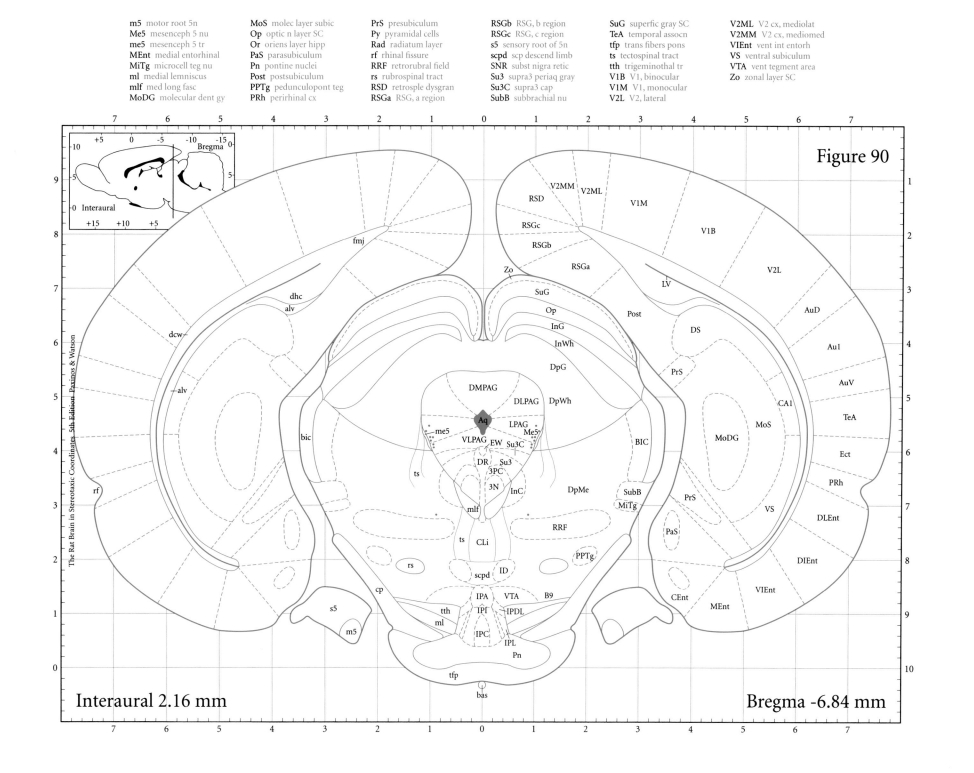

Figure 90

Interaural 2.16 mm

Bregma -6.84 mm

m5 motor root 5n
Me5 mesenceph 5 nu
me5 mesenceph 5 tr
MEnt medial entorhinal
MiTg microcell teg nu
ml medial lemniscus
mlf med long fasc
MoDG molecular dent gy

MoS molec layer subic
Op optic n layer SC
Or oriens layer hipp
PaS parasubiculum
Pn pontine nuclei
Post postsubiculum
PPTg pedunculopont teg
PRh perirhinal cx

PrS presubiculum
Py pyramidal cells
Rad radiatum layer
rf rhinal fissure
RRF retrorubral field
rs rubrospinal tract
RSD retrosple dysgran
RSGa RSG, a region

RSGb RSG, b region
RSGc RSG, c region
s5 sensory root of 5n
scpd scp descend limb
SNR subst nigra retic
Su3 supra3 periaq gray
Su3C supra3 cap
SubB subbrachial nu

SuG superfic gray SC
TeA temporal assocn
tfp trans fibers pons
ts tectospinal tract
tth trigeminothal tr
V1B V1, binocular
V1M V1, monocular
V2L V2, lateral

V2ML V2 cx, mediolat
V2MM V2 cx, mediomed
VIEnt vent int entorh
VS ventral subiculum
VTA vent tegment area
Zo zonal layer SC

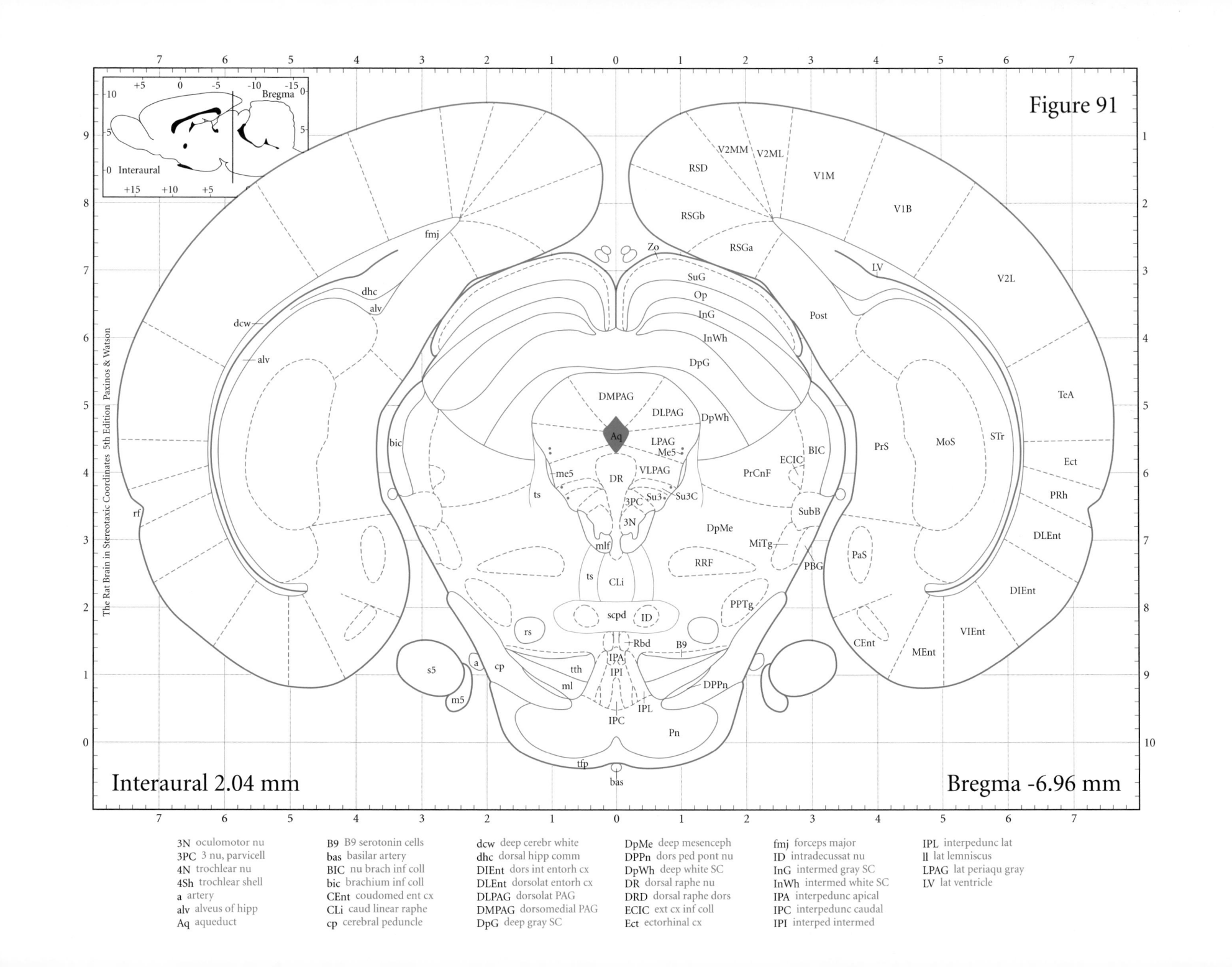

Figure 91

Interaural 2.04 mm

Bregma -6.96 mm

3N oculomotor nu
3PC 3 nu, parvicell
4N trochlear nu
4Sh trochlear shell
a artery
alv alveus of hipp
Aq aqueduct

B9 B9 serotonin cells
bas basilar artery
BIC nu brach inf coll
bic brachium inf coll
CEnt coudomed ent cx
CLi caud linear raphe
cp cerebral peduncle

dcw deep cerebr white
dhc dorsal hipp comm
DIEnt dors int entorh cx
DLEnt dorsolat entorh cx
DLPAG dorsolat PAG
DMPAG dorsomedial PAG
DpG deep gray SC

DpMe deep mesenceph
DPPn dors ped pont nu
DpWh deep white SC
DR dorsal raphe nu
DRD dorsal raphe dors
ECIC ext cx inf coll
Ect ectorhinal cx

fmj forceps major
ID intradecussat nu
InG intermed gray SC
InWh intermed white SC
IPA interpedunc apical
IPC interpedunc caudal
IPI interped intermed

IPL interpedunc lat
ll lat lemniscus
LPAG lat periaqu gray
LV lat ventricle

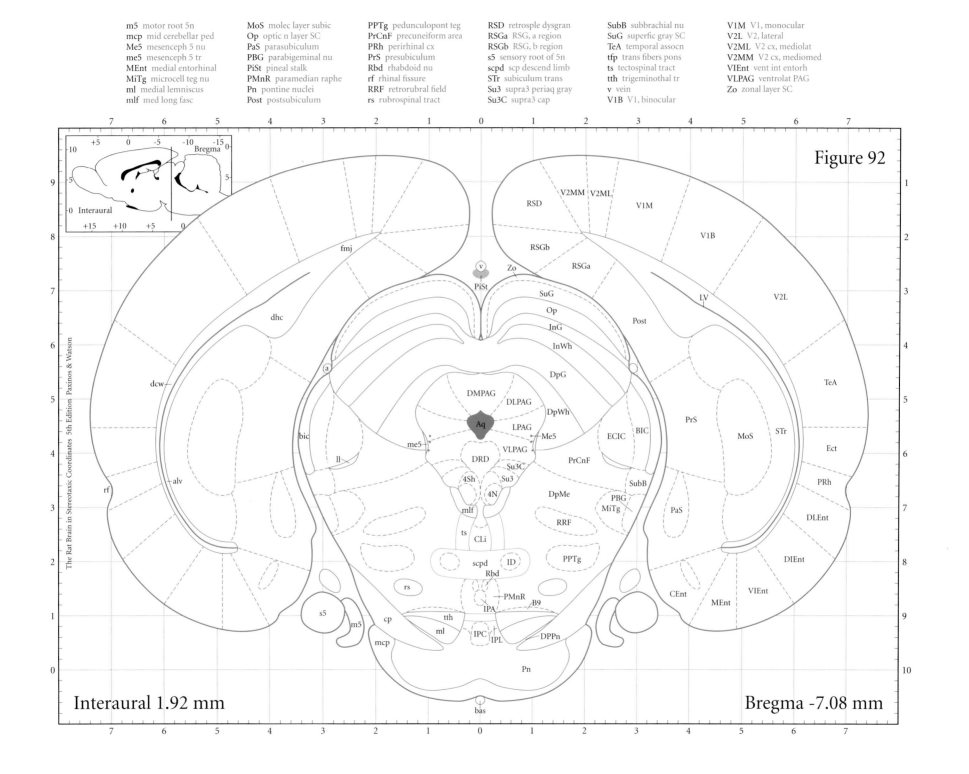

The Rat Brain in Stereotaxic Coordinates 5th Edition Paxinos & Watson

Figure 92

Interaural 1.92 mm

Bregma -7.08 mm

m5 motor root 5n
mcp mid cerebellar ped
Me5 mesenceph 5 nu
me5 mesenceph 5 tr
MEnt medial entorhinal
MiTg microcell teg nu
ml medial lemniscus
mlf med long fasc

MoS molec layer subic
Op optic n layer SC
PaS parasubiculum
PBG parabigeminal nu
PiSt pineal stalk
PMnR paramedian raphe
Pn pontine nuclei
Post postsubiculum

PPTg pedunculopont teg
PrCnF precuneiform area
PRh perirhinal cx
PrS presubiculum
Rbd rhabdoid nu
rf rhinal fissure
RRF retrorubral field
rs rubrospinal tract

RSD retrosple dysgran
RSGa RSG, a region
RSGb RSG, b region
s5 sensory root of 5n
scpd scp descend limb
STr subiculum trans
Su3 supra3 periaq gray
Su3C supra3 cap

SubB subbrachial nu
SuG superfic gray SC
TeA temporal assocn
tfp trans fibers pons
ts tectospinal tract
tth trigeminothal tr
v vein
V1B V1, binocular

V1M V1, monocular
V2L V2, lateral
V2ML V2 cx, mediolat
V2MM V2 cx, mediomed
VIEnt vent int entorh
VLPAG ventrolat PAG
Zo zonal layer SC

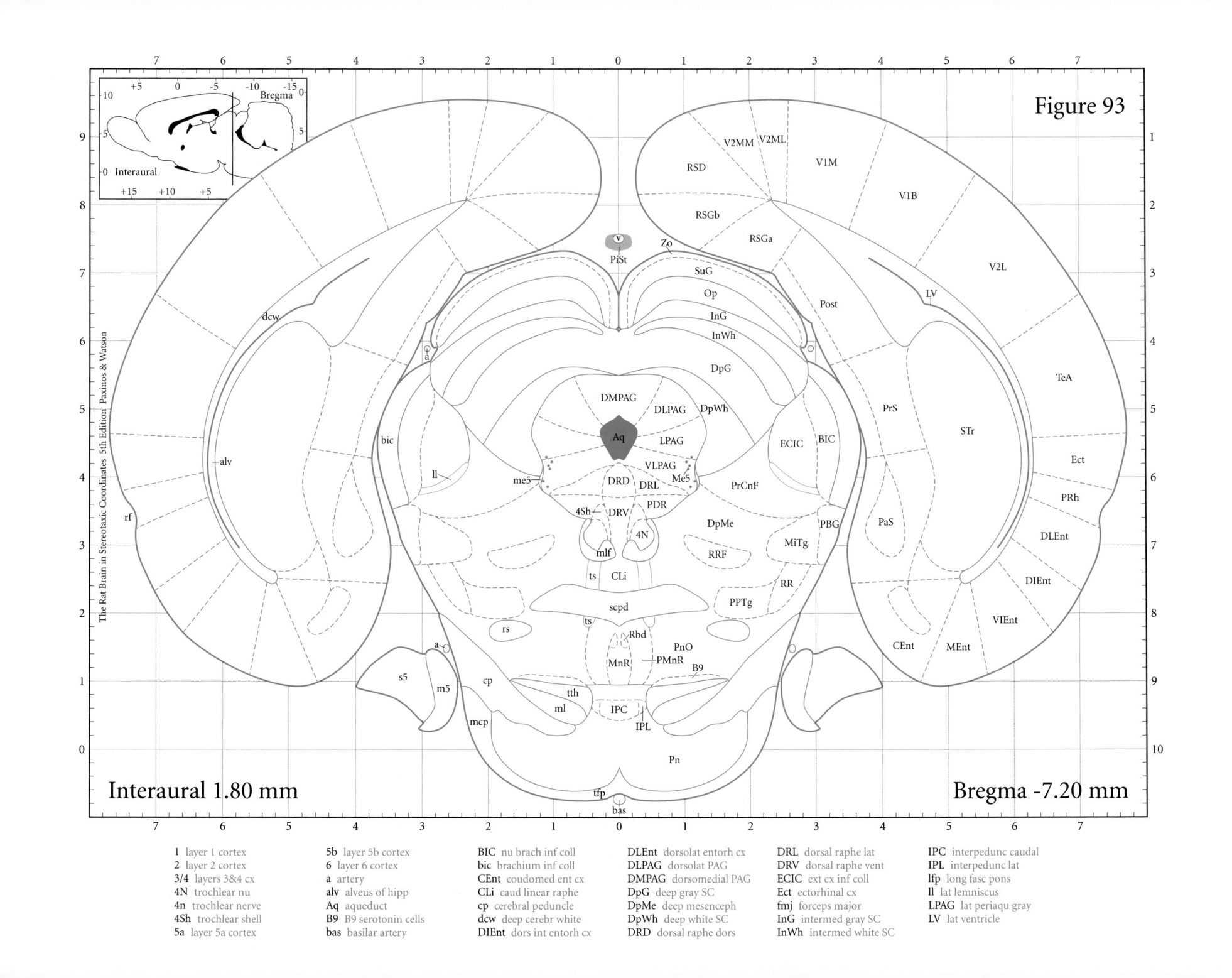

Figure 93

Interaural 1.80 mm

Bregma -7.20 mm

1	layer 1 cortex	5b	layer 5b cortex	BIC	nu brach inf coll
2	layer 2 cortex	6	layer 6 cortex	bic	brachium inf coll
3/4	layers 3&4 cx	a	artery	CEnt	coudomed ent cx
4N	trochlear nu	alv	alveus of hipp	CLi	caud linear raphe
4n	trochlear nerve	Aq	aqueduct	cp	cerebral peduncle
4Sh	trochlear shell	B9	B9 serotonin cells	dcw	deep cerebr white
5a	layer 5a cortex	bas	basilar artery	DIEnt	dors int entorh cx

DLEnt	dorsolat entorh cx	DRL	dorsal raphe lat
DLPAG	dorsolat PAG	DRV	dorsal raphe vent
DMPAG	dorsomedial PAG	ECIC	ext cx inf coll
DpG	deep gray SC	Ect	ectorhinal cx
DpMe	deep mesenceph	fmj	forceps major
DpWh	deep white SC	InG	intermed gray SC
DRD	dorsal raphe dors	InWh	intermed white SC

IPC	interpedunc caudal
IPL	interpedunc lat
lfp	long fasc pons
ll	lat lemniscus
LPAG	lat periaqu gray
LV	lat ventricle

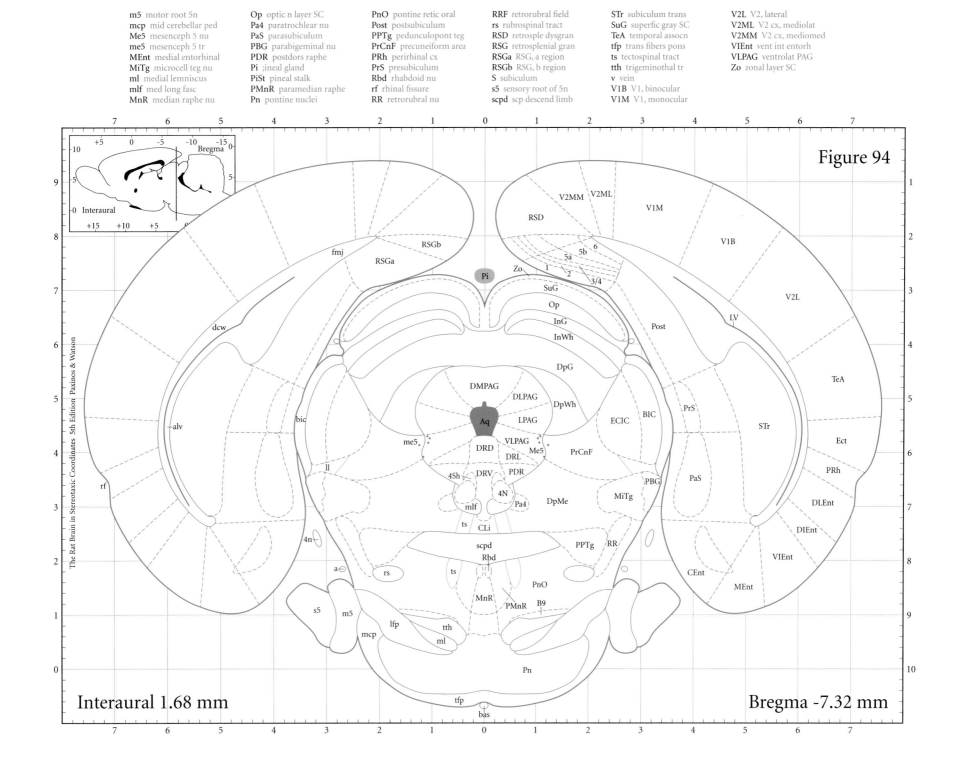

Figure 94

Interaural 1.68 mm

Bregma -7.32 mm

m5 motor root 5n
mcp mid cerebellar ped
Me5 mesenceph 5 nu
me5 mesenceph 5 tr
MEnt medial entorhinal
MiTg microcell teg nu
ml medial lemniscus
mlf med long fasc
MnR median raphe nu

Op optic n layer SC
Pa4 paratrochlear nu
PaS parasubiculum
PBG parabigeminal nu
PDR postdors raphe
Pi ;ineal gland
PiSt pineal stalk
PMnR paramedian raphe
Pn pontine nuclei

PnO pontine retic oral
Post postsubiculum
PPTg pedunculopont teg
PrCnF precuneiform area
PRh perirhinal cx
PrS presubiculum
Rbd rhabdoid nu
rf rhinal fissure
RR retrorubral nu

RRF retrorubral field
rs rubrospinal tract
RSD retrosple dysgran
RSG retrosplenial gran
RSGa RSG, a region
RSGb RSG, b region
S subiculum
s5 sensory root of 5n
scpd scp descend limb

STr subiculum trans
SuG superfic gray SC
TeA temporal assocn
tfp trans fibers pons
ts tectospinal tract
tth trigeminothal tr
v vein
V1B V1, binocular
V1M V1, monocular

V2L V2, lateral
V2ML V2 cx, mediolat
V2MM V2 cx, mediomed
VIEnt vent int entorh
VLPAG ventrolat PAG
Zo zonal layer SC

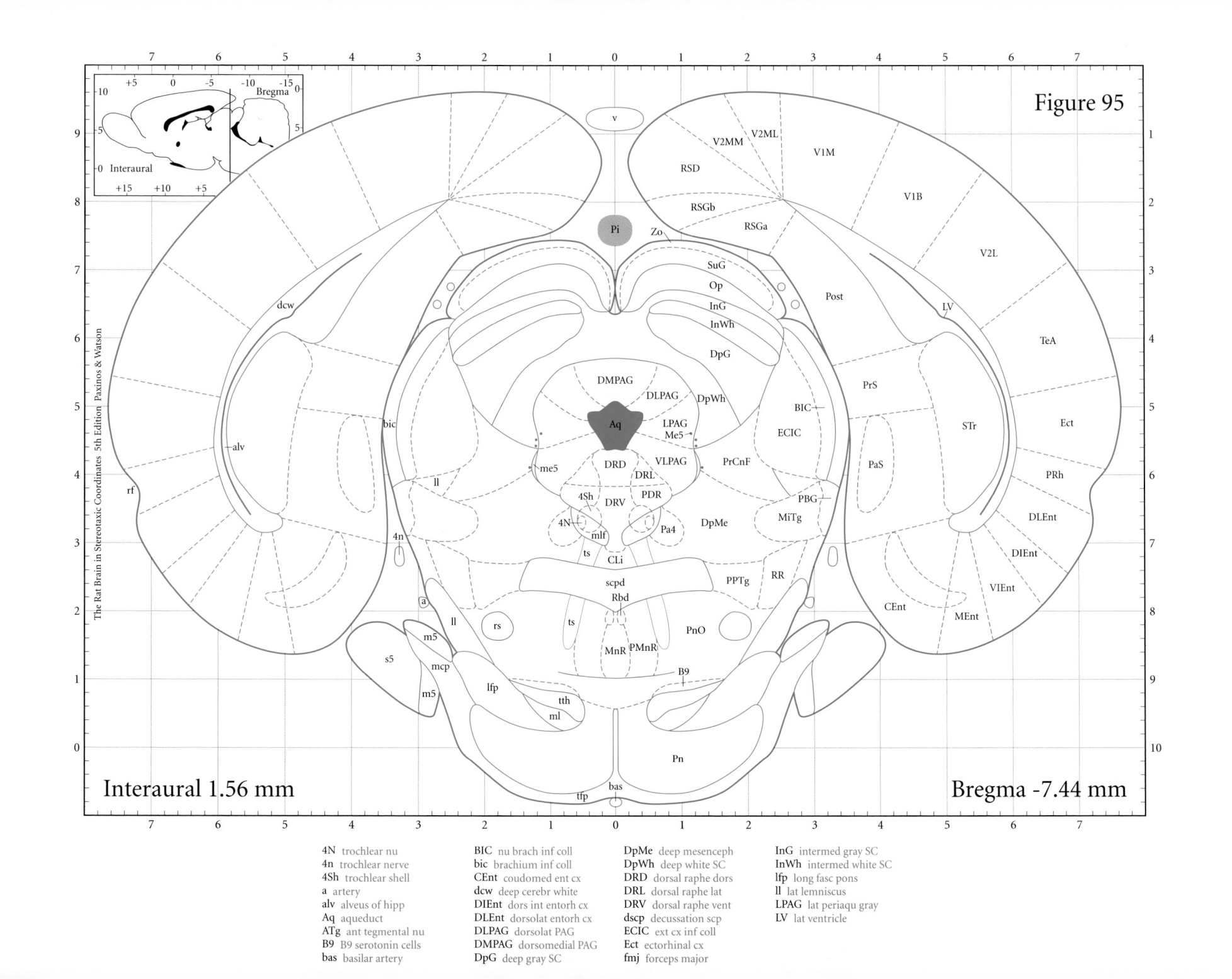

Figure 95

Interaural 1.56 mm

Bregma -7.44 mm

4N   trochlear nu
4n   trochlear nerve
4Sh trochlear shell
a   artery
alv alveus of hipp
Aq   aqueduct
ATg ant tegmental nu
B9   B9 serotonin cells
bas basilar artery

BIC nu brach inf coll
bic brachium inf coll
CEnt coudomed ent cx
dcw deep cerebr white
DIEnt dors int entorh cx
DLEnt dorsolat entorh cx
DLPAG dorsolat PAG
DMPAG dorsomedial PAG
DpG deep gray SC

DpMe deep mesenceph
DpWh deep white SC
DRD dorsal raphe dors
DRL dorsal raphe lat
DRV dorsal raphe vent
dscp decussation scp
ECIC ext cx inf coll
Ect ectorhinal cx
fmj forceps major

InG intermed gray SC
InWh intermed white SC
lfp long fasc pons
ll   lat lemniscus
LPAG lat periaqu gray
LV   lat ventricle

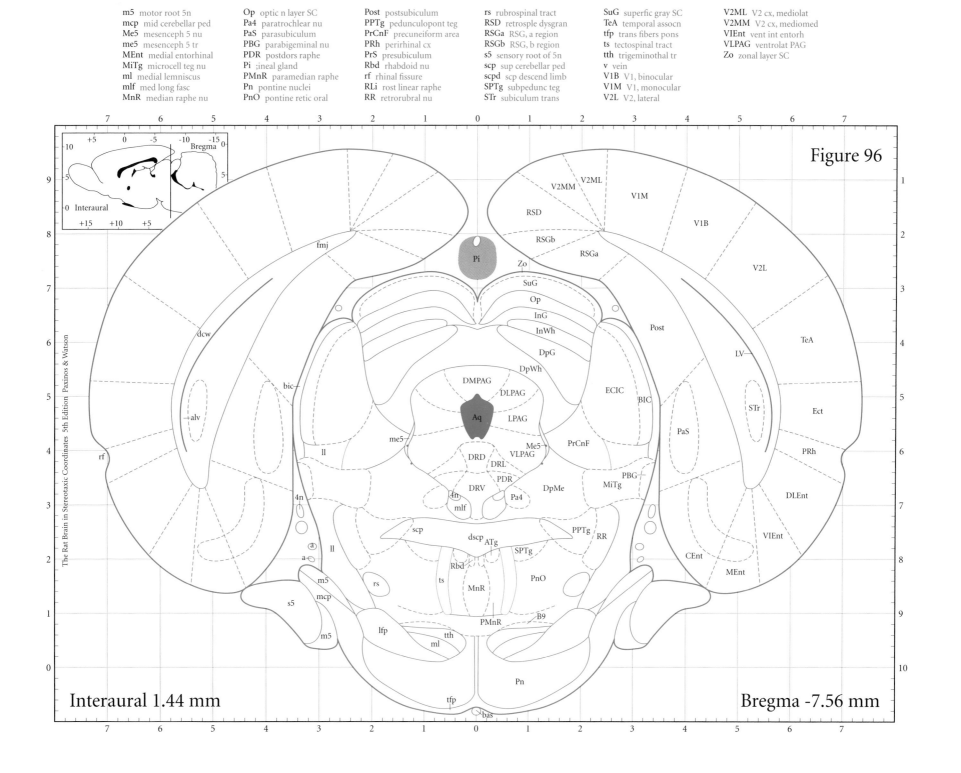

m5 motor root 5n
mcp mid cerebellar ped
Me5 mesenceph 5 nu
me5 mesenceph 5 tr
MEnt medial entorhinal
MiTg microcell teg nu
ml medial lemniscus
mlf med long fasc
MnR median raphe nu

Op optic n layer SC
Pa4 paratrochlear nu
PaS parasubiculum
PBG parabigeminal nu
PDR postdors raphe
Pi ;ineal gland
PMnR paramedian raphe
Pn pontine nuclei
PnO pontine retic oral

Post postsubiculum
PPTg pedunculopont teg
PrCnF precuneiform area
PRh perirhinal cx
PrS presubiculum
Rbd rhabdoid nu
rf rhinal fissure
RLi rost linear raphe
RR retrorubral nu

rs rubrospinal tract
RSD retrosple dysgran
RSGa RSG, a region
RSGb RSG, b region
s5 sensory root of 5n
scp sup cerebellar ped
scpd scp descend limb
SPTg subpedunc teg
STr subiculum trans

SuG superfic gray SC
TeA temporal assocn
tfp trans fibers pons
ts tectospinal tract
tth trigeminothal tr
v vein
V1B V1, binocular
V1M V1, monocular
V2L V2, lateral

V2ML V2 cx, mediolat
V2MM V2 cx, mediomed
VIEnt vent int entorh
VLPAG ventrolat PAG
Zo zonal layer SC

Figure 96

Interaural 1.44 mm

Bregma -7.56 mm

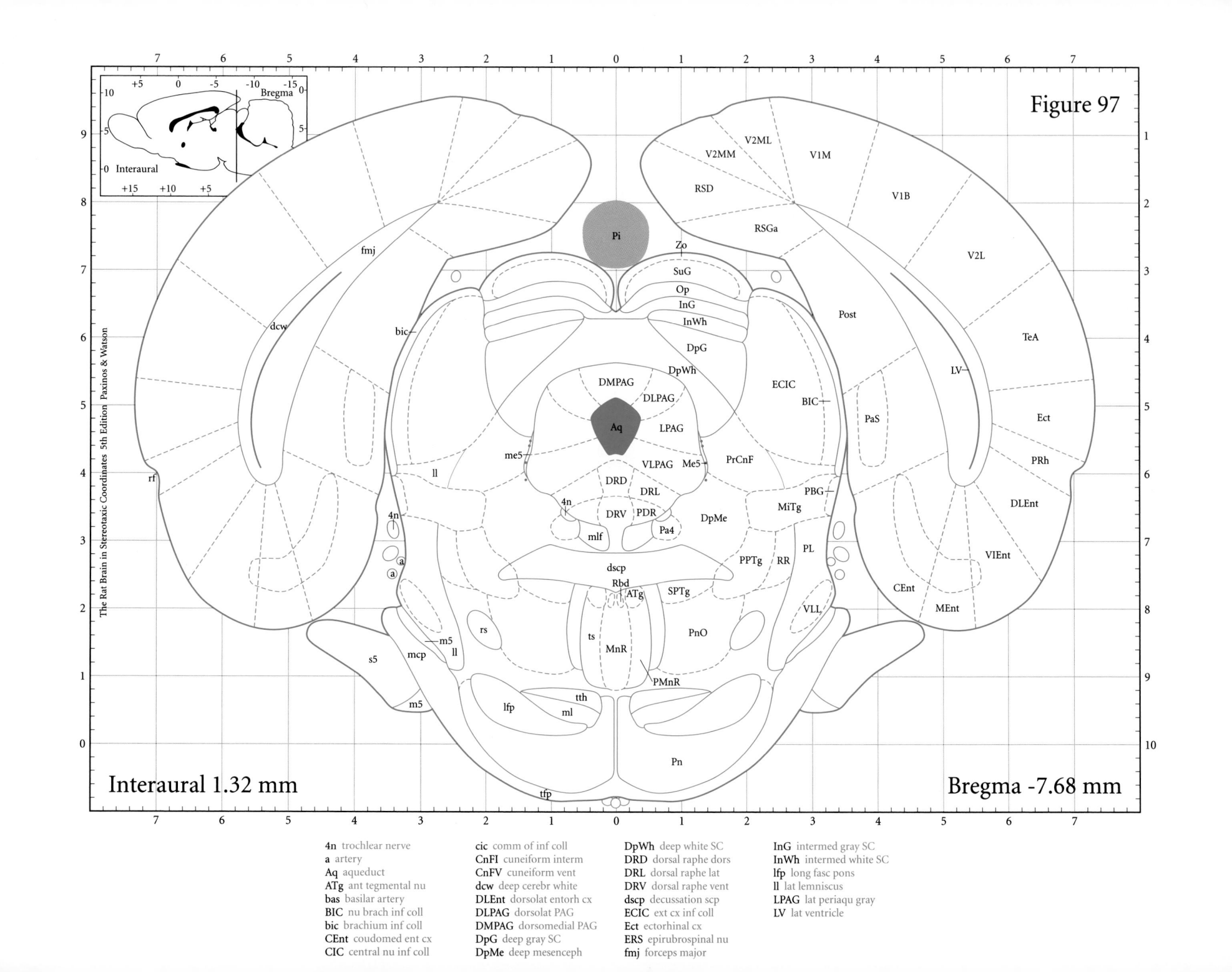

Figure 97

Interaural 1.32 mm

Bregma -7.68 mm

4n trochlear nerve
a artery
Aq aqueduct
ATg ant tegmental nu
bas basilar artery
BIC nu brach inf coll
bic brachium inf coll
CEnt coudomed ent cx
CIC central nu inf coll

cic comm of inf coll
CnFI cuneiform interm
CnFV cuneiform vent
dcw deep cerebr white
DLEnt dorsolat entorh cx
DLPAG dorsolat PAG
DMPAG dorsomedial PAG
DpG deep gray SC
DpMe deep mesenceph

DpWh deep white SC
DRD dorsal raphe dors
DRL dorsal raphe lat
DRV dorsal raphe vent
dscp decussation scp
ECIC ext cx inf coll
Ect ectorhinal cx
ERS epirubrospinal nu
fmj forceps major

InG intermed gray SC
InWh intermed white SC
lfp long fasc pons
ll lat lemniscus
LPAG lat periaqu gray
LV lat ventricle

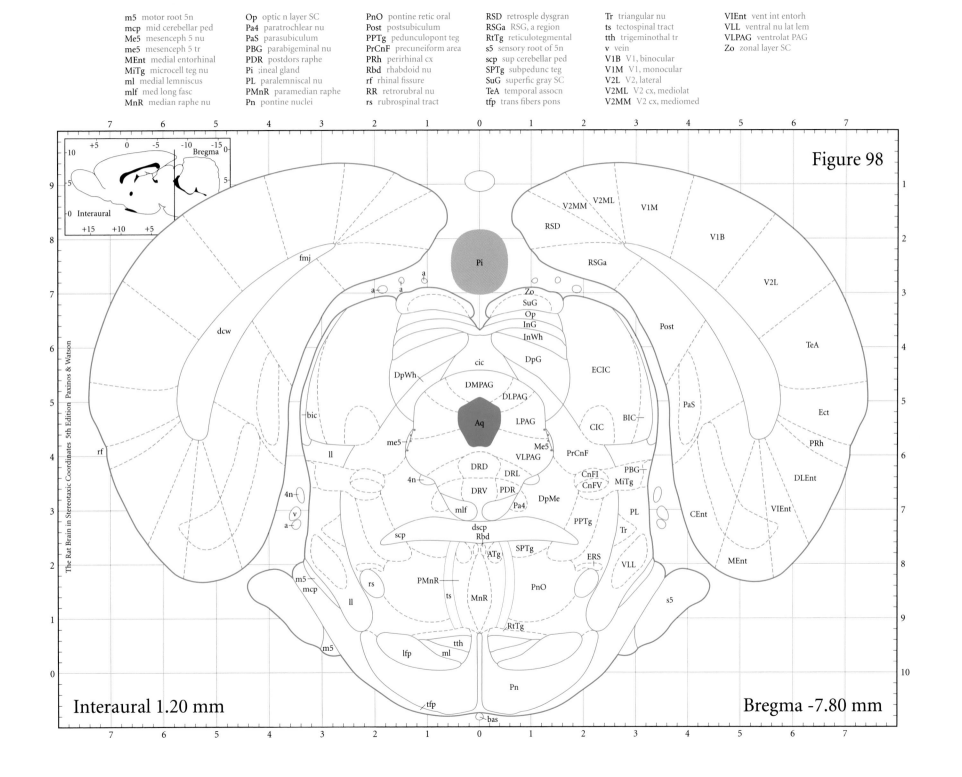

The Rat Brain in Stereotaxic Coordinates 5th Edition Paxinos & Watson

Figure 98

Interaural 1.20 mm

Bregma -7.80 mm

m5 motor root 5n
mcp mid cerebellar ped
Me5 mesenceph 5 nu
me5 mesenceph 5 tr
MEnt medial entorhinal
MiTg microcell teg nu
ml medial lemniscus
mlf med long fasc
MnR median raphe nu

Op optic n layer SC
Pa4 paratrochlear nu
PaS parasubiculum
PBG parabigeminal nu
PDR postdors raphe
Pi ;ineal gland
PL paralemniscal nu
PMnR paramedian raphe
Pn pontine nuclei

PnO pontine retic oral
Post postsubiculum
PPTg pedunculopont teg
PrCnF precuneiform area
PRh perirhinal cx
Rbd rhabdoid nu
rf rhinal fissure
RR retrorubral nu
rs rubrospinal tract

RSD retrosple dysgran
RSGa RSG, a region
RtTg reticulotegmental
s5 sensory root of 5n
scp sup cerebellar ped
SPTg subpedunc teg
SuG superfic gray SC
TeA temporal assocn
tfp trans fibers pons

Tr triangular nu
ts tectospinal tract
tth trigeminothal tr
v vein
V1B V1, binocular
V1M V1, monocular
V2L V2, lateral
V2ML V2 cx, mediolat
V2MM V2 cx, mediomed

VIEnt vent int entorh
VLL ventral nu lat lem
VLPAG ventrolat PAG
Zo zonal layer SC

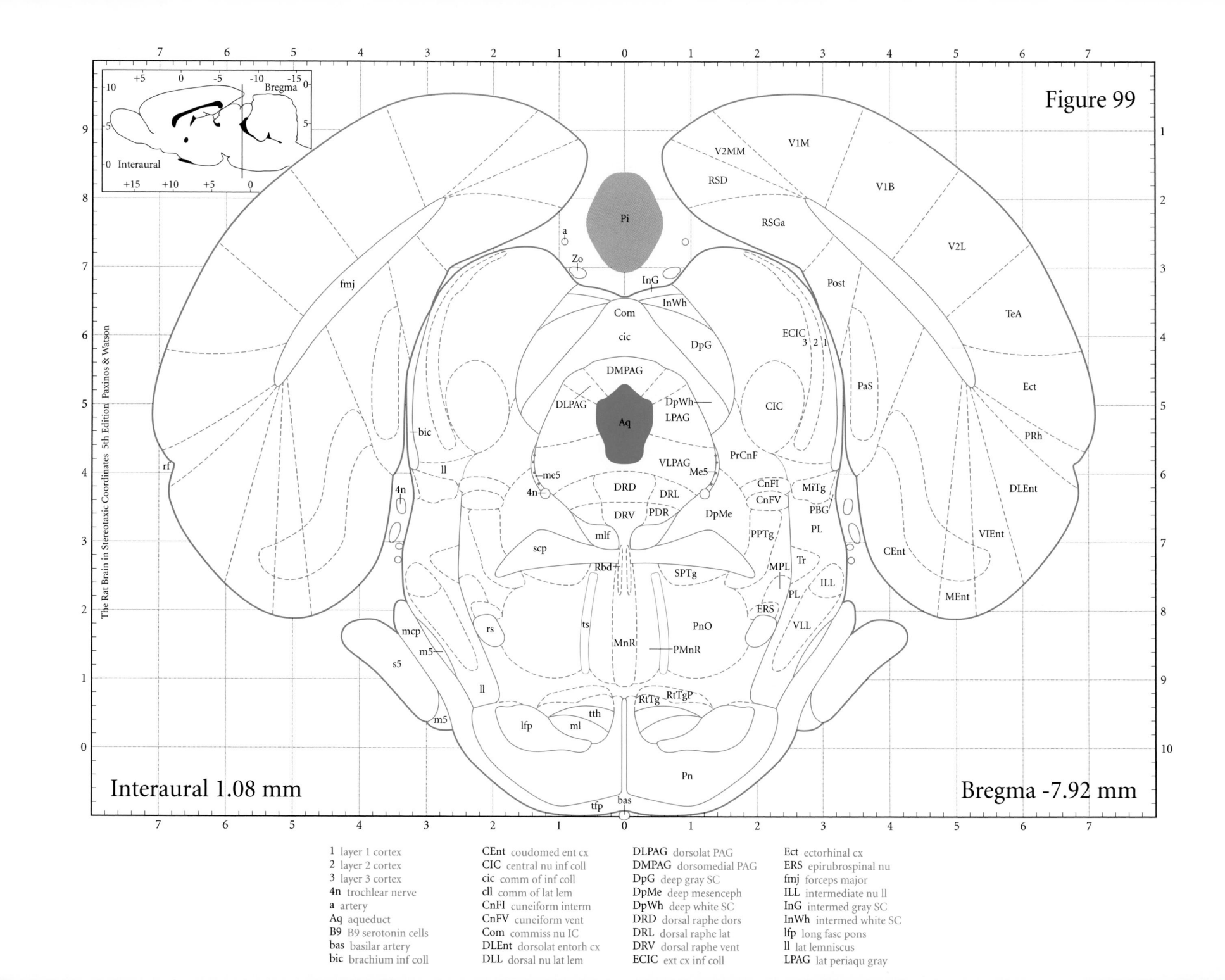

Figure 99

Interaural 1.08 mm

Bregma -7.92 mm

1   layer 1 cortex
2   layer 2 cortex
3   layer 3 cortex
4n trochlear nerve
a   artery
Aq aqueduct
B9 B9 serotonin cells
bas basilar artery
bic brachium inf coll

CEnt coudomed ent cx
CIC   central nu inf coll
cic   comm of inf coll
cll   comm of lat lem
CnFI cuneiform interm
CnFV cuneiform vent
Com   commiss nu IC
DLEnt dorsolat entorh cx
DLL   dorsal nu lat lem

DLPAG dorsolat PAG
DMPAG dorsomedial PAG
DpG   deep gray SC
DpMe deep mesenceph
DpWh deep white SC
DRD   dorsal raphe dors
DRL   dorsal raphe lat
DRV   dorsal raphe vent
ECIC ext cx inf coll

Ect   ectorhinal cx
ERS   epirubrospinal nu
fmj   forceps major
ILL   intermediate nu ll
InG   intermed gray SC
InWh intermed white SC
lfp   long fasc pons
ll   lat lemniscus
LPAG lat periaqu gray

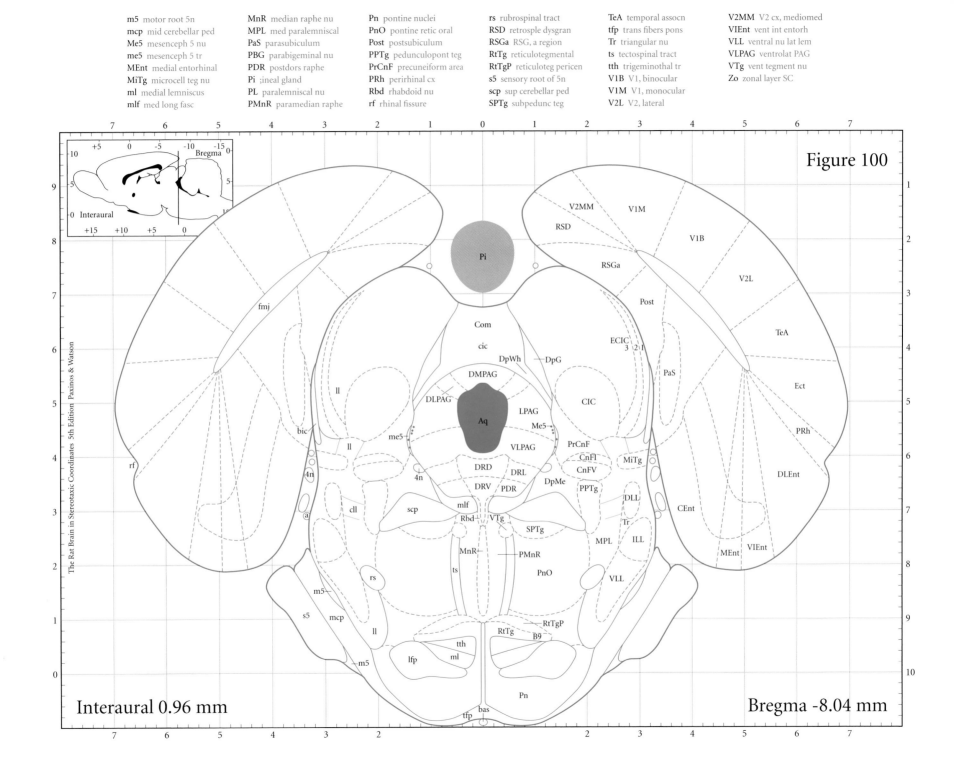

m5 motor root 5n
mcp mid cerebellar ped
Me5 mesenceph 5 nu
me5 mesenceph 5 tr
MEnt medial entorhinal
MiTg microcell teg nu
ml medial lemniscus
mlf med long fasc

MnR median raphe nu
MPL med paralemniscal
PaS parasubiculum
PBG parabigeminal nu
PDR postdors raphe
Pi ;ineal gland
PL paralemniscal nu
PMnR paramedian raphe

Pn pontine nuclei
PnO pontine retic oral
Post postsubiculum
PPTg pedunculopont teg
PrCnF precuneiform area
PRh perirhinal cx
Rbd rhabdoid nu
rf rhinal fissure

rs rubrospinal tract
RSD retrosple dysgran
RSGa RSG, a region
RtTg reticulotegmental
RtTgP reticuloteg pericen
s5 sensory root of 5n
scp sup cerebellar ped
SPTg subpedunc teg

TeA temporal assocn
tfp trans fibers pons
Tr triangular nu
ts tectospinal tract
tth trigeminothal tr
V1B V1, binocular
V1M V1, monocular
V2L V2, lateral

V2MM V2 cx, mediomed
VIEnt vent int entorh
VLL ventral nu lat lem
VLPAG ventrolat PAG
VTg vent tegment nu
Zo zonal layer SC

Figure 100

Interaural 0.96 mm

Bregma -8.04 mm

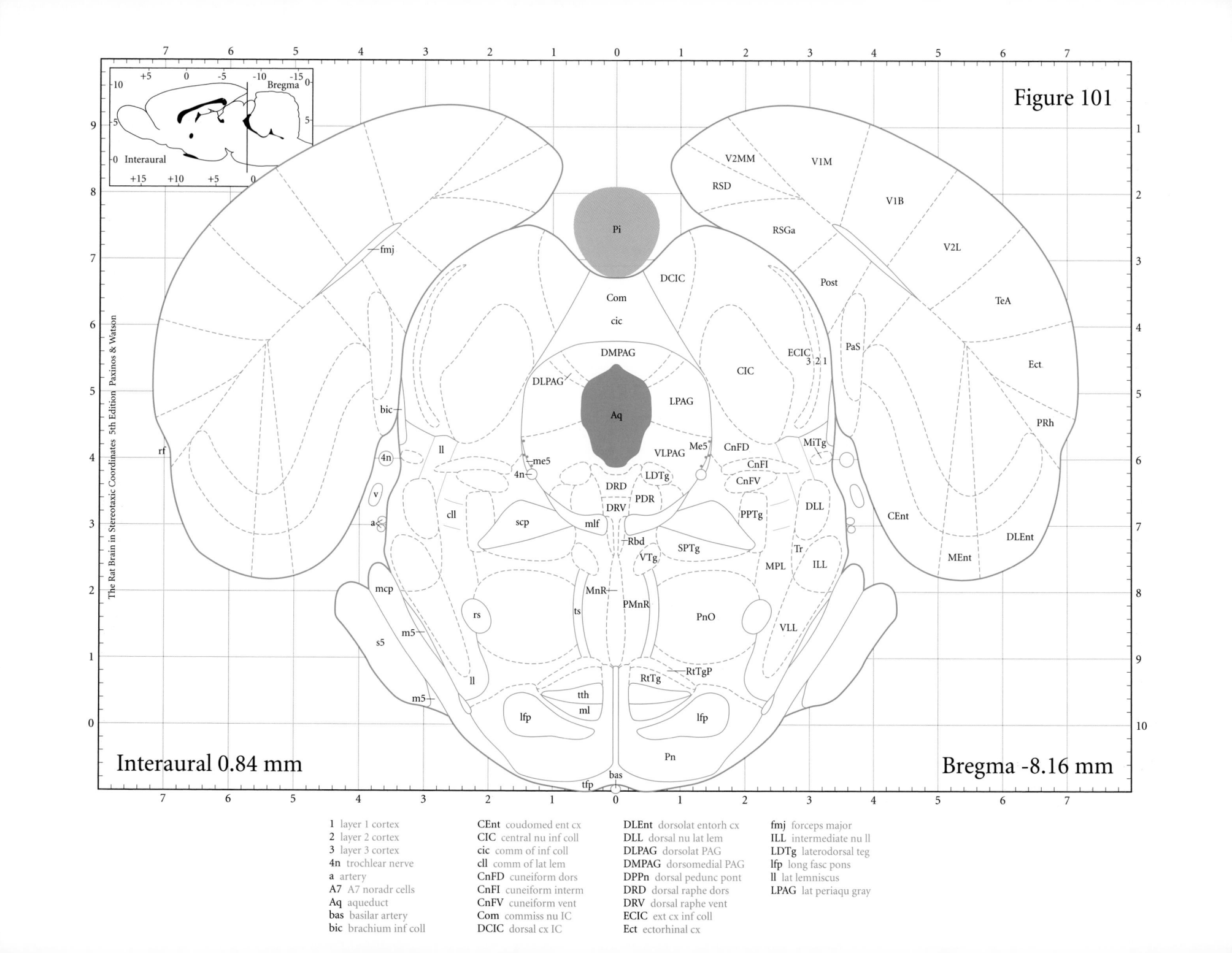

Figure 101

Interaural 0.84 mm

Bregma -8.16 mm

The Rat Brain in Stereotaxic Coordinates 5th Edition Paxinos & Watson

1 layer 1 cortex	CEnt coudomed ent cx	DLEnt dorsolat entorh cx	fmj forceps major
2 layer 2 cortex	CIC central nu inf coll	DLL dorsal nu lat lem	ILL intermediate nu ll
3 layer 3 cortex	cic comm of inf coll	DLPAG dorsolat PAG	LDTg laterodorsal teg
4n trochlear nerve	cll comm of lat lem	DMPAG dorsomedial PAG	lfp long fasc pons
a artery	CnFD cuneiform dors	DPPn dorsal pedunc pont	ll lat lemniscus
A7 A7 noradr cells	CnFI cuneiform interm	DRD dorsal raphe dors	LPAG lat periaqu gray
Aq aqueduct	CnFV cuneiform vent	DRV dorsal raphe vent	
bas basilar artery	Com commiss nu IC	ECIC ext cx inf coll	
bic brachium inf coll	DCIC dorsal cx IC	Ect ectorhinal cx	

Figure 102

m5 motor root 5n
mcp mid cerebellar ped
Me5 mesenceph 5 nu
me5 mesenceph 5 tr
MEnt medial entorhinal
MiTg microcell teg nu
ml medial lemniscus
mlf med long fasc

MnR median raphe nu
MPL med paralemniscal
PaS parasubiculum
PDR postdors raphe
Pi ;ineal gland
PMnR paramedian raphe
Pn pontine nuclei
PnO pontine retic oral

Post postsubiculum
PPTg pedunculopont teg
PRh perirhinal cx
Rbd rhabdoid nu
rf rhinal fissure
rs rubrospinal tract
RSD retrosple dysgran
RSGa RSG, a region

RtTg reticulotegmental
RtTgP reticuloteg pericen
s5 sensory root of 5n
Sag sagulum nu
scp sup cerebellar ped
SPTg subpedunc teg
TeA temporal assocn
tfp trans fibers pons

Tr triangular nu
ts tectospinal tract
tth trigeminothal tr
v vein
V1B V1, binocular
V1M V1, monocular
V2L V2, lateral
V2MM V2 cx, mediomed

VLL ventral nu lat lem
VLPAG ventrolat PAG
VTg vent tegment nu

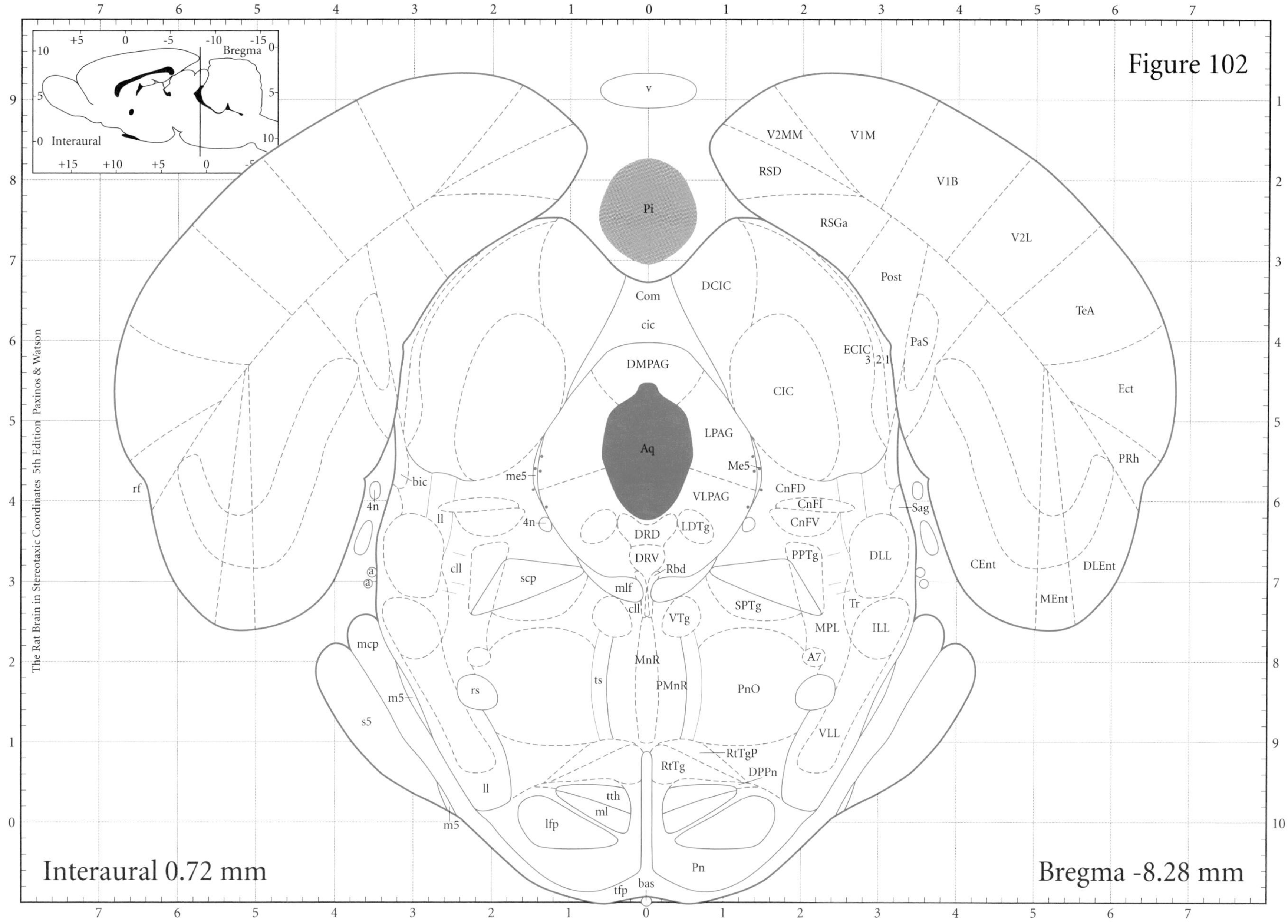

Interaural 0.72 mm

Bregma -8.28 mm

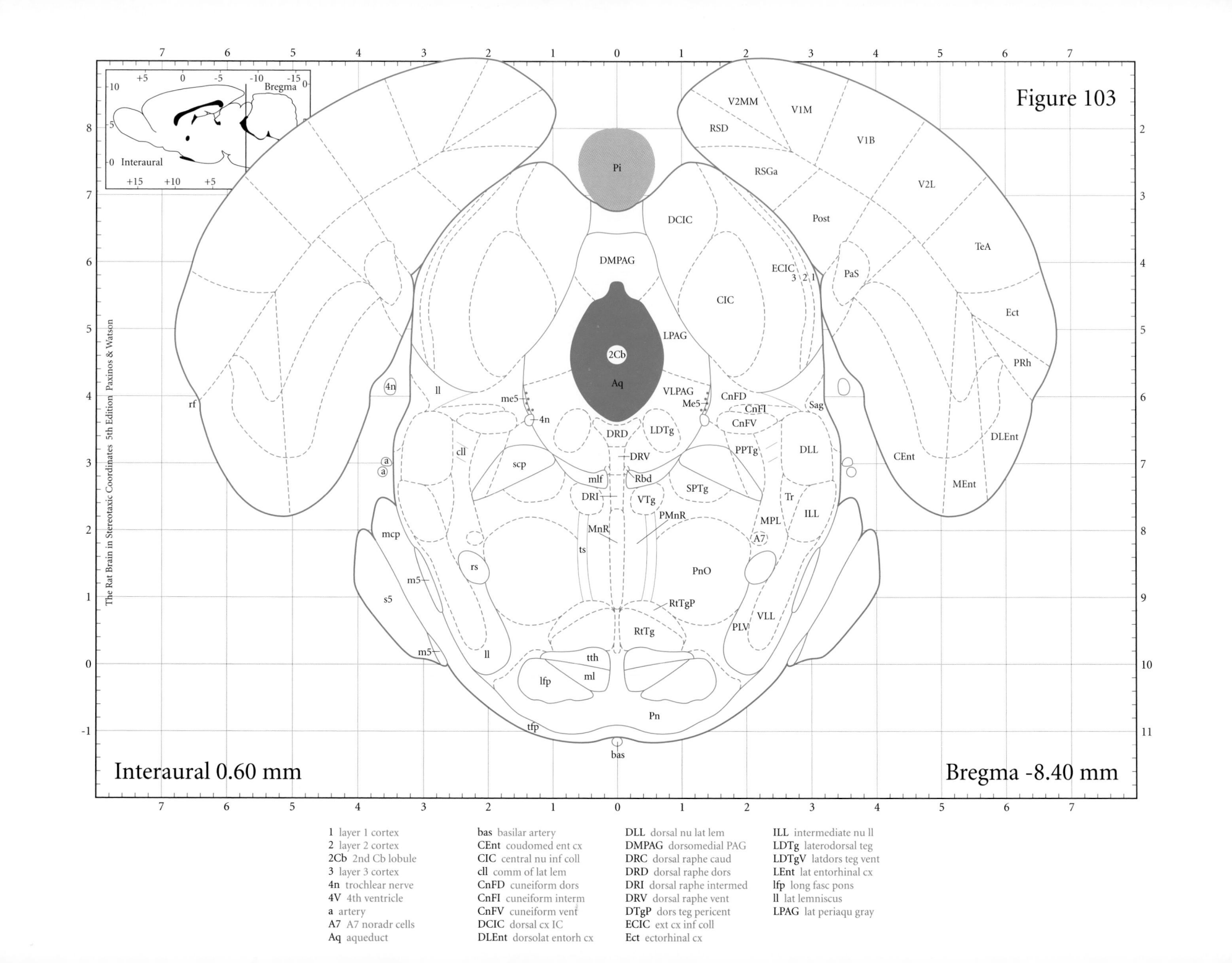

Figure 103

Interaural 0.60 mm

Bregma -8.40 mm

The Rat Brain in Stereotaxic Coordinates 5th Edition Paxinos & Watson

1 layer 1 cortex
2 layer 2 cortex
2Cb 2nd Cb lobule
3 layer 3 cortex
4n trochlear nerve
4V 4th ventricle
a artery
A7 A7 noradr cells
Aq aqueduct

bas basilar artery
CEnt coudomed ent cx
CIC central nu inf coll
cll comm of lat lem
CnFD cuneiform dors
CnFI cuneiform interm
CnFV cuneiform vent
DCIC dorsal cx IC
DLEnt dorsolat entorh cx

DLL dorsal nu lat lem
DMPAG dorsomedial PAG
DRC dorsal raphe caud
DRD dorsal raphe dors
DRI dorsal raphe intermed
DRV dorsal raphe vent
DTgP dors teg pericent
ECIC ext cx inf coll
Ect ectorhinal cx

ILL intermediate nu ll
LDTg laterodorsal teg
LDTgV latdors teg vent
LEnt lat entorhinal cx
lfp long fasc pons
ll lat lemniscus
LPAG lat periaqu gray

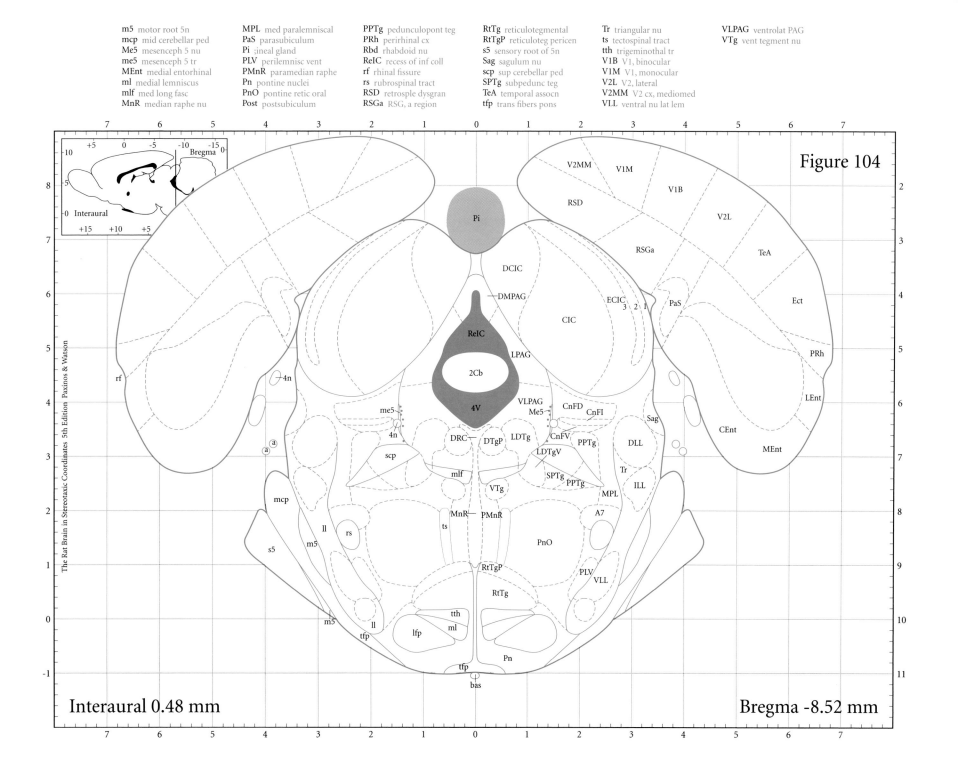

Figure 104

Interaural 0.48 mm

Bregma -8.52 mm

The Rat Brain in Stereotaxic Coordinates 5th Edition Paxinos & Watson

m5 motor root 5n
mcp mid cerebellar ped
Me5 mesenceph 5 nu
me5 mesenceph 5 tr
MEnt medial entorhinal
ml medial lemniscus
mlf med long fasc
MnR median raphe nu

MPL med paralemniscal
PaS parasubiculum
Pi ;ineal gland
PLV perilemnisc vent
PMnR paramedian raphe
Pn pontine nuclei
PnO pontine retic oral
Post postsubiculum

PPTg pedunculopont teg
PRh perirhinal cx
Rbd rhabdoid nu
ReIC recess of inf coll
rf rhinal fissure
rs rubrospinal tract
RSD retrosple dysgran
RSGa RSG, a region

RtTg reticulotegmental
RtTgP reticuloteg pericen
s5 sensory root of 5n
Sag sagulum nu
scp sup cerebellar ped
SPTg subpedunc teg
TeA temporal associn
tfp trans fibers pons

Tr triangular nu
ts tectospinal tract
tth trigeminothal tr
V1B V1, binocular
V1M V1, monocular
V2L V2, lateral
V2MM V2 cx, mediomed
VLL ventral nu lat lem

VLPAG ventrolat PAG
VTg vent tegment nu

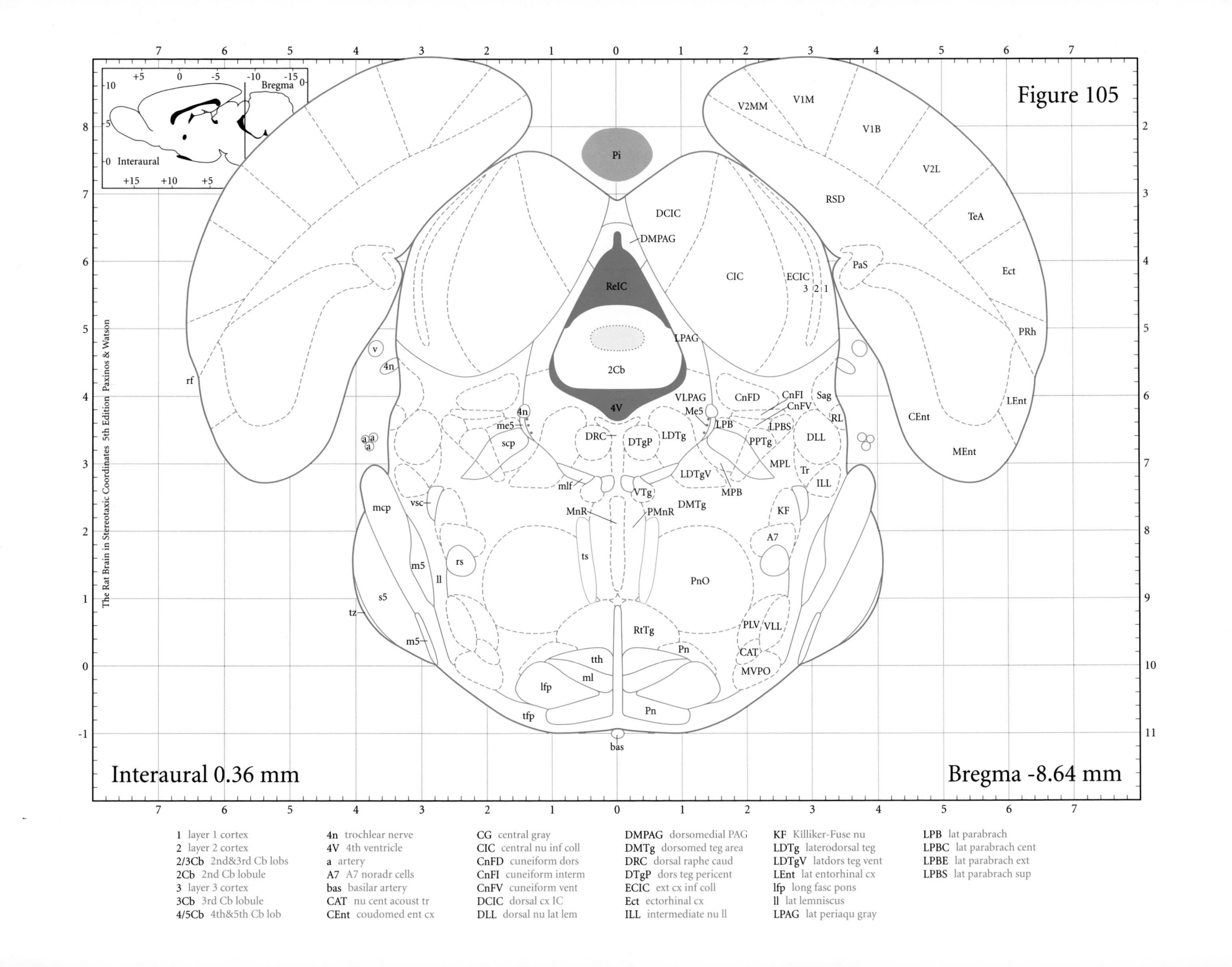

Figure 105

Interaural 0.36 mm

Bregma -8.64 mm

The Rat Brain in Stereotaxic Coordinates 5th Edition Paxinos & Watson

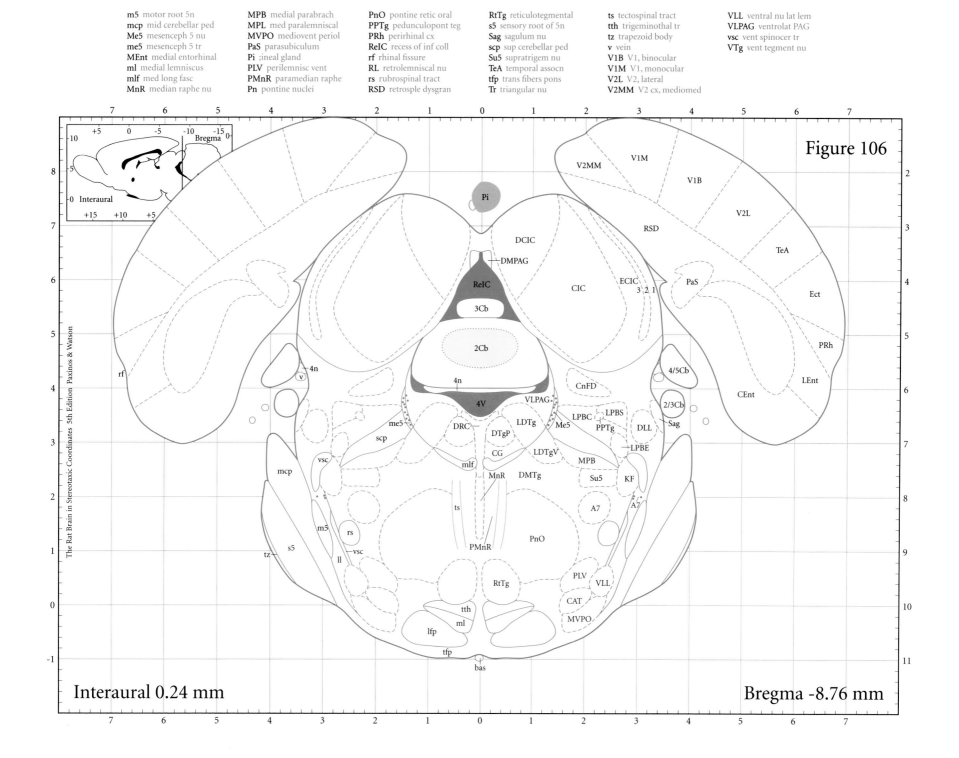

The Rat Brain in Stereotaxic Coordinates 5th Edition Paxinos & Watson

Figure 106

Interaural 0.24 mm

Bregma -8.76 mm

m5 motor root 5n
mcp mid cerebellar ped
Me5 mesenceph 5 nu
me5 mesenceph 5 tr
MEnt medial entorhinal
ml medial lemniscus
mlf med long fasc
MnR median raphe nu

MPB medial parabrach
MPL med paralemniscal
MVPO mediovent periol
PaS parasubiculum
Pi ;ineal gland
PLV perilemnisc vent
PMnR paramedian raphe
Pn pontine nuclei

PnO pontine retic oral
PPTg pedunculopont teg
PRh perirhinal cx
ReIC recess of inf coll
rf rhinal fissure
RL retrolemniscal nu
rs rubrospinal tract
RSD retrosple dysgran

RtTg reticulotegmental
s5 sensory root of 5n
Sag sagulum nu
scp sup cerebellar ped
Su5 supratrigem nu
TeA temporal assocn
tfp trans fibers pons
Tr triangular nu

ts tectospinal tract
tth trigeminothal tr
tz trapezoid body
v vein
V1B V1, binocular
V1M V1, monocular
V2L V2, lateral
V2MM V2 cx, mediomed

VLL ventral nu lat lem
VLPAG ventrolat PAG
vsc vent spinocer tr
VTg vent tegment nu

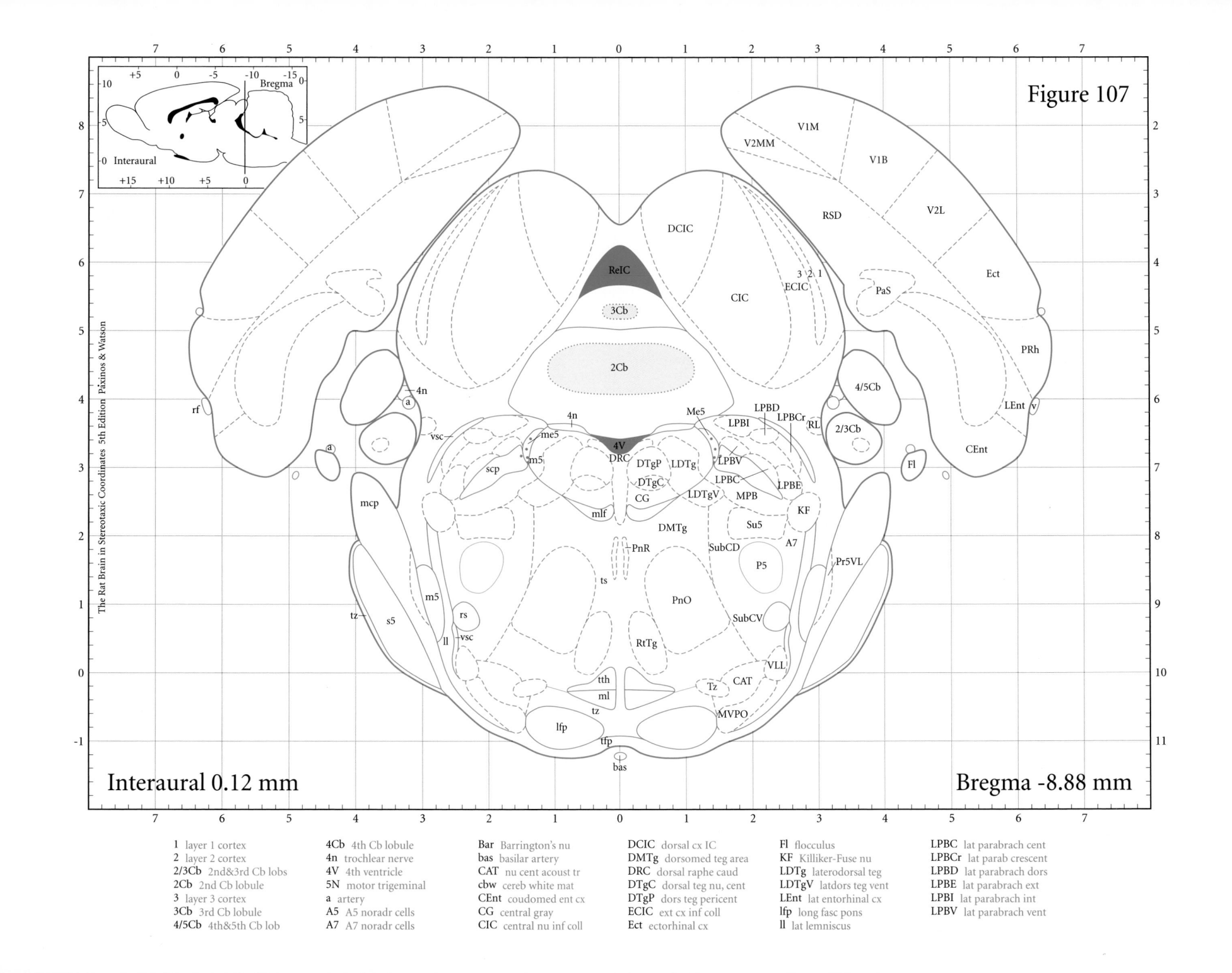

Figure 107

Interaural 0.12 mm

Bregma -8.88 mm

The Rat Brain in Stereotaxic Coordinates 5th Edition Paxinos & Watson

1 layer 1 cortex	4Cb 4th Cb lobule	Bar Barrington's nu	DCIC dorsal cx IC	Fl flocculus	LPBC lat parabrach cent
2 layer 2 cortex	4n trochlear nerve	bas basilar artery	DMTg dorsomed teg area	KF Killiker-Fuse nu	LPBCr lat parab crescent
2/3Cb 2nd&3rd Cb lobs	4V 4th ventricle	CAT nu cent acoust tr	DRC dorsal raphe caud	LDTg laterodorsal teg	LPBD lat parabrach dors
2Cb 2nd Cb lobule	5N motor trigeminal	cbw cereb white mat	DTgC dorsal teg nu, cent	LDTgV latdors teg vent	LPBE lat parabrach ext
3 layer 3 cortex	a artery	CEnt coudomed ent cx	DTgP dors teg pericent	LEnt lat entorhinal cx	LPBI lat parabrach int
3Cb 3rd Cb lobule	A5 A5 noradr cells	CG central gray	ECIC ext cx inf coll	lfp long fasc pons	LPBV lat parabrach vent
4/5Cb 4th&5th Cb lob	A7 A7 noradr cells	CIC central nu inf coll	Ect ectorhinal cx	ll lat lemniscus	

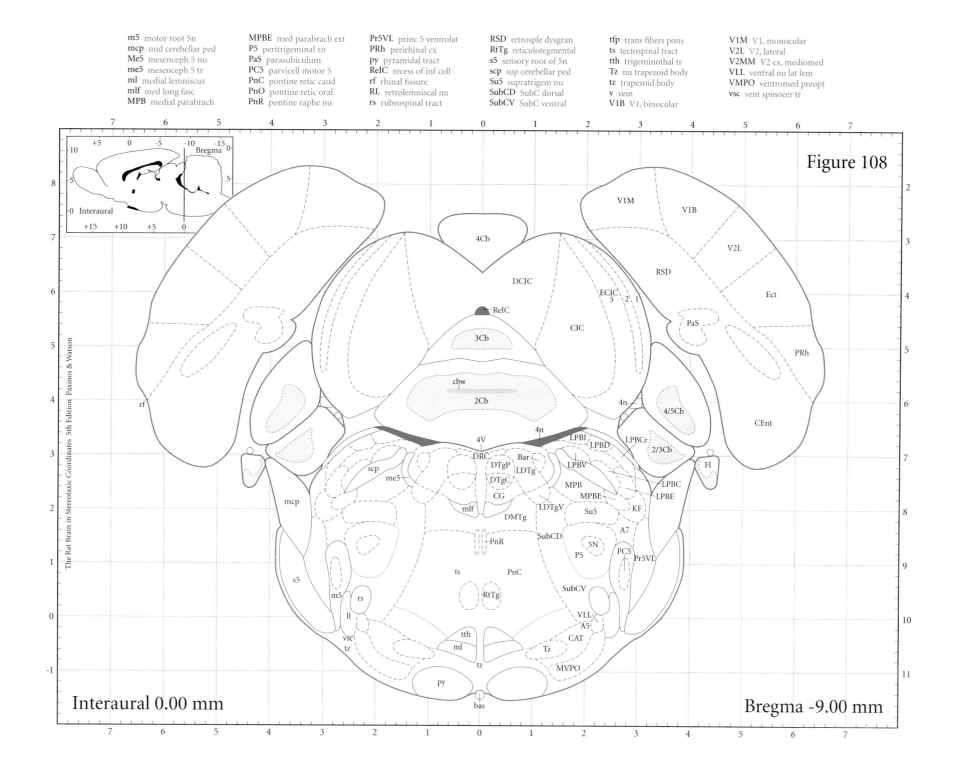

Figure 108

Interaural 0.00 mm

Bregma -9.00 mm

m5 motor root 5n
mcp mid cerebellar ped
Me5 mesenceph 5 nu
me5 mesenceph 5 tr
ml medial lemniscus
mlf med long fasc
MPB medial parabrach

MPBE med parabrach ext
P5 peritrigeminal zn
PaS parasubiculum
PC5 parvicell motor 5
PnC pontine retic caud
PnO pontine retic oral
PnR pontine raphe nu

Pr5VL princ 5 ventrolat
PRh perirhinal cx
py pyramidal tract
ReIC recess of inf coll
RL retrolemniscal nu
rs rubrospinal tract

RSD retrosple dysgran
RtTg reticulotegmental
s5 sensory root of 5n
scp sup cerebellar ped
Su5 supratrigem nu
SubCD SubC dorsal
SubCV SubC ventral

tfp trans fibers pons
ts tectospinal tract
tth trigeminothal tr
Tz nu trapezoid body
tz trapezoid body
v vein
V1B V1, binocular

V1M V1, monocular
V2L V2, lateral
V2MM V2 cx, mediomed
VLL ventral nu lat lem
VMPO ventromed preopt
vsc vent spinocer tr

The Rat Brain in Stereotaxic Coordinates 5th Edition Paxinos & Watson

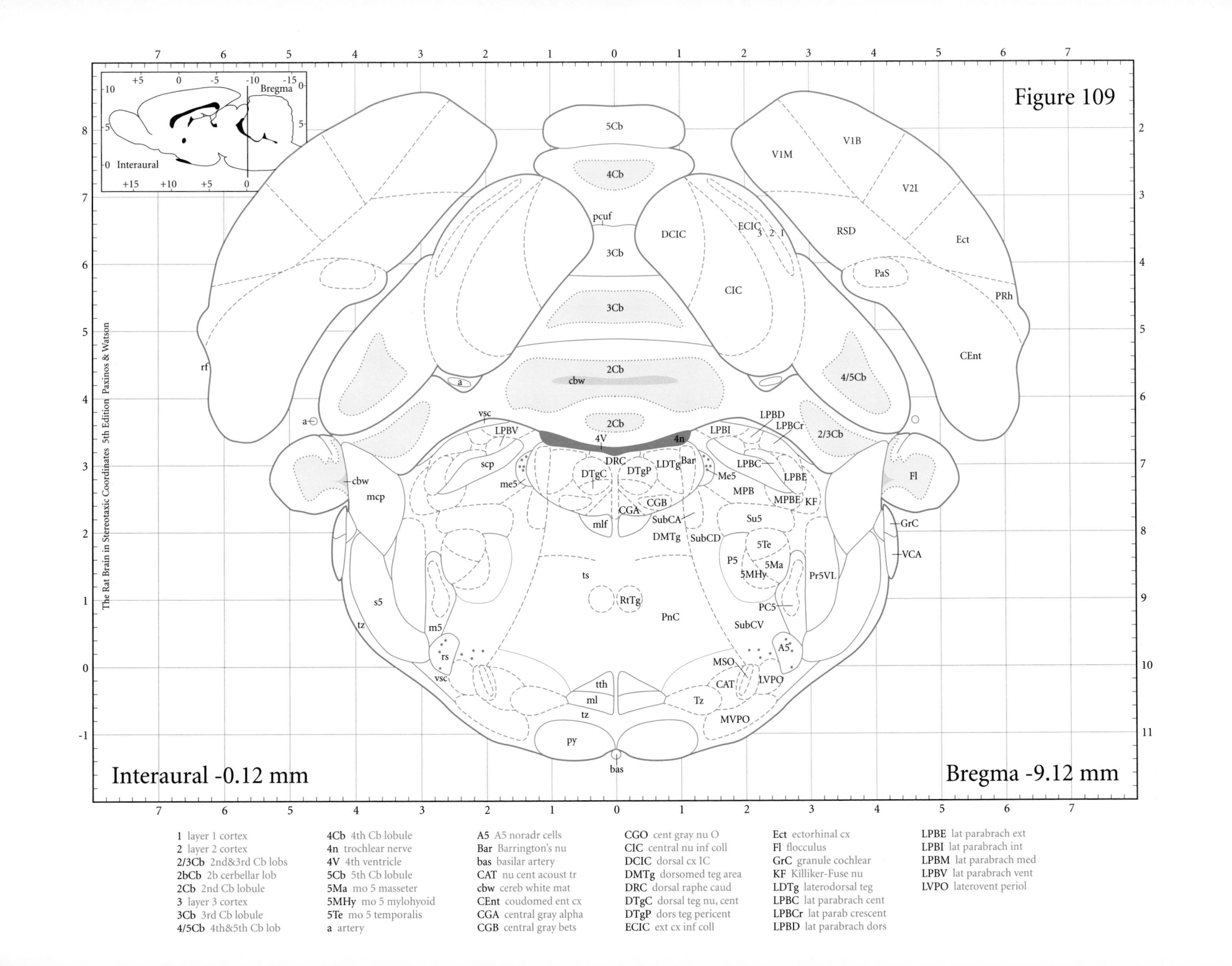

Figure 109

The Rat Brain in Stereotaxic Coordinates 5th Edition Paxinos & Watson

Interaural -0.12 mm

Bregma -9.12 mm

1 layer 1 cortex
2 layer 2 cortex
2/3Cb 2nd&3rd Cb lobs
2bCb 2b cerbellar lob
2Cb 2nd Cb lobule
3 layer 3 cortex
3Cb 3rd Cb lobule
4/5Cb 4th&5th Cb lob

4Cb 4th Cb lobule
4n trochlear nerve
4V 4th ventricle
5Cb 5th Cb lobule
5Ma mo 5 masseter
5MHy mo 5 mylohyoid
5Te mo 5 temporalis
a artery

A5 A5 noradr cells
Bar Barrington's nu
bas basilar artery
CAT nu cent acoust tr
cbw cereb white mat
CEnt coudomed ent cx
CGA central gray alpha
CGB central gray bets

CGO cent gray nu O
CIC central nu inf coll
DCIC dorsal cx IC
DMTg dorsomed teg area
DRC dorsal raphe caud
DTgC dorsal teg nu, cent
DTgP dors teg pericent
ECIC ext cx inf coll

Ect ectorhinal cx
Fl flocculus
GrC granule cochlear
KF Killiker-Fuse nu
LDTg laterodorsal teg
LPBC lat parabrach cent
LPBCr lat parab crescent
LPBD lat parabrach dors

LPBE lat parabrach ext
LPBI lat parabrach int
LPBM lat parabrach med
LPBV lat parabrach vent
LVPO laterovent periol

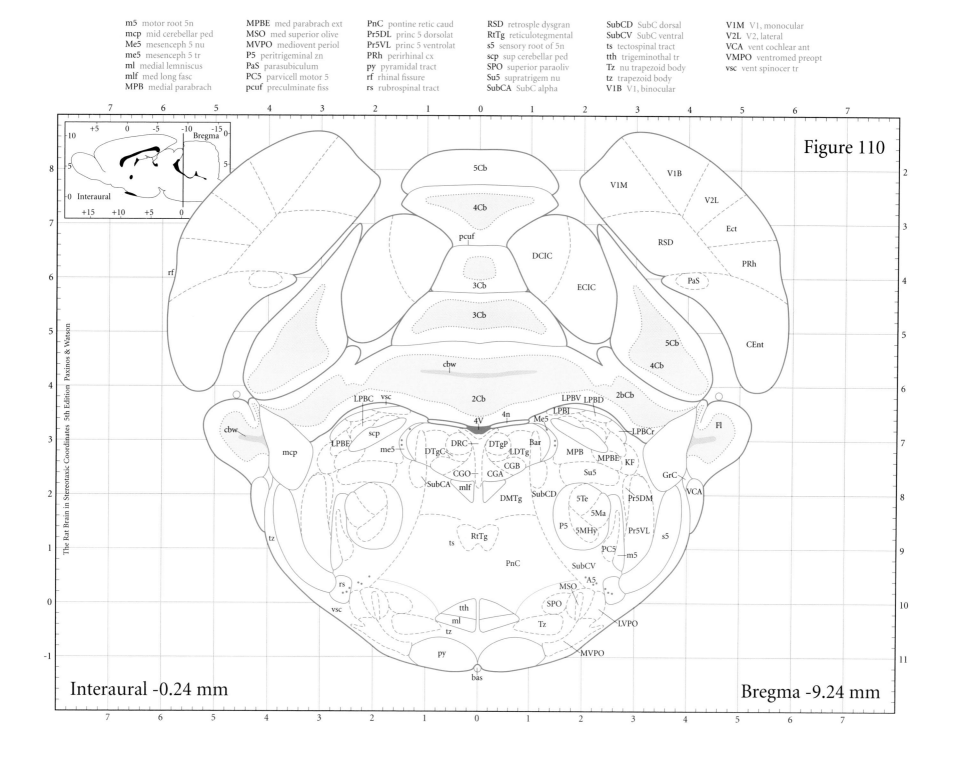

m5 motor root 5n
mcp mid cerebellar ped
Me5 mesenceph 5 nu
me5 mesenceph 5 tr
ml medial lemniscus
mlf med long fasc
MPB medial parabrach

MPBE med parabrach ext
MSO med superior olive
MVPO mediovent periol
P5 peritrigeminal zn
PaS parasubiculum
PC5 parvicell motor 5
pcuf preculminate fiss

PnC pontine retic caud
Pr5DL princ 5 dorsolat
Pr5VL princ 5 ventrolat
PRh perirhinal cx
py pyramidal tract
rf rhinal fissure
rs rubrospinal tract

RSD retrosple dysgran
RtTg reticulotegmental
s5 sensory root of 5n
scp sup cerebellar ped
SPO superior paraoliv
Su5 supratrigem nu
SubCA SubC alpha

SubCD SubC dorsal
SubCV SubC ventral
ts tectospinal tract
tth trigeminothal tr
Tz nu trapezoid body
tz trapezoid body
V1B V1, binocular

V1M V1, monocular
V2L V2, lateral
VCA vent cochlear ant
VMPO ventromed preopt
vsc vent spinocer tr

Figure 110

Interaural -0.24 mm

Bregma -9.24 mm

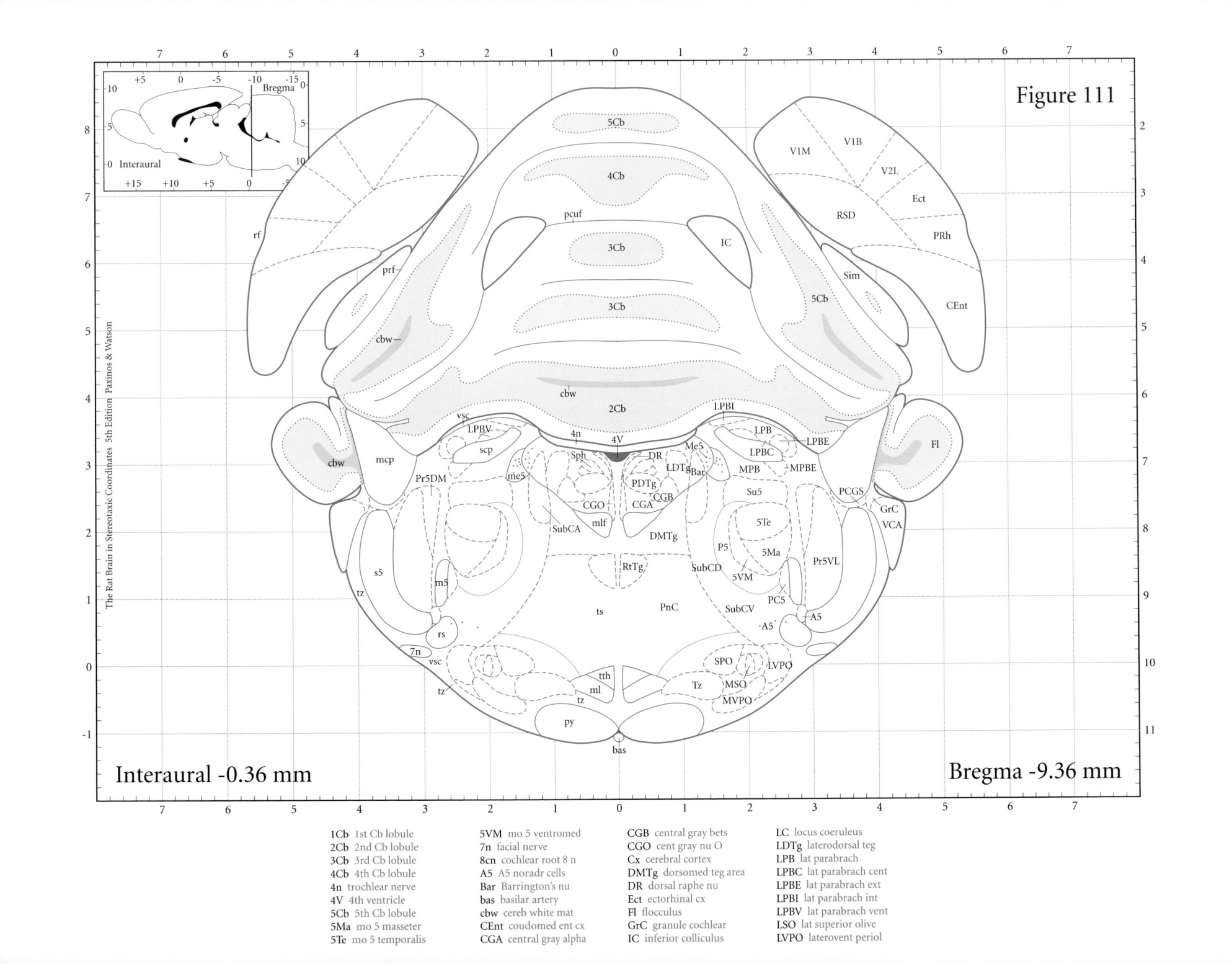

Figure 111

Interaural -0.36 mm

Bregma -9.36 mm

The Rat Brain in Stereotaxic Coordinates 5th Edition Paxinos & Watson

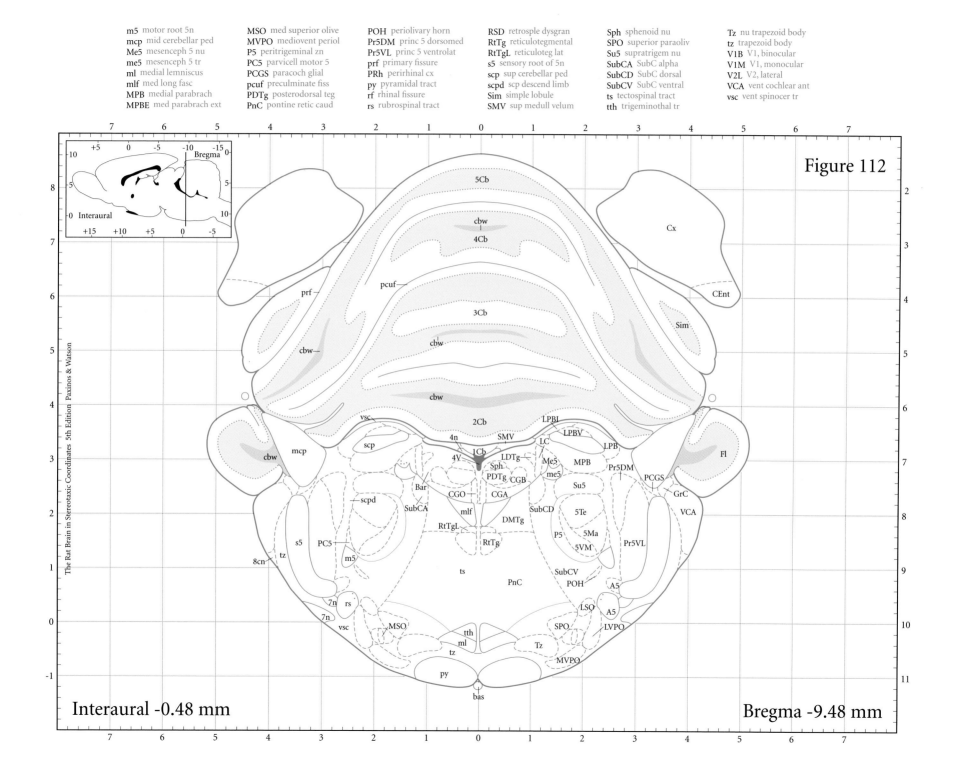

m5 motor root 5n
mcp mid cerebellar ped
Me5 mesenceph 5 nu
me5 mesenceph 5 tr
ml medial lemniscus
mlf med long fasc
MPB medial parabrach
MPBE med parabrach ext

MSO med superior olive
MVPO medovent periol
P5 peritrigeminal zn
PC5 parvicell motor 5
PCGS paracoch glial
pcuf preculminate fiss
PDTg posterodorsal teg
PnC pontine retic caud

POH periolivary horn
Pr5DM princ 5 dorsomed
Pr5VL princ 5 ventrolat
prf primary fissure
PRh perirhinal cx
py pyramidal tract
rf rhinal fissure
rs rubrospinal tract

RSD retrosple dysgran
RtTg reticulotegmental
RtTgL reticuloteg lat
s5 sensory root of 5n
scp sup cerebellar ped
scpd scp descend limb
Sim simple lobule
SMV sup medull velum

Sph sphenoid nu
SPO superior paraoliv
Su5 supratrigem nu
SubCA SubC alpha
SubCD SubC dorsal
SubCV SubC ventral
ts tectospinal tract
tth trigeminothal tr

Tz nu trapezoid body
tz trapezoid body
V1B V1, binocular
V1M V1, monocular
V2L V2, lateral
VCA vent cochlear ant
vsc vent spinocer tr

Figure 112

Interaural -0.48 mm

Bregma -9.48 mm

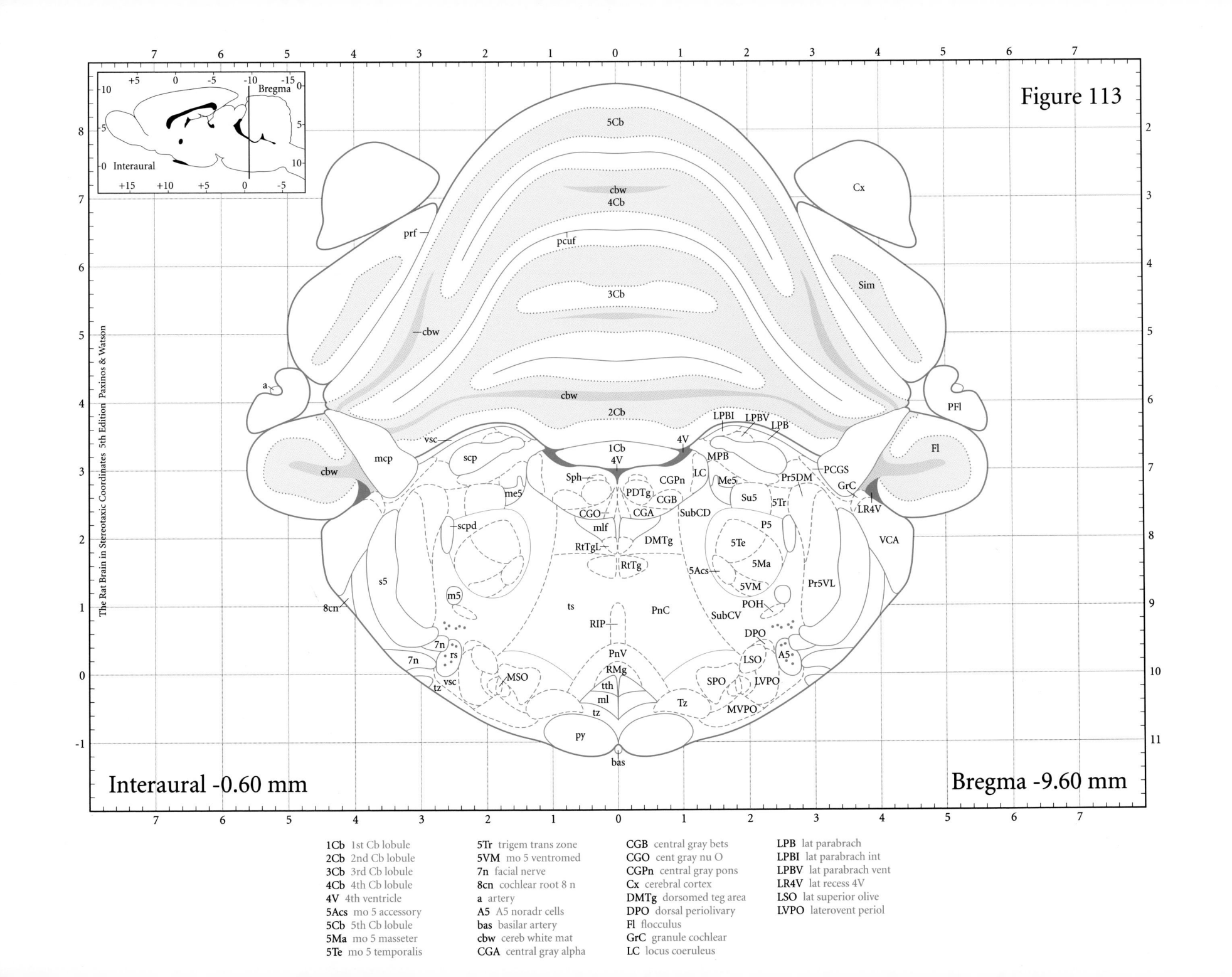

Figure 113

Interaural -0.60 mm

Bregma -9.60 mm

1Cb	1st Cb lobule	**5Tr** trigem trans zone	**CGB** central gray bets	**LPB** lat parabrach
2Cb	2nd Cb lobule	**5VM** mo 5 ventromed	**CGO** cent gray nu O	**LPBI** lat parabrach int
3Cb	3rd Cb lobule	**7n** facial nerve	**CGPn** central gray pons	**LPBV** lat parabrach vent
4Cb	4th Cb lobule	**8cn** cochlear root 8 n	**Cx** cerebral cortex	**LR4V** lat recess 4V
4V	4th ventricle	**a** artery	**DMTg** dorsomed teg area	**LSO** lat superior olive
5Acs	mo 5 accessory	**A5** A5 noradr cells	**DPO** dorsal periolivary	**LVPO** laterovent periol
5Cb	5th Cb lobule	**bas** basilar artery	**Fl** flocculus	
5Ma	mo 5 masseter	**cbw** cereb white mat	**GrC** granule cochlear	
5Te	mo 5 temporalis	**CGA** central gray alpha	**LC** locus coeruleus	

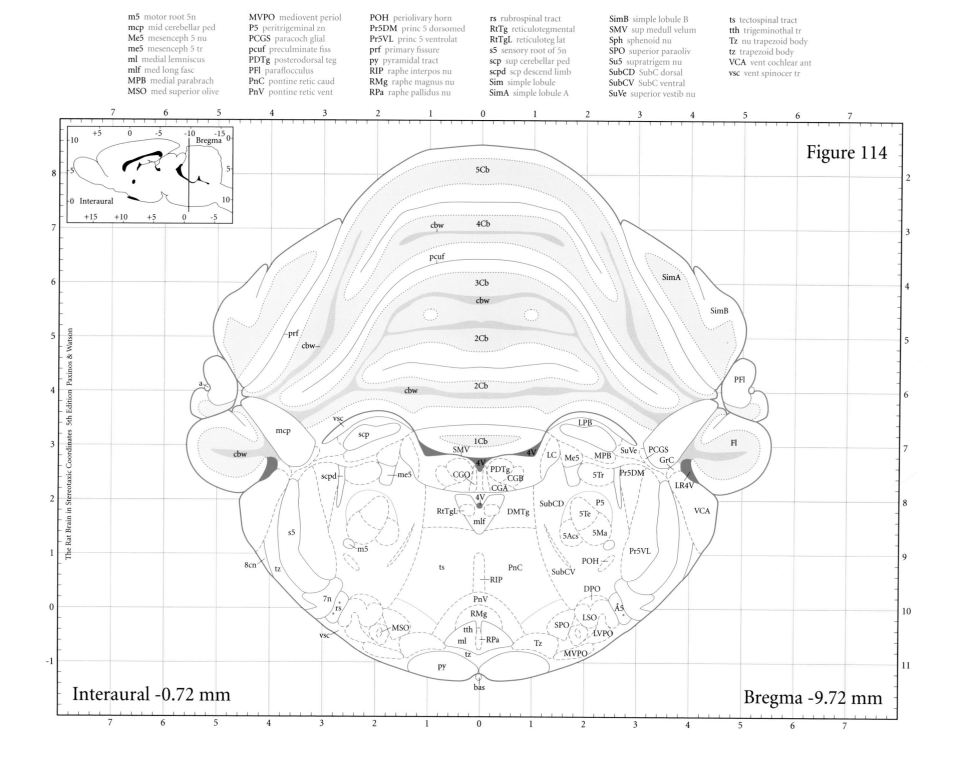

m5   motor root 5n
mcp   mid cerebellar ped
Me5   mesenceph 5 nu
me5   mesenceph 5 tr
ml   medial lemniscus
mlf   med long fasc
MPB   medial parabrach
MSO   med superior olive

MVPO mediovent periol
P5   peritrigeminal zn
PCGS paracoch glial
pcuf preculminate fiss
PDTg posterodorsal teg
PFl   paraflocculus
PnC   pontine retic caud
PnV   pontine retic vent

POH   periolivary horn
Pr5DM princ 5 dorsomed
Pr5VL princ 5 ventrolat
prf   primary fissure
py   pyramidal tract
RIP   raphe interpos nu
RMg   raphe magnus nu
RPa   raphe pallidus nu

rs   rubrospinal tract
RtTg reticulotegmental
RtTgL reticuloteg lat
s5   sensory root of 5n
scp   sup cerebellar ped
scpd scp descend limb
Sim   simple lobule
SimA simple lobule A

SimB simple lobule B
SMV   sup medull velum
Sph   sphenoid nu
SPO   superior paraoliv
Su5   supratrigem nu
SubCD SubC dorsal
SubCV SubC ventral
SuVe superior vestib nu

ts   tectospinal tract
tth   trigeminothal tr
Tz   nu trapezoid body
tz   trapezoid body
VCA   vent cochlear ant
vsc   vent spinocer tr

Figure 114

Interaural -0.72 mm

Bregma -9.72 mm

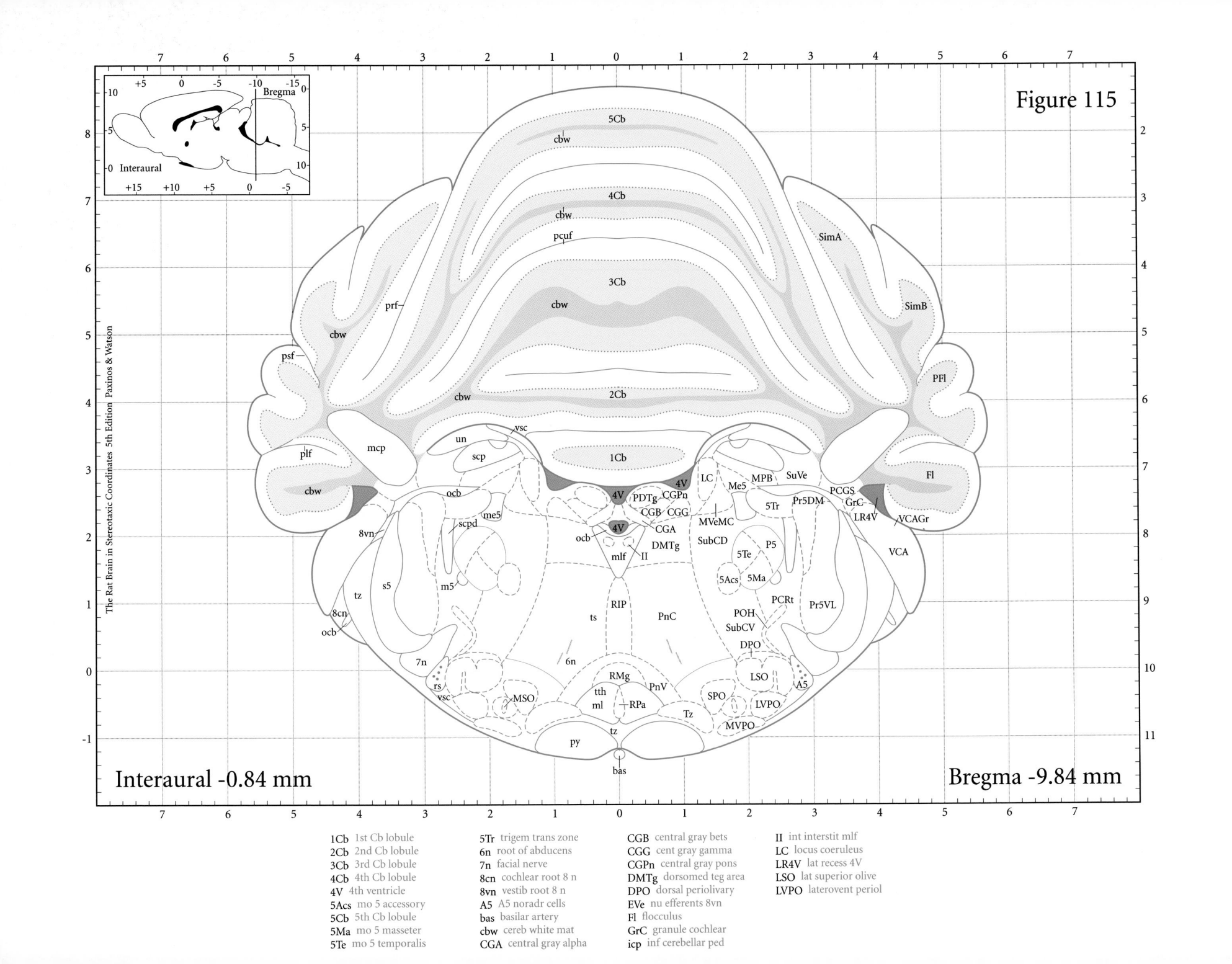

Figure 115

Interaural -0.84 mm

Bregma -9.84 mm

The Rat Brain in Stereotaxic Coordinates 5th Edition Paxinos & Watson

1Cb 1st Cb lobule	5Tr trigem trans zone	CGB central gray bets	II int interstit mlf
2Cb 2nd Cb lobule	6n root of abducens	CGG cent gray gamma	LC locus coeruleus
3Cb 3rd Cb lobule	7n facial nerve	CGPn central gray pons	LR4V lat recess 4V
4Cb 4th Cb lobule	8cn cochlear root 8 n	DMTg dorsomed teg area	LSO lat superior olive
4V 4th ventricle	8vn vestib root 8 n	DPO dorsal periolivary	LVPO laterovent periol
5Acs mo 5 accessory	A5 A5 noradr cells	EVe nu efferents 8vn	
5Cb 5th Cb lobule	bas basilar artery	Fl flocculus	
5Ma mo 5 masseter	cbw cereb white mat	GrC granule cochlear	
5Te mo 5 temporalis	CGA central gray alpha	icp inf cerebellar ped	

Figure 116

The Rat Brain in Stereotaxic Coordinates 5th Edition Paxinos & Watson

m5   motor root 5n
mcp   mid cerebellar ped
Me5   mesenceph 5 nu
me5   mesenceph 5 tr
ml   medial lemniscus
mlf   med long fasc
MPB   medial parabrach
MSO   med superior olive
MVeMC   med vestib magno

MVPO   mediovent periol
ocb   olivocochl bundle
P5   peritrigeminal zn
PCGS   paracoch glial
PCRt   parvicell ret nu
pcuf   preculminate fiss
PDTg   posterodorsal teg
PFl   paraflocculus
plf   posterolat fissure

PnC   pontine retic caud
PnV   pontine retic vent
POH   periolivary horn
Pr5DM   princ 5 dorsomed
Pr5VL   princ 5 ventrolat
prf   primary fissure
psf   post superior fiss
py   pyramidal tract
RIP   raphe interpos nu

RMg   raphe magnus nu
RPa   raphe pallidus nu
rs   rubrospinal tract
s5   sensory root of 5n
scp   sup cerebellar ped
scpd   scp descend limb
SGe   supragenual nu
SimA   simple lobule A
SimB   simple lobule B

SMV   sup medull velum
SPO   superior paraoliv
SubCD   SubC dorsal
SubCV   SubC ventral
SuVe   superior vestib nu
ts   tectospinal tract
tth   trigeminothal tr
Tz   nu trapezoid body
tz   trapezoid body

un   uncinate fascic
VCA   vent cochlear ant
VCAGr   VCA granule layer
veme   vestibulomes tr
vsc   vent spinocer tr

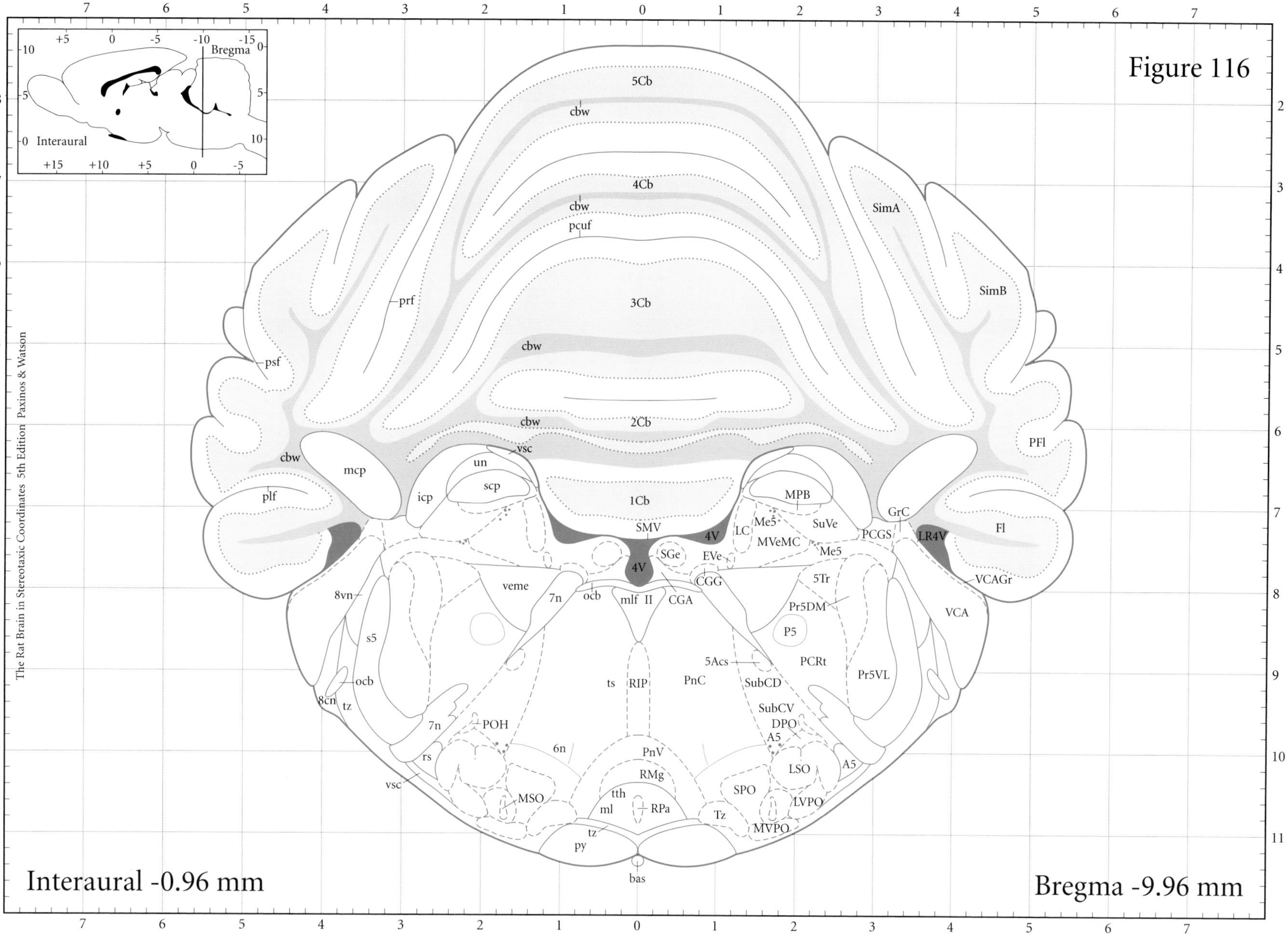

Interaural -0.96 mm

Bregma -9.96 mm

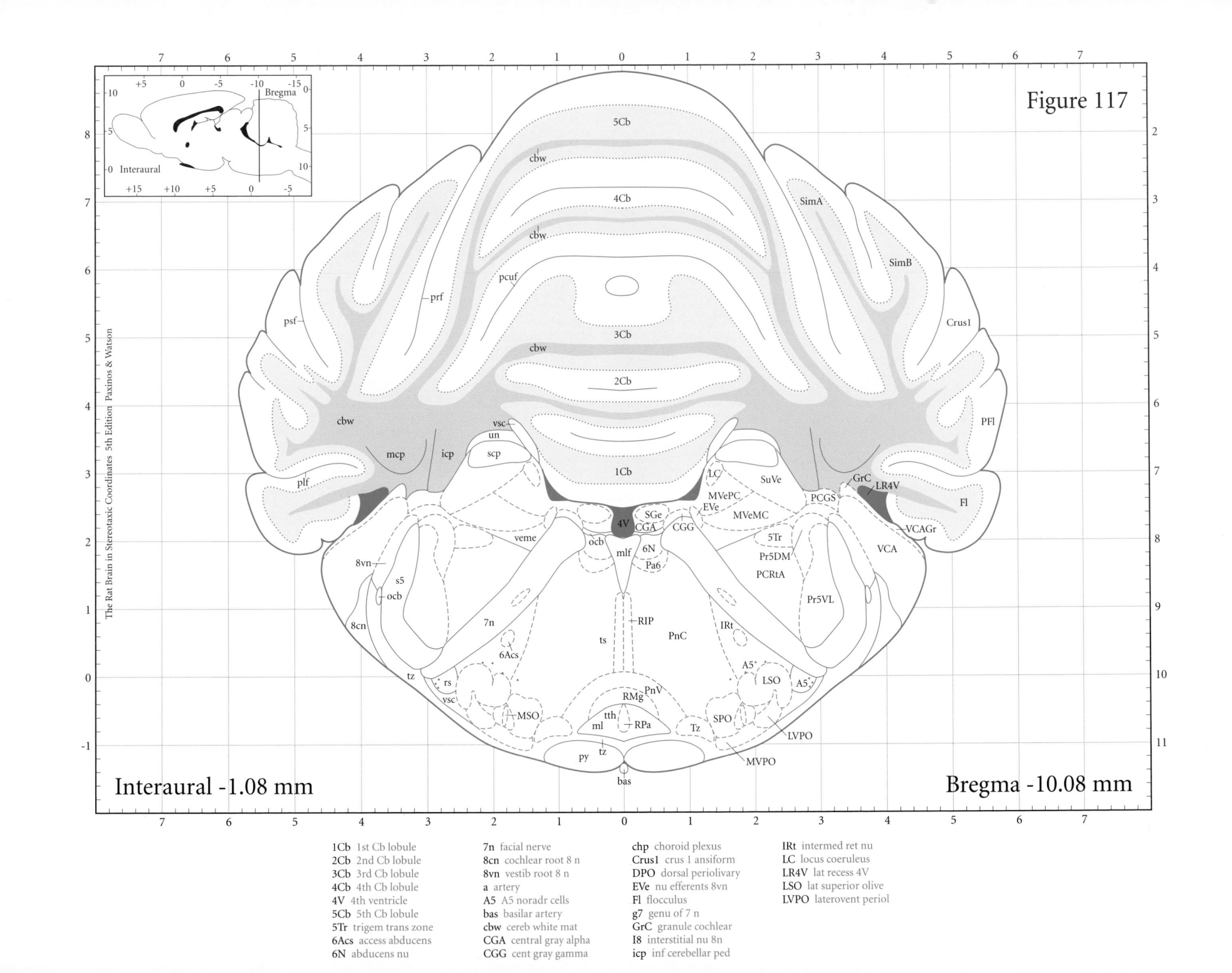

Figure 117

Interaural -1.08 mm

Bregma -10.08 mm

1Cb 1st Cb lobule
2Cb 2nd Cb lobule
3Cb 3rd Cb lobule
4Cb 4th Cb lobule
4V 4th ventricle
5Cb 5th Cb lobule
5Tr trigem trans zone
6Acs access abducens
6N abducens nu

7n facial nerve
8cn cochlear root 8 n
8vn vestib root 8 n
a artery
A5 A5 noradr cells
bas basilar artery
cbw cereb white mat
CGA central gray alpha
CGG cent gray gamma

chp choroid plexus
Crus1 crus 1 ansiform
DPO dorsal periolivary
EVe nu efferents 8vn
Fl flocculus
g7 genu of 7 n
GrC granule cochlear
I8 interstitial nu 8n
icp inf cerebellar ped

IRt intermed ret nu
LC locus coeruleus
LR4V lat recess 4V
LSO lat superior olive
LVPO laterovent periol

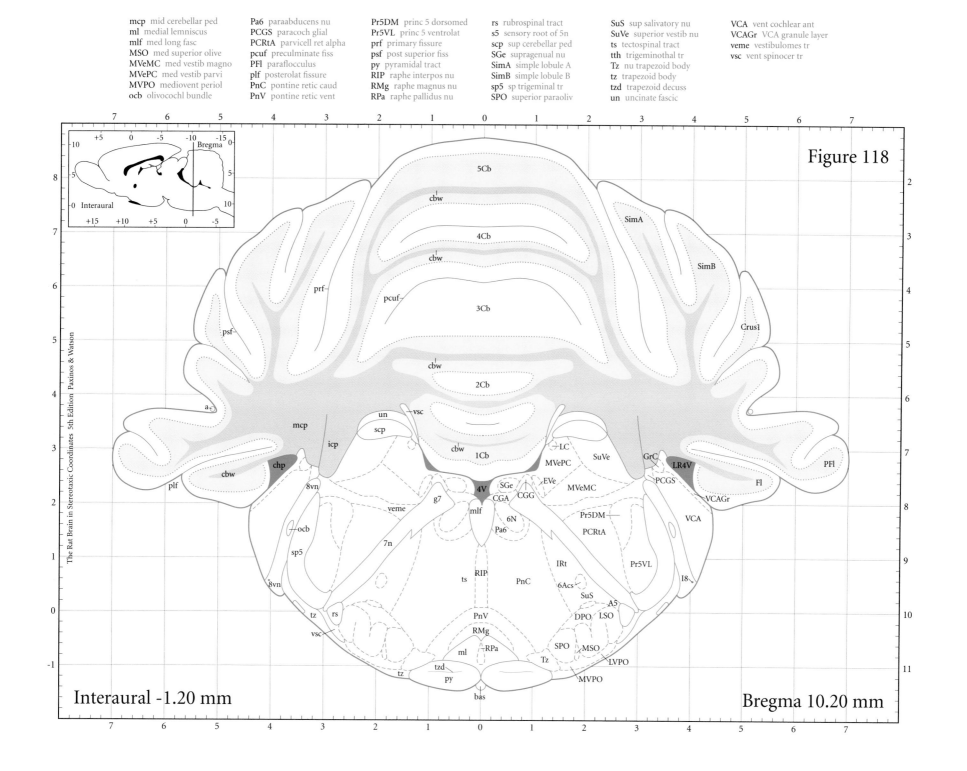

Figure 118

Interaural -1.20 mm

Bregma 10.20 mm

mcp mid cerebellar ped
ml medial lemniscus
mlf med long fasc
MSO med superior olive
MVeMC med vestib magno
MVePC med vestib parvi
MVPO mediovent periol
ocb olivocochl bundle

Pa6 paraabducens nu
PCGS paracoch glial
PCRtA parvicell ret alpha
pcuf preculminate fiss
PFl paraflocculus
plf posterolat fissure
PnC pontine retic caud
PnV pontine retic vent

Pr5DM princ 5 dorsomed
Pr5VL princ 5 ventrolat
prf primary fissure
psf post superior fiss
py pyramidal tract
RIP raphe interpos nu
RMg raphe magnus nu
RPa raphe pallidus nu

rs rubrospinal tract
s5 sensory root of 5n
scp sup cerebellar ped
SGe supragenual nu
SimA simple lobule A
SimB simple lobule B
sp5 sp trigeminal tr
SPO superior paraoliv

SuS sup salivatory nu
SuVe superior vestib nu
ts tectospinal tract
tth trigeminothal tr
Tz nu trapezoid body
tz trapezoid body
tzd trapezoid decuss
un uncinate fascic

VCA vent cochlear ant
VCAGr VCA granule layer
veme vestibulomes tr
vsc vent spinocer tr

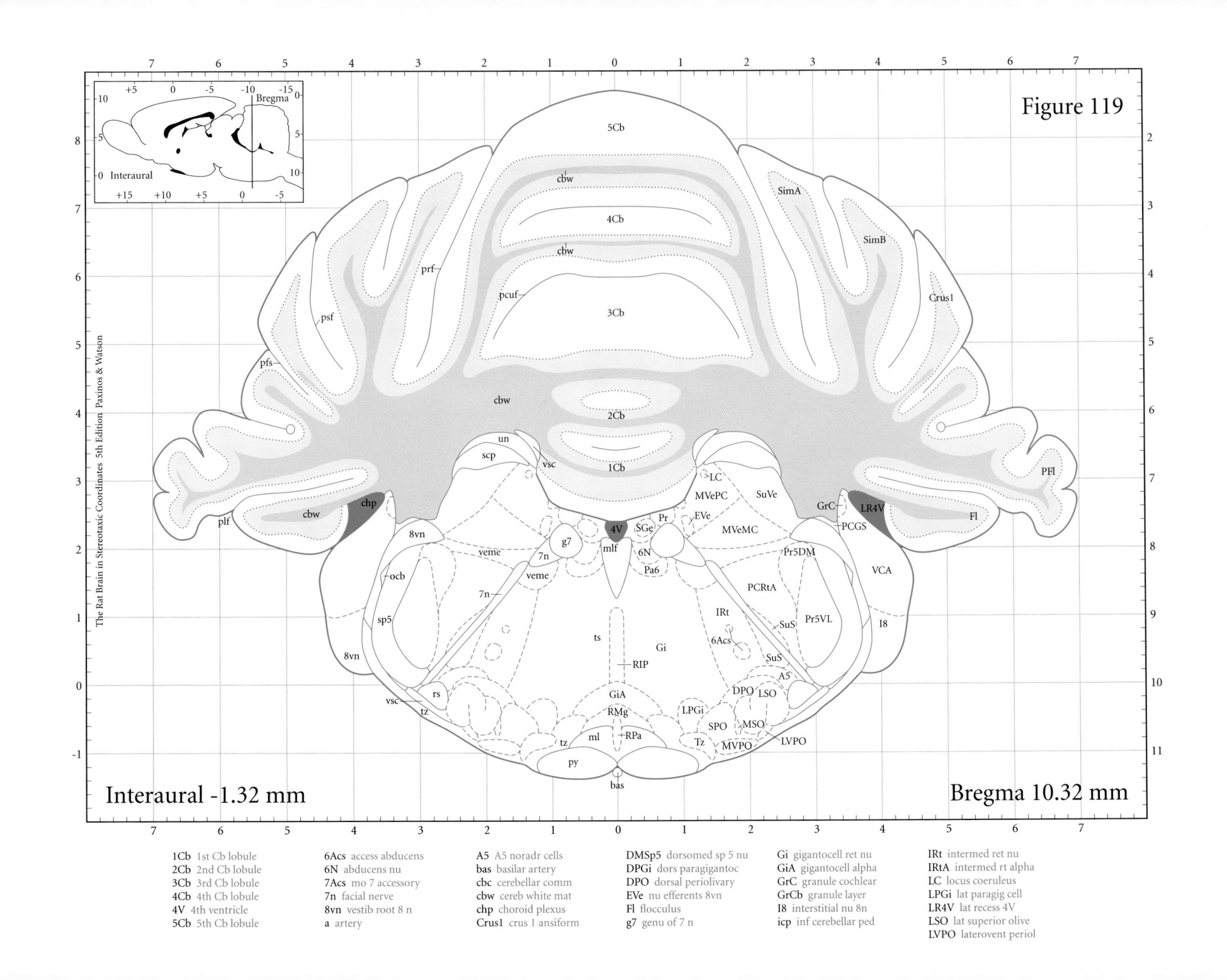

Figure 119

Interaural -1.32 mm

Bregma 10.32 mm

The Rat Brain in Stereotaxic Coordinates 5th Edition Paxinos & Watson

1Cb 1st Cb lobule	6Acs access abducens	A5 A5 noradr cells	DMSp5 dorsomed sp 5 nu	Gi gigantocell ret nu	IRt intermed ret nu
2Cb 2nd Cb lobule	6N abducens nu	bas basilar artery	DPGi dors paragigantoc	GiA gigantocell alpha	IRtA intermed rt alpha
3Cb 3rd Cb lobule	7Acs mo 7 accessory	cbc cerebellar comm	DPO dorsal periolivary	GrC granule cochlear	LC locus coeruleus
4Cb 4th Cb lobule	7n facial nerve	cbw cereb white mat	EVe nu efferents 8vn	GrCb granule layer	LPGi lat paragig cell
4V 4th ventricle	8vn vestib root 8 n	chp choroid plexus	Fl flocculus	I8 interstitial nu 8n	LR4V lat recess 4V
5Cb 5th Cb lobule	a artery	Crus1 crus 1 ansiform	g7 genu of 7 n	icp inf cerebellar ped	LSO lat superior olive
					LVPO laterovent periol

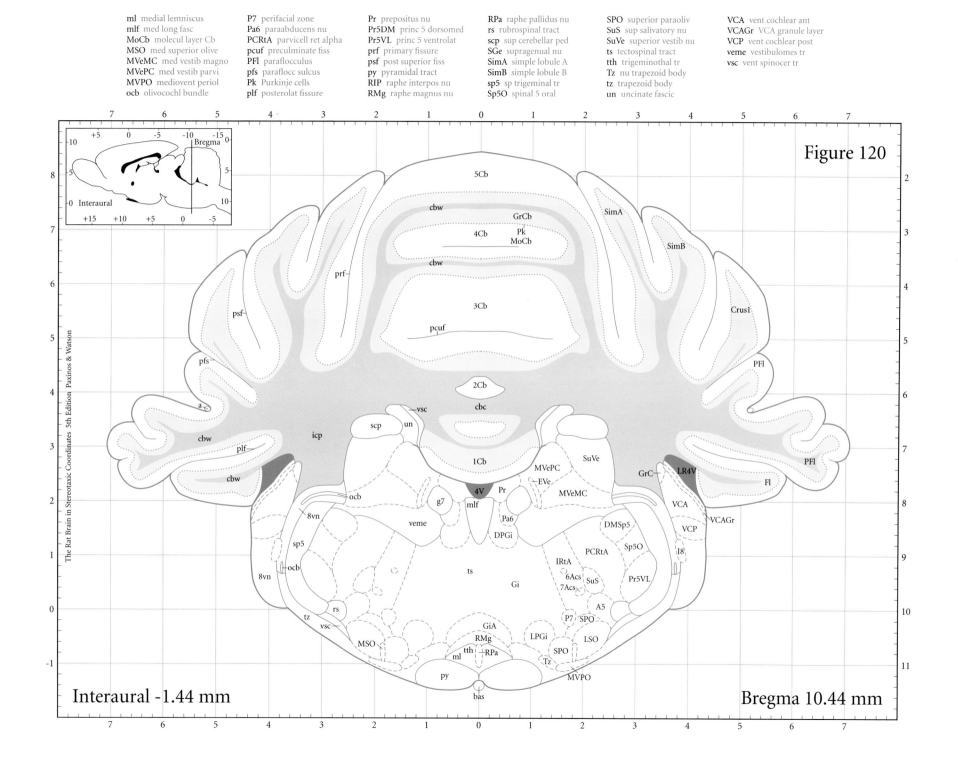

Figure 120

Interaural -1.44 mm

Bregma 10.44 mm

ml medial lemniscus
mlf med long fasc
MoCb molecul layer Cb
MSO med superior olive
MVeMC med vestib magno
MVePC med vestib parvi
MVPO mediovent periol
ocb olivocochl bundle

P7 perifacial zone
Pa6 paraabducens nu
PCRtA parvicell ret alpha
pcuf preculminate fiss
PFl paraflocculus
pfs paraflocc sulcus
Pk Purkinje cells
plf posterolat fissure

Pr prepositus nu
Pr5DM princ 5 dorsomed
Pr5VL princ 5 ventrolat
prf primary fissure
psf post superior fiss
py pyramidal tract
RIP raphe interpos nu
RMg raphe magnus nu

RPa raphe pallidus nu
rs rubrospinal tract
scp sup cerebellar ped
SGe supragenual nu
SimA simple lobule A
SimB simple lobule B
sp5 sp trigeminal tr
Sp5O spinal 5 oral

SPO superior paraoliv
SuS sup salivatory nu
SuVe superior vestib nu
ts tectospinal tract
tth trigeminothal tr
Tz nu trapezoid body
tz trapezoid body
un uncinate fascic

VCA vent cochlear ant
VCAGr VCA granule layer
VCP vent cochlear post
veme vestibulomes tr
vsc vent spinocer tr

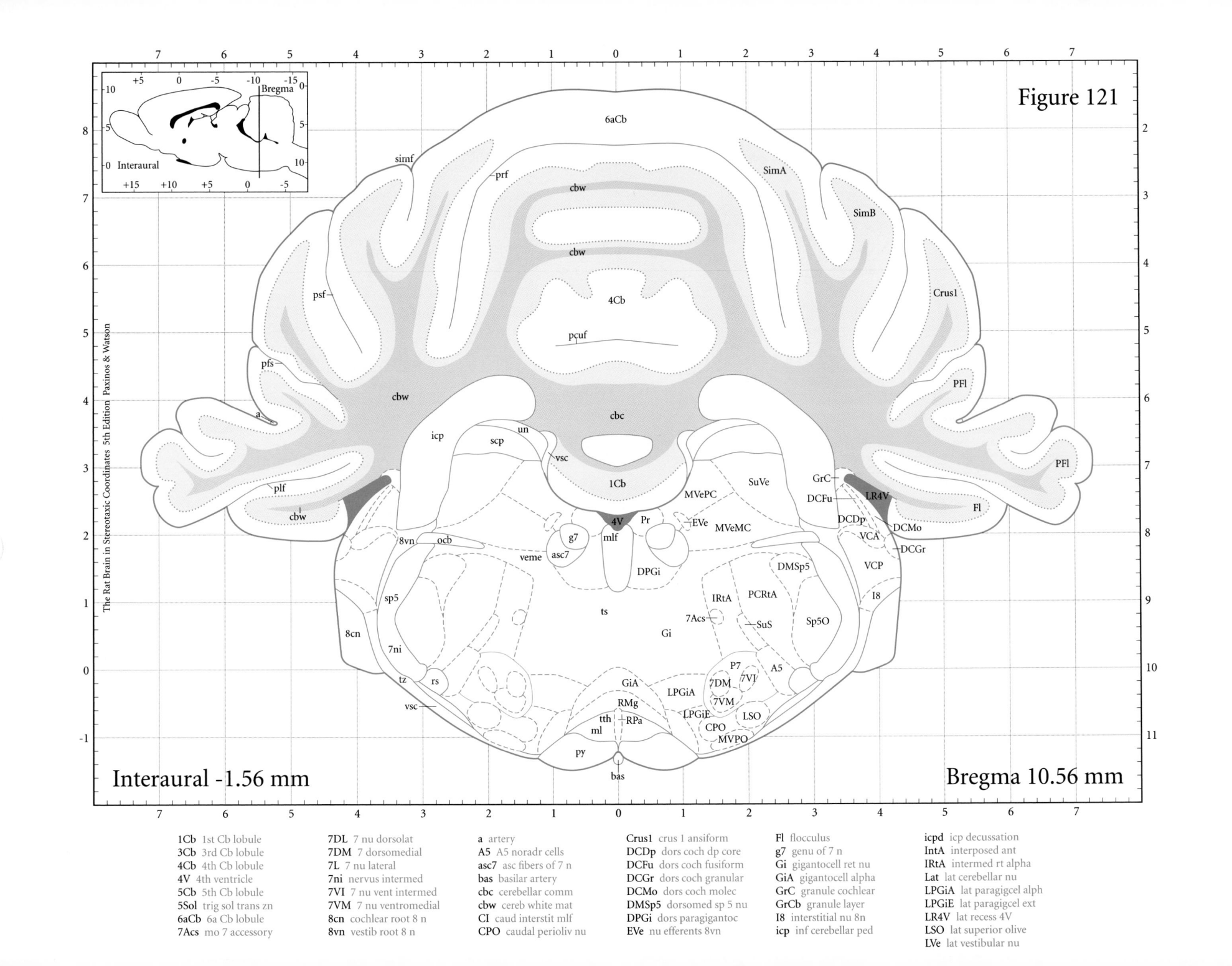

Figure 121

Interaural -1.56 mm

Bregma 10.56 mm

The Rat Brain in Stereotaxic Coordinates 5th Edition Paxinos & Watson

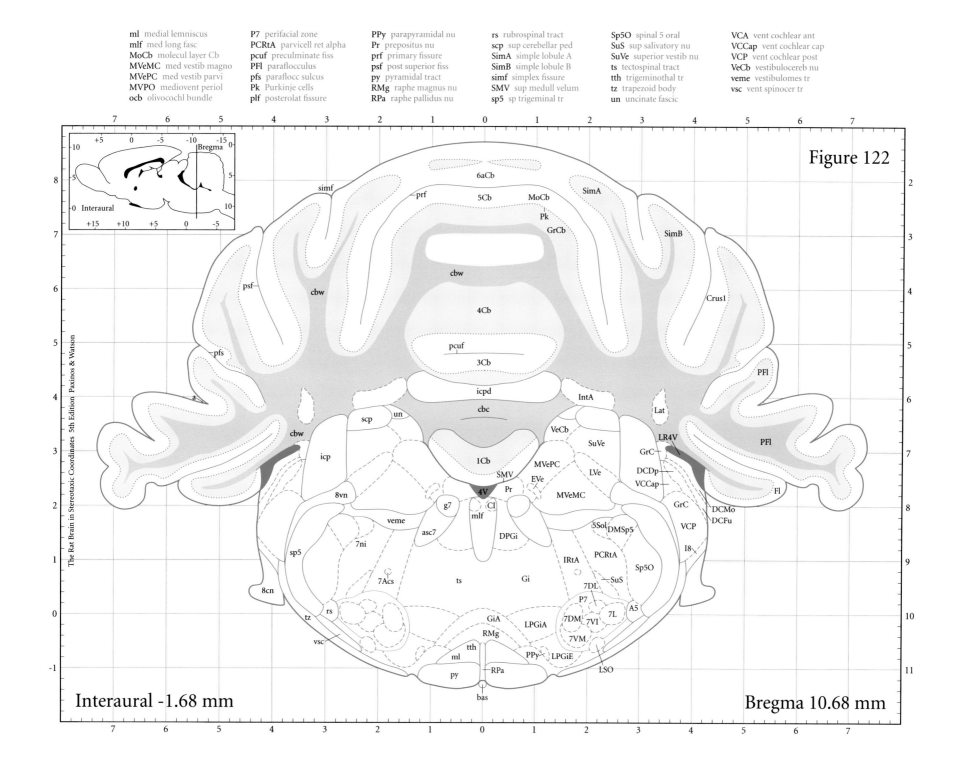

The Rat Brain in Stereotaxic Coordinates 5th Edition Paxinos & Watson

ml medial lemniscus
mlf med long fasc
MoCb molecul layer Cb
MVeMC med vestib magno
MVePC med vestib parvi
MVPO mediovent periol
ocb olivocochl bundle

P7 perifacial zone
PCRtA parvicell ret alpha
pcuf preculminate fiss
PFl paraflocculus
pfs paraflocc sulcus
Pk Purkinje cells
plf posterolat fissure

PPy parapyramidal nu
Pr prepositus nu
prf primary fissure
psf post superior fiss
py pyramidal tract
RMg raphe magnus nu
RPa raphe pallidus nu

rs rubrospinal tract
scp sup cerebellar ped
SimA simple lobule A
SimB simple lobule B
simf simplex fissure
SMV sup medull velum
sp5 sp trigeminal tr

Sp5O spinal 5 oral
SuS sup salivatory nu
SuVe superior vestib nu
ts tectospinal tract
tth trigeminothal tr
tz trapezoid body
un uncinate fascic

VCA vent cochlear ant
VCCap vent cochlear cap
VCP vent cochlear post
VeCb vestibulocereb nu
veme vestibulomes tr
vsc vent spinocer tr

Figure 122

Interaural -1.68 mm

Bregma 10.68 mm

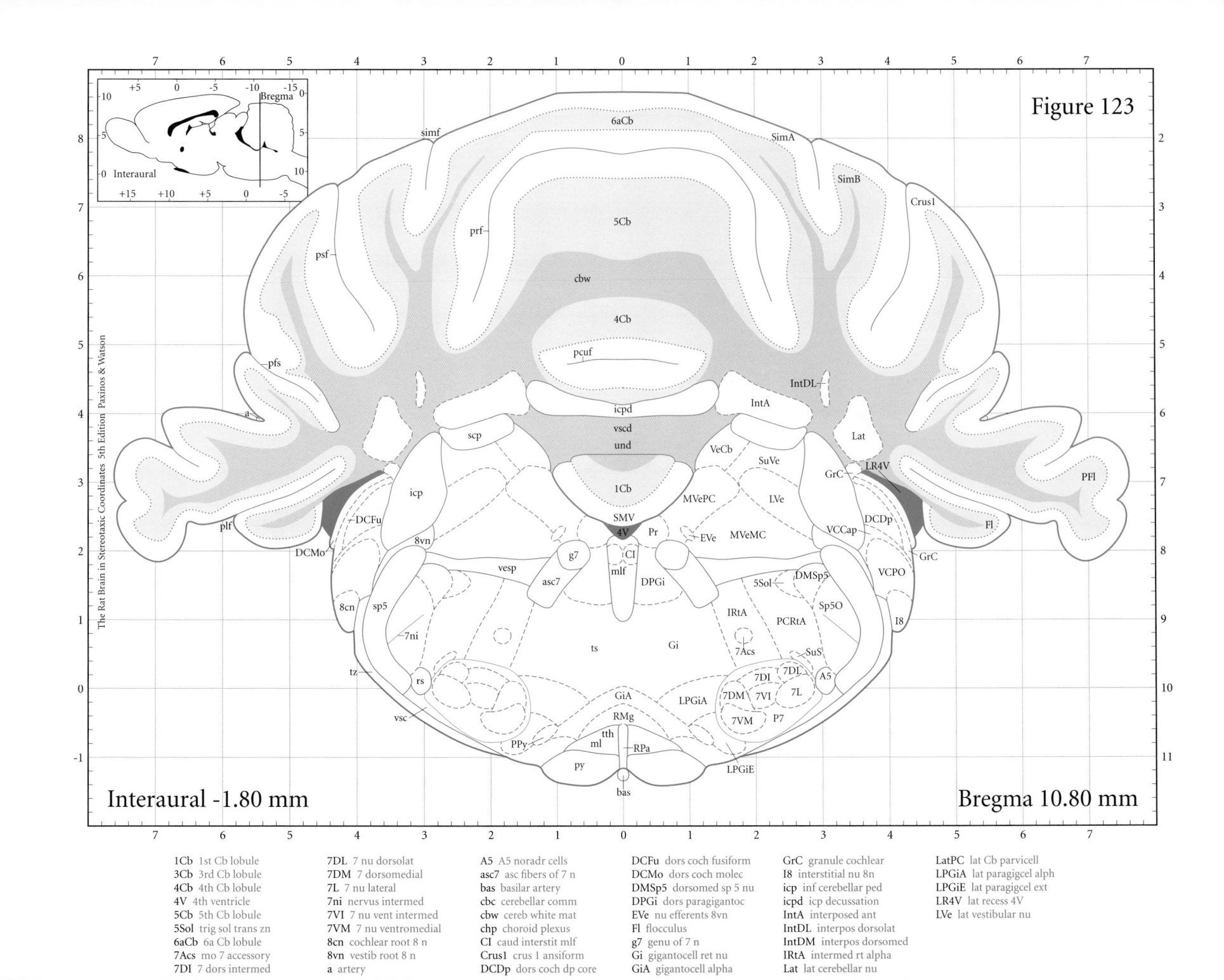

Figure 123

Interaural -1.80 mm

Bregma 10.80 mm

1Cb 1st Cb lobule	7DL 7 nu dorsolat	A5 A5 noradr cells	DCFu dors coch fusiform	GrC granule cochlear	LatPC lat Cb parvicell
3Cb 3rd Cb lobule	7DM 7 dorsomedial	asc7 asc fibers of 7 n	DCMo dors coch molec	I8 interstitial nu 8n	LPGiA lat paragigcel alph
4Cb 4th Cb lobule	7L 7 nu lateral	bas basilar artery	DMSp5 dorsomed sp 5 nu	icp inf cerebellar ped	LPGiE lat paragigcel ext
4V 4th ventricle	7ni nervus intermed	cbc cerebellar comm	DPGi dors paragigantoc	icpd icp decussation	LR4V lat recess 4V
5Cb 5th Cb lobule	7VI 7 nu vent intermed	cbw cereb white mat	EVe nu efferents 8vn	IntA interposed ant	LVe lat vestibular nu
5Sol trig sol trans zn	7VM 7 nu ventromedial	chp choroid plexus	Fl flocculus	IntDL interpos dorsolat	
6aCb 6a Cb lobule	8cn cochlear root 8 n	CI caud interstit mlf	g7 genu of 7 n	IntDM interpos dorsomed	
7Acs mo 7 accessory	8vn vestib root 8 n	Crus1 crus 1 ansiform	Gi gigantocell ret nu	IRtA intermed rt alpha	
7DI 7 dors intermed	a artery	DCDp dors coch dp core	GiA gigantocell alpha	Lat lat cerebellar nu	

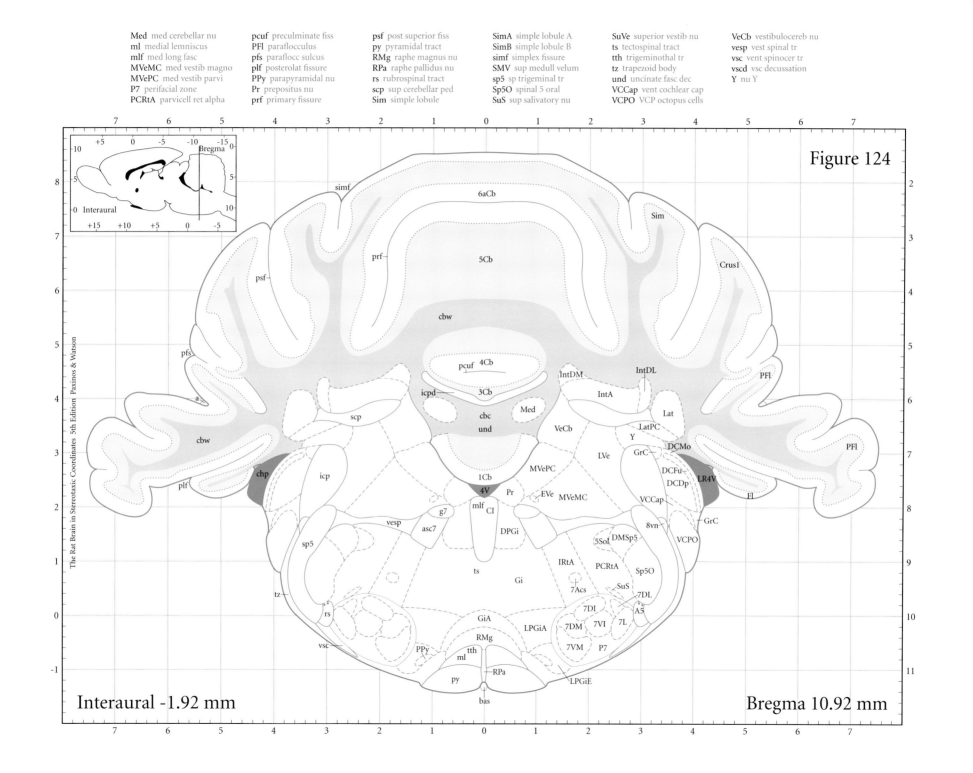

The Rat Brain in Stereotaxic Coordinates 5th Edition Paxinos & Watson

Figure 124

Interaural -1.92 mm

Bregma 10.92 mm

Med med cerebellar nu
ml medial lemniscus
mlf med long fasc
MVeMC med vestib magno
MVePC med vestib parvi
P7 perifacial zone
PCRtA parvicell ret alpha

pcuf preculminate fiss
PFl paraflocculus
pfs paraflocc sulcus
plf posterolat fissure
PPy parapyramidal nu
Pr prepositus nu
prf primary fissure

psf post superior fiss
py pyramidal tract
RMg raphe magnus nu
RPa raphe pallidus nu
rs rubrospinal tract
scp sup cerebellar ped
Sim simple lobule

SimA simple lobule A
SimB simple lobule B
simf simplex fissure
SMV sup medull velum
sp5 sp trigeminal tr
Sp5O spinal 5 oral
SuS sup salivatory nu

SuVe superior vestib nu
ts tectospinal tract
tth trigeminothal tr
tz trapezoid body
und uncinate fasc dec
VCCap vent cochlear cap
VCPO VCP octopus cells

VeCb vestibulocereb nu
vesp vest spinal tr
vsc vent spinocer tr
vscd vsc decussation
Y nu Y

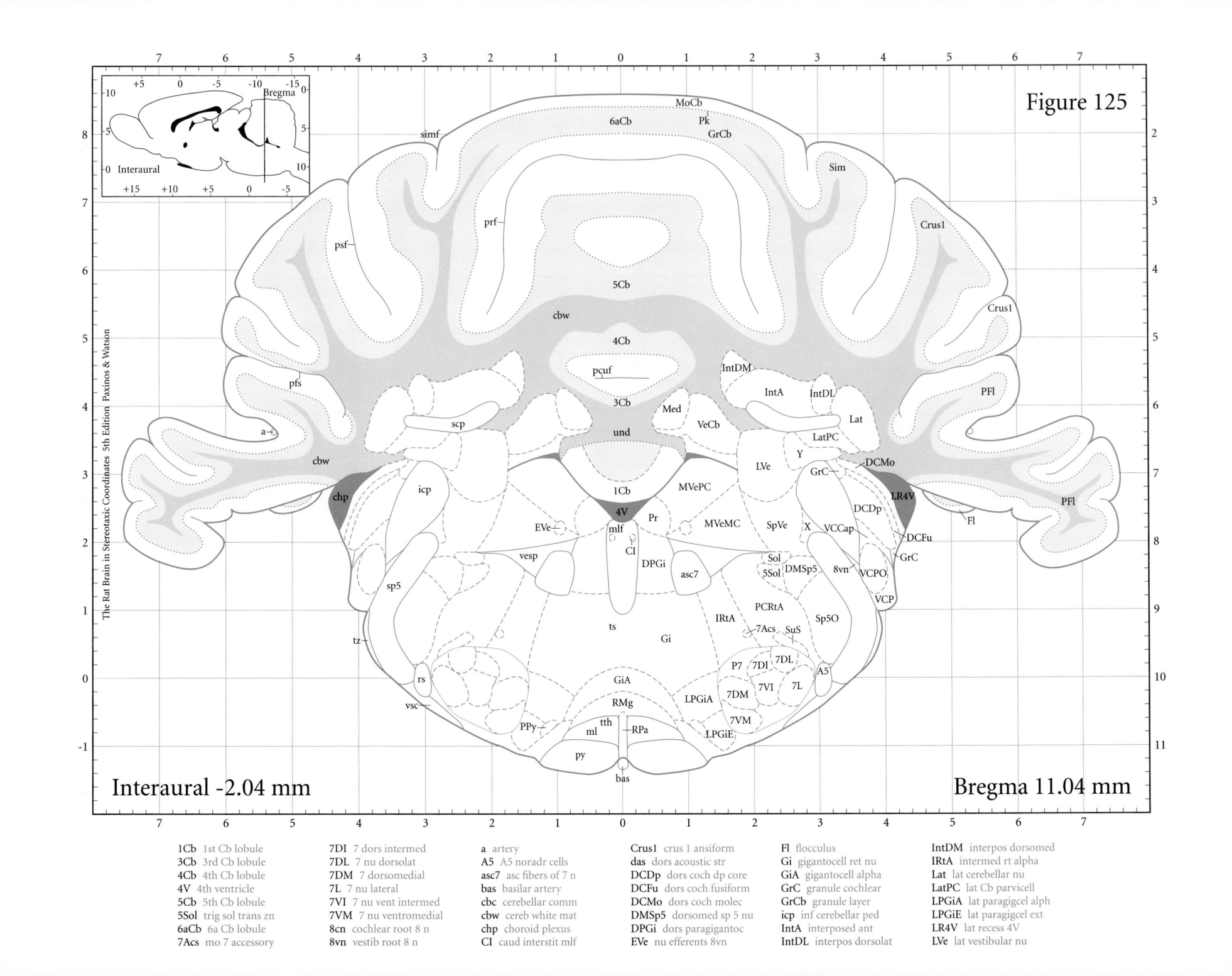

Figure 125

Interaural -2.04 mm

Bregma 11.04 mm

1Cb 1st Cb lobule	7DI 7 dors intermed	a artery	Crus1 crus 1 ansiform	Fl flocculus	IntDM interpos dorsomed
3Cb 3rd Cb lobule	7DL 7 nu dorsolat	A5 A5 noradr cells	das dors acoustic str	Gi gigantocell ret nu	IRtA intermed rt alpha
4Cb 4th Cb lobule	7DM 7 dorsomedial	asc7 asc fibers of 7 n	DCDp dors coch dp core	GiA gigantocell alpha	Lat lat cerebellar nu
4V 4th ventricle	7L 7 nu lateral	bas basilar artery	DCFu dors coch fusiform	GrC granule cochlear	LatPC lat Cb parvicell
5Cb 5th Cb lobule	7VI 7 nu vent intermed	cbc cerebellar comm	DCMo dors coch molec	GrCb granule layer	LPGiA lat paragigcel alph
5Sol trig sol trans zn	7VM 7 nu ventromedial	cbw cereb white mat	DMSp5 dorsomed sp 5 nu	icp inf cerebellar ped	LPGiE lat paragigcel ext
6aCb 6a Cb lobule	8cn cochlear root 8 n	chp choroid plexus	DPGi dors paragigantoc	IntA interposed ant	LR4V lat recess 4V
7Acs mo 7 accessory	8vn vestib root 8 n	CI caud interstit mlf	EVe nu efferents 8vn	IntDL interpos dorsolat	LVe lat vestibular nu

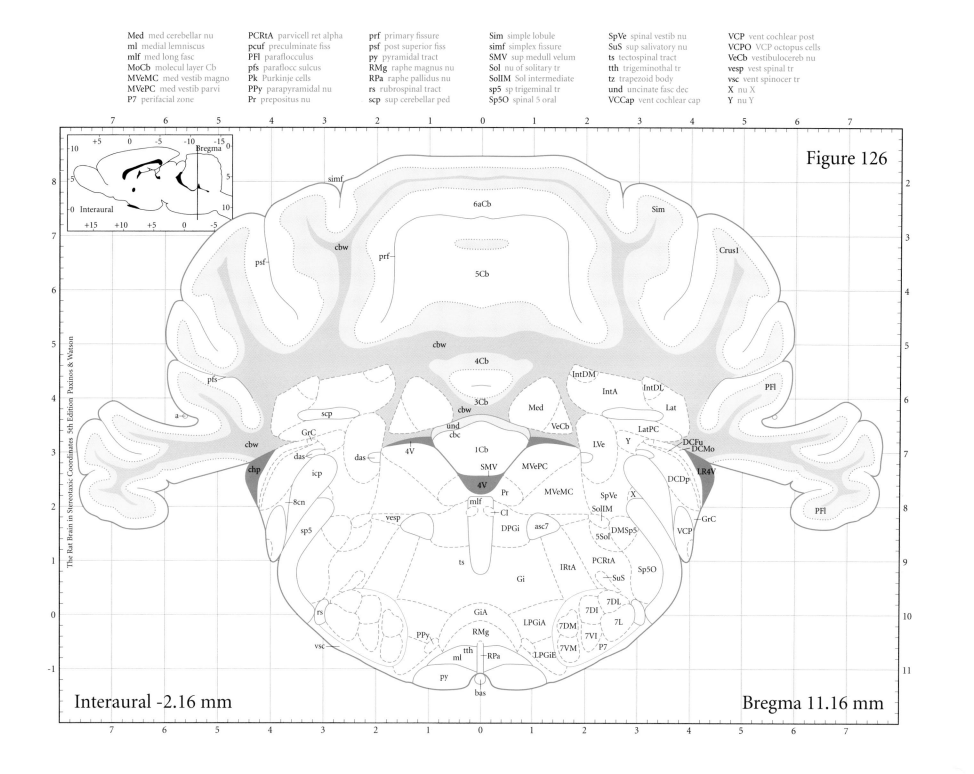

Figure 126

Med   med cerebellar nu
ml   medial lemniscus
mlf   med long fasc
MoCb   molecul layer Cb
MVeMC   med vestib magno
MVePC   med vestib parvi
P7   perifacial zone

PCRtA   parvicell ret alpha
pcuf   preculminate fiss
PFl   paraflocculus
pfs   paraflocc sulcus
Pk   Purkinje cells
PPy   parapyramidal nu
Pr   prepositus nu

prf   primary fissure
psf   post superior fiss
py   pyramidal tract
RMg   raphe magnus nu
RPa   raphe pallidus nu
rs   rubrospinal tract
scp   sup cerebellar ped

Sim   simple lobule
simf   simplex fissure
SMV   sup medull velum
Sol   nu of solitary tr
SolIM   Sol intermediate
sp5   sp trigeminal tr
Sp5O   spinal 5 oral

SpVe   spinal vestib nu
SuS   sup salivatory nu
ts   tectospinal tract
tth   trigeminothal tr
tz   trapezoid body
und   uncinate fasc dec
VCCap   vent cochlear cap

VCP   vent cochlear post
VCPO   VCP octopus cells
VeCb   vestibulocereb nu
vesp   vest spinal tr
vsc   vent spinocer tr
X   nu X
Y   nu Y

Interaural -2.16 mm

Bregma 11.16 mm

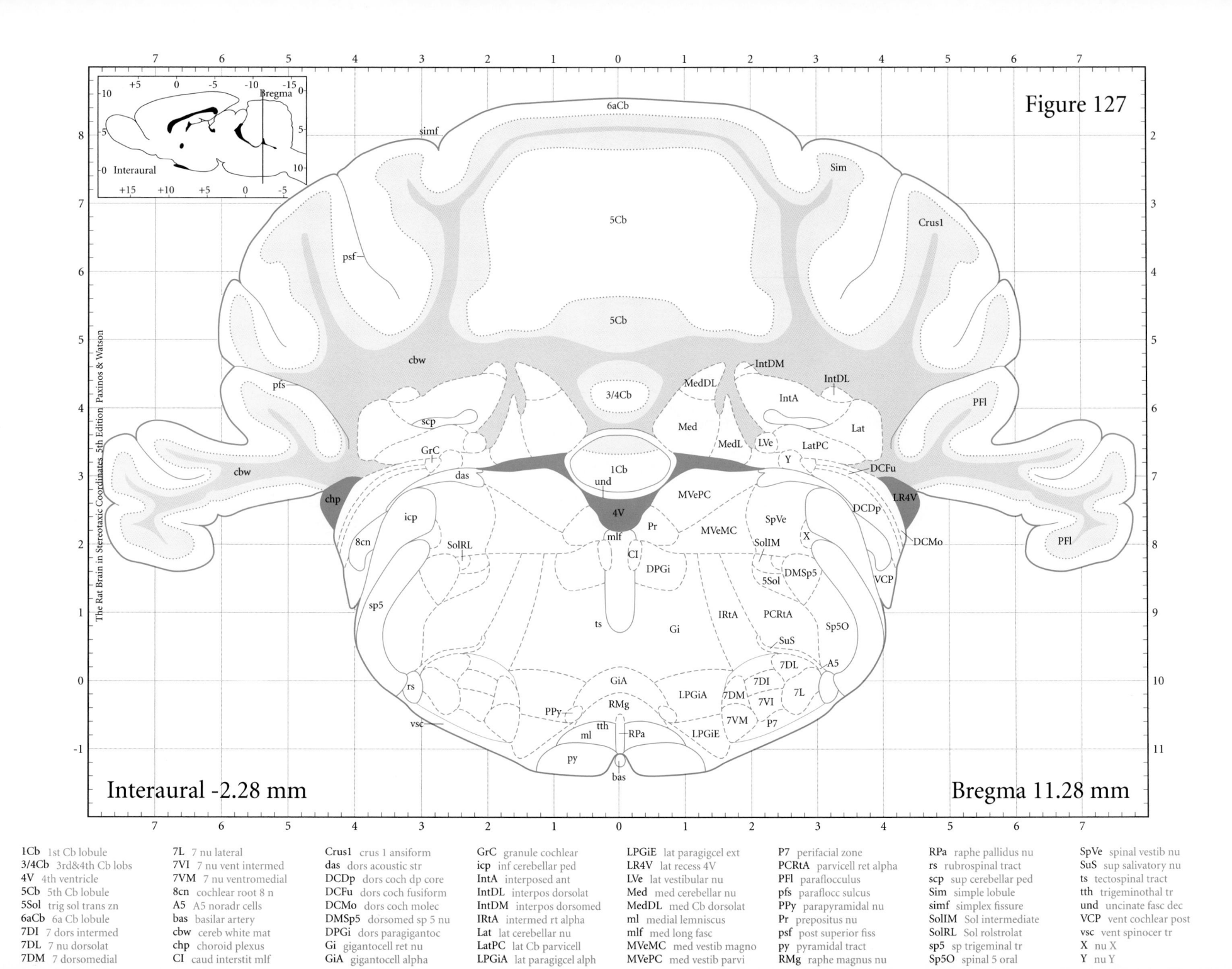

Figure 127

Interaural -2.28 mm

Bregma 11.28 mm

The Rat Brain in Stereotaxic Coordinates, 5th Edition Paxinos & Watson

1Cb 1st Cb lobule	7L 7 nu lateral	Crus1 crus 1 ansiform	GrC granule cochlear	LPGiE lat paragigcel ext	P7 perifacial zone
3/4Cb 3rd&4th Cb lobs	7VI 7 nu vent intermed	das dors acoustic str	icp inf cerebellar ped	LR4V lat recess 4V	PCRtA parvicell ret alpha
4V 4th ventricle	7VM 7 nu ventromedial	DCDp dors coch dp core	IntA interposed ant	LVe lat vestibular nu	PFl paraflocculus
5Cb 5th Cb lobule	8cn cochlear root 8 n	DCFu dors coch fusiform	IntDL interpos dorsolat	Med med cerebellar nu	pfs paraflocc sulcus
5Sol trig sol trans zn	A5 A5 noradr cells	DCMo dors coch molec	IntDM interpos dorsomed	MedDL med Cb dorsolat	PPy parapyramidal nu
6aCb 6a Cb lobule	bas basilar artery	DMSp5 dorsomed sp 5 nu	IRtA intermed rt alpha	ml medial lemniscus	Pr prepositus nu
7DI 7 dors intermed	cbw cereb white mat	DPGi dors paragigantoc	Lat lat cerebellar nu	mlf med long fasc	psf post superior fiss
7DL 7 nu dorsolat	chp choroid plexus	Gi gigantocell ret nu	LatPC lat Cb parvicell	MVeMC med vestib magno	py pyramidal tract
7DM 7 dorsomedial	CI caud interstit mlf	GiA gigantocell alpha	LPGiA lat paragigcel alph	MVePC med vestib parvi	RMg raphe magnus nu

RPa raphe pallidus nu	SpVe spinal vestib nu
rs rubrospinal tract	SuS sup salivatory nu
scp sup cerebellar ped	ts tectospinal tract
Sim simple lobule	tth trigeminothal tr
simf simplex fissure	und uncinate fasc dec
SolIM Sol intermediate	VCP vent cochlear post
SolRL Sol rolstrolat	vsc vent spinocer tr
sp5 sp trigeminal tr	X nu X
Sp5O spinal 5 oral	Y nu Y

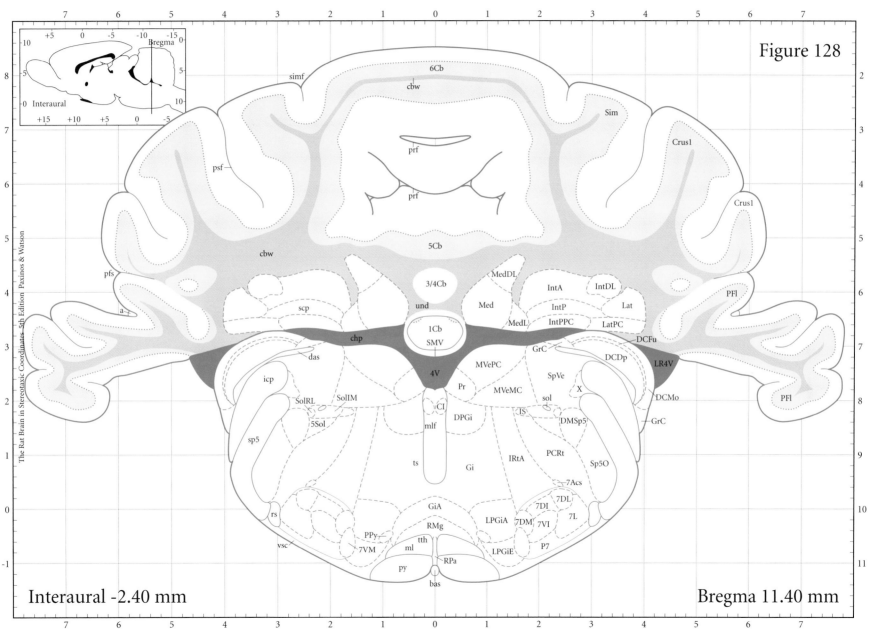

Figure 128

Interaural -2.40 mm

Bregma 11.40 mm

1Cb 1st Cb lobule
3/4Cb 3rd&4th Cb lobs
4V 4th ventricle
5Cb 5th Cb lobule
5Sol trig sol trans zn
6Cb 6th Cb lobule
7Acs mo t accessory
7DI 7 dors intermed
7DL 7 nu dorsolat

7DM 7 dorsomedial
7L 7 nu lateral
7VI 7 nu vent intermed
7VM 7 nu ventromedial
bas basilar artery
cbw cereb white mat
chp choroid plexus
CI caud interstit mlf
Crus1 crus 1 ansiform

das dors acoustic str
DCDp dors coch dp core
DCFu dors coch fusiform
DCMo dors coch molec
DMSp5 dorsomed sp 5 nu
DPGi dors paragigantoc
Gi gigantocell ret nu
GiA gigantocell alpha
GrC granule cochlear

icp inf cerebellar ped
IntA interposed ant
IntDL interpos dorsolat
IntP interpos post
IntPPC interpos pos parv
IRtA intermed rt alpha
IS inf salivatory nu
Lat lat cerebellar nu
LatPC lat Cb parvicell

LPGiA lat paragigcel alph
LPGiE lat paragigcel ext
LR4V lat recess 4V
Med med cerebellar nu
MedDL med Cb dorsolat
MedL med Cb lateral
ml medial lemniscus
mlf med long fasc
MVeMC med vestib magno

MVePC med vestib parvi
PCRt parvicell ret nu
PFl paraflocculus
pfs paraflocc sulcus
PPy parapyramidal nu
Pr prepositus nu
prf primary fissure
psf post superior fiss
py pyramidal tract

RMg raphe magnus nu
RPa raphe pallidus nu
rs rubrospinal tract
scp sup cerebellar ped
Sim simple lobule
simf simplex fissure
SMV sup medull velum
sol solitary tract
SolIM Sol intermediate

SolRL Sol rolstrolat
sp5 sp trigeminal tr
Sp5O spinal 5 oral
SpVe spinal vestib nu
ts tectospinal tract
tth trigeminothal tr
und uncinate fasc dec
vsc vent spinocer tr
X nu X

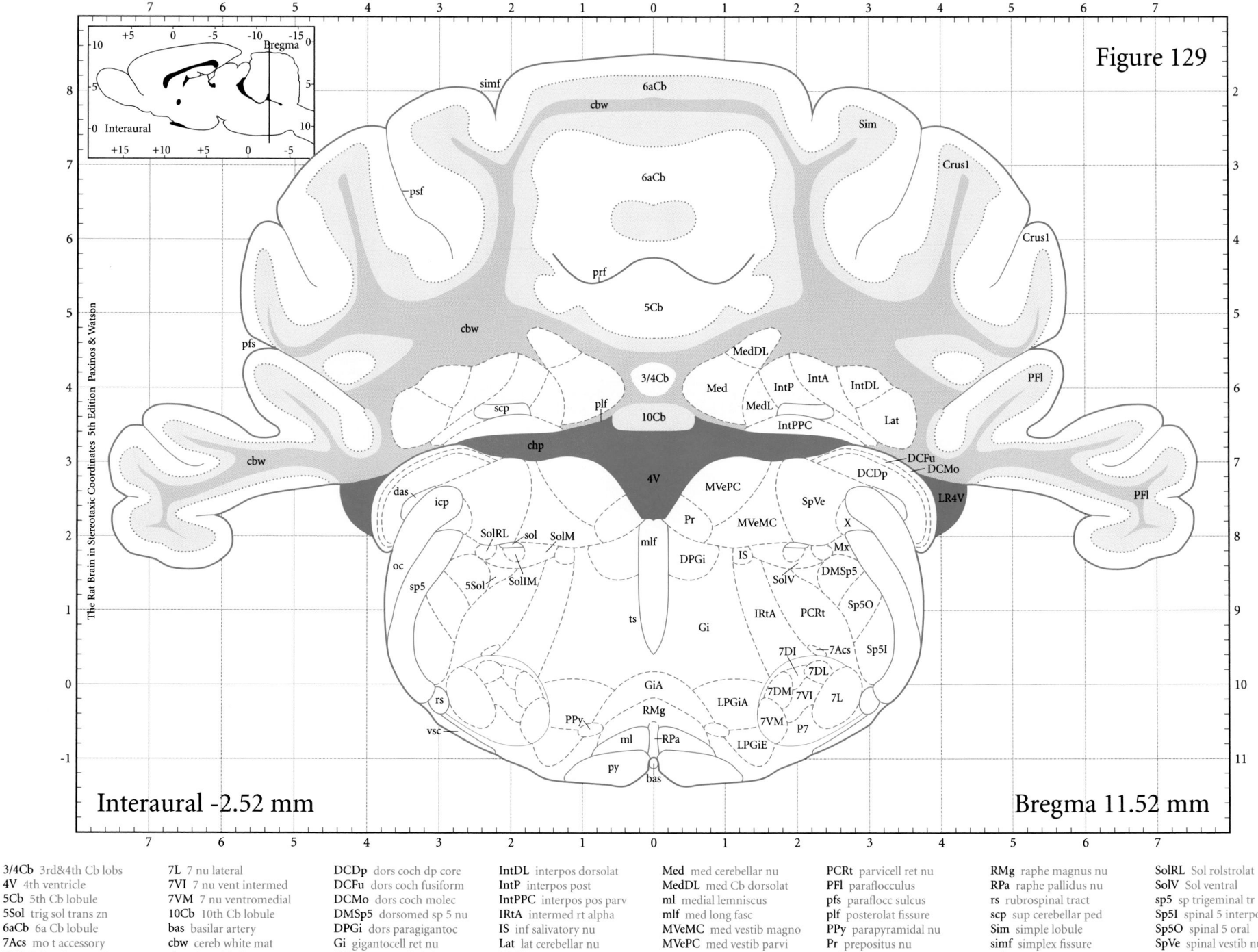

Figure 129

Interaural -2.52 mm

Bregma 11.52 mm

3/4Cb 3rd&4th Cb lobs
4V 4th ventricle
5Cb 5th Cb lobule
5Sol trig sol trans zn
6aCb 6a Cb lobule
7Acs mo t accessory
7DI 7 dors intermed
7DL 7 nu dorsolat
7DM 7 dorsomedial

7L 7 nu lateral
7VI 7 nu vent intermed
7VM 7 nu ventromedial
10Cb 10th Cb lobule
bas basilar artery
cbw cereb white mat
chp choroid plexus
Crus1 crus 1 ansiform
das dors acoustic str

DCDp dors coch dp core
DCFu dors coch fusiform
DCMo dors coch molec
DMSp5 dorsomed sp 5 nu
DPGi dors paragigantoc
Gi gigantocell ret nu
GiA gigantocell alpha
icp inf cerebellar ped
IntA interposed ant

IntDL interpos dorsolat
IntP interpos post
IntPPC interpos pos parv
IRtA intermed rt alpha
IS inf salivatory nu
Lat lat cerebellar nu
LPGiA lat paragigcel alph
LPGiE lat paragigcel ext
LR4V lat recess 4V

Med med cerebellar nu
MedDL med Cb dorsolat
ml medial lemniscus
mlf med long fasc
MVeMC med vestib magno
MVePC med vestib parvi
Mx matrix region
oc olivocerebellar tr
P7 perifacial zone

PCRt parvicell ret nu
PFl paraflocculus
pfs paraflocc sulcus
plf posterolat fissure
PPy parapyramidal nu
Pr prepositus nu
prf primary fissure
psf post superior fiss
py pyramidal tract

RMg raphe magnus nu
RPa raphe pallidus nu
rs rubrospinal tract
scp sup cerebellar ped
Sim simple lobule
simf simplex fissure
sol solitary tract
SolIM Sol intermediate
SolM Sol medial

SolRL Sol rolstrolat
SolV Sol ventral
sp5 sp trigeminal tr
Sp5I spinal 5 interpolar
Sp5O spinal 5 oral
SpVe spinal vestib nu
ts tectospinal tract
vsc vent spinocer tr
X nu X

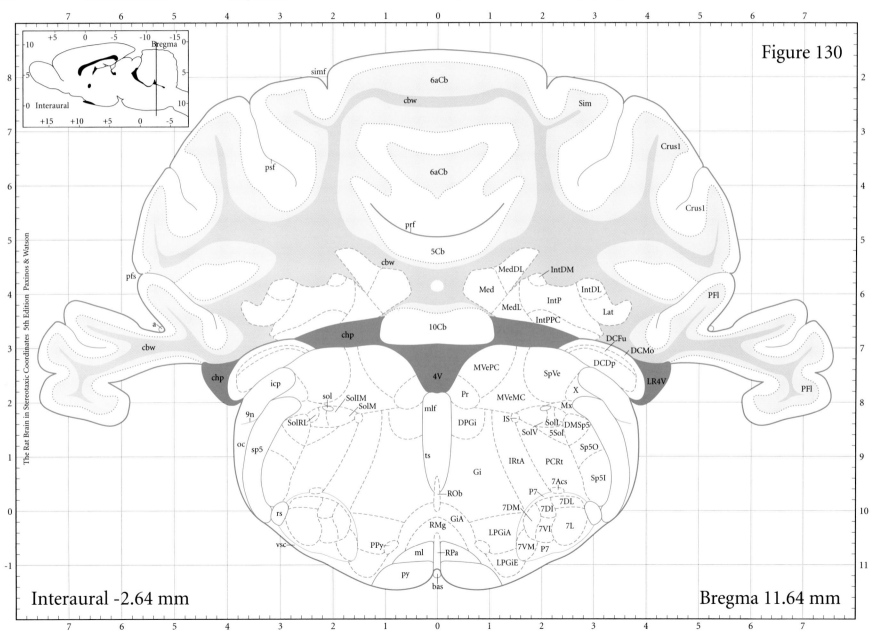

Figure 130

The Rat Brain in Stereotaxic Coordinates 5th Edition Paxinos & Watson

Interaural -2.64 mm

Bregma 11.64 mm

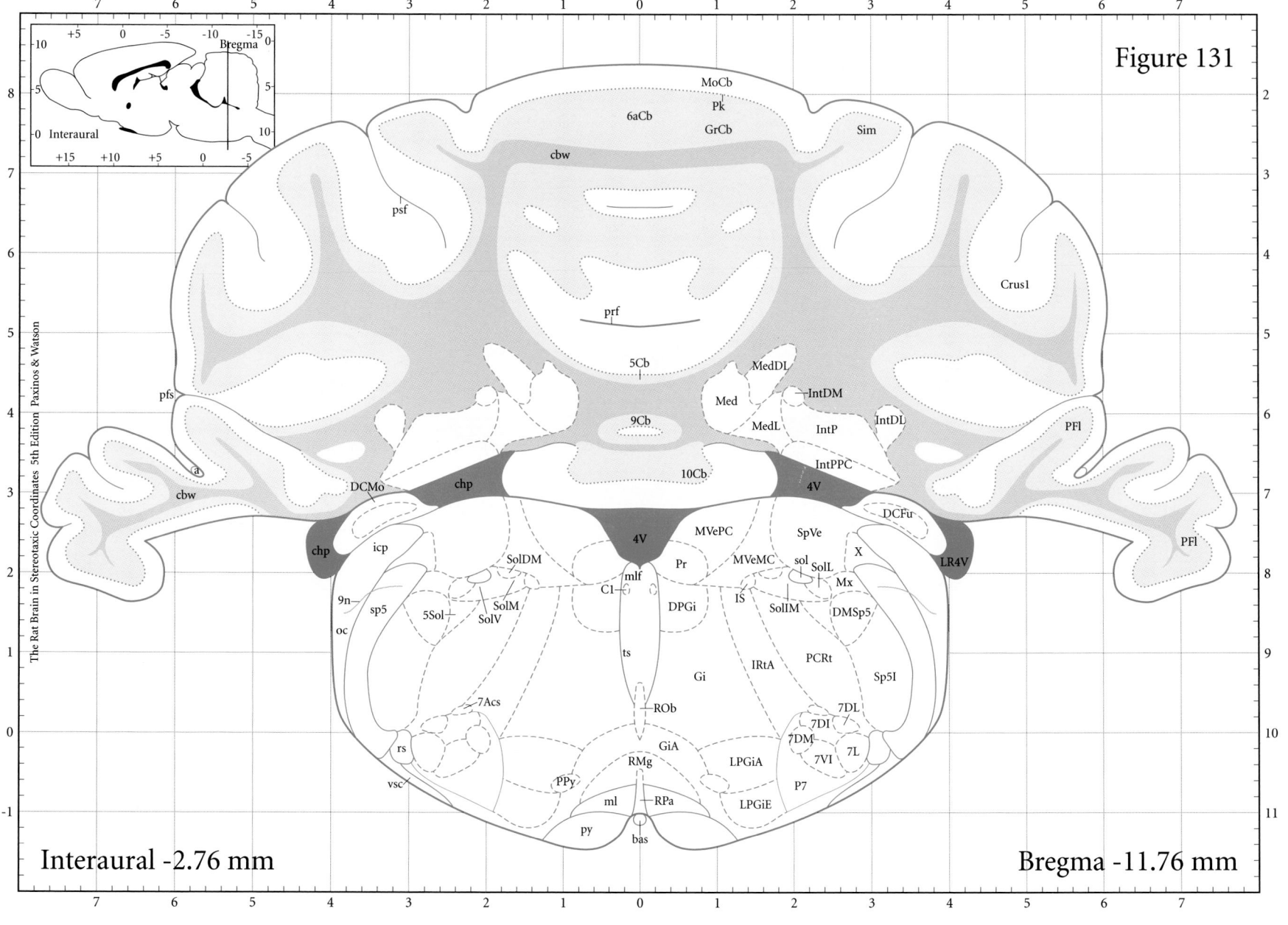

Figure 131

Interaural -2.76 mm

Bregma -11.76 mm

The Rat Brain in Stereotaxic Coordinates 5th Edition Paxinos & Watson

4V 4th ventricle	7VI 7 nu vent intermed	Crus1 crus 1 ansiform	IntDM interpos dorsomed	ml medial lemniscus	PFl paraflocculus	ROb raphe obscurus nu	SolV Sol ventral
5Cb 5th Cb lobule	9Cb 9th Cb lobule	DCFu dors coch fusiform	IntP interpos post	mlf med long fasc	pfs paraflocc sulcus	RPa raphe pallidus nu	sp5 sp trigeminal tr
5Sol trig sol trans zn	9n glossopharyngeal n	DCMo dors coch molec	IRtA intermed rt alpha	MoCb molecul layer Cb	Pk Purkinje cells	rs rubrospinal tract	Sp5I spinal 5 interpolar
6aCb 6a Cb lobule	10Cb 10th Cb lobule	DMSp5 dorsomed sp 5 nu	IS inf salivatory nu	MVeMC med vestib magno	PPy parapyramidal nu	Sim simple lobule	SpVe spinal vestib nu
7Acs mo t accessory	a artery	DPGi dors paragigantoc	LPGiA lat paragigcel alph	MVePC med vestib parvi	Pr prepositus nu	sol solitary tract	ts tectospinal tract
7DI 7 dors intermed	bas basilar artery	Gi gigantocell ret nu	LPGiE lat paragigcel ext	Mx matrix region	prf primary fissure	SolDM Sol dorsomedial	vsc vent spinocer tr
7DL 7 nu dorsolat	C1 C1 adren cells	GiA gigantocell alpha	LR4V lat recess 4V	oc olivocerebellar tr	psf post superior fiss	SolIM Sol intermediate	X nu X
7DM 7 dorsomedial	cbw cereb white mat	icp inf cerebellar ped	Med med cerebellar nu	P7 perifacial zone	py pyramidal tract	SolL Sol lateral	
7L 7 nu lateral	chp choroid plexus	IntDL interpos dorsolat	MedDL med Cb dorsolat	PCRt parvicell ret nu	RMg raphe magnus nu	SolM Sol medial	

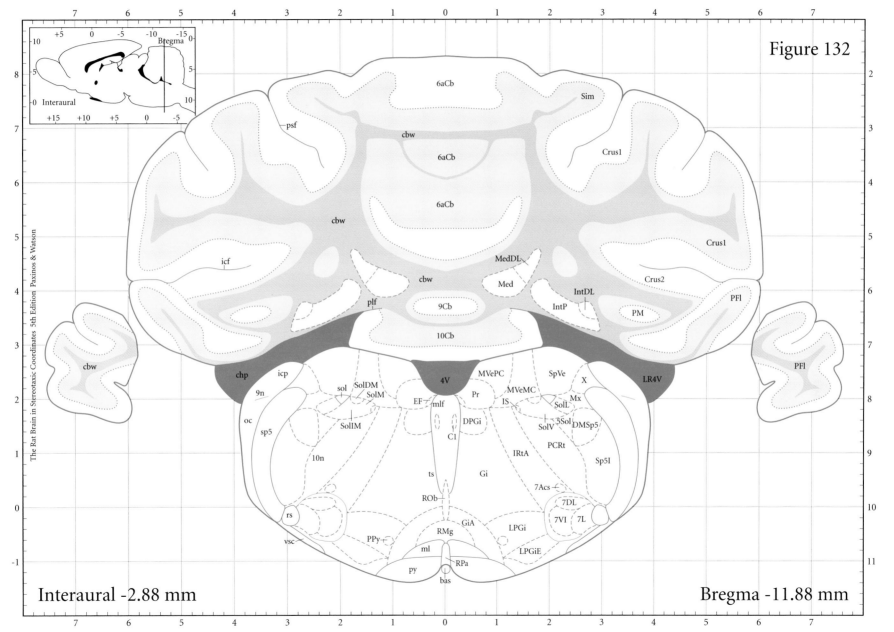

4V   4th ventricle
5Sol   trig sol trans zn
6aCb   6a Cb lobule
7Acs   mo t accessory
7DL   7 nu dorsolat
7L   7 nu lateral
7VI   7 nu vent intermed
9Cb   9th Cb lobule

9n   glossopharyngeal n
10Cb   10th Cb lobule
10n   vagus n
bas   basilar artery
C1   C1 adren cells
cbw   cereb white mat
chp   choroid plexus
Crus1   crus 1 ansiform

Crus2   crus 2 ansiform
DMSp5   dorsomed sp 5 nu
DPGi   dors paragigantoc
EF   epifascicular nu
Gi   gigantocell ret nu
GiA   gigantocell alpha
icf   intercrural fissure
icp   inf cerebellar ped

IntDL   interpos dorsolat
IntP   interpos post
IRtA   intermed rt alpha
IS   inf salivatory nu
LPGi   lat paragig cell
LPGiE   lat paragigcel ext
LR4V   lat recess 4V
Med   med cerebellar nu

MedDL   med Cb dorsolat
ml   medial lemniscus
mlf   med long fasc
MVeMC   med vestib magno
MVePC   med vestib parvi
Mx   matrix region
oc   olivocerebellar tr
PCRt   parvicell ret nu

PFl   paraflocculus
plf   posterolat fissure
PM   paramedian lobule
PPy   parapyramidal nu
Pr   prepositus nu
psf   post superior fiss
RMg   raphe magnus nu

ROb   raphe obscurus nu
RPa   raphe pallidus nu
rs   rubrospinal tract
Sim   simple lobule
sol   solitary tract
SolDM   Sol dorsomedial
SolIM   Sol intermediate
SolL   Sol lateral

SolM   Sol medial
SolV   Sol ventral
sp5   sp trigeminal tr
Sp5I   spinal 5 interpolar
SpVe   spinal vestib nu
ts   tectospinal tract
vsc   vent spinocer tr
X   nu X

Figure 132

Interaural -2.88 mm

Bregma -11.88 mm

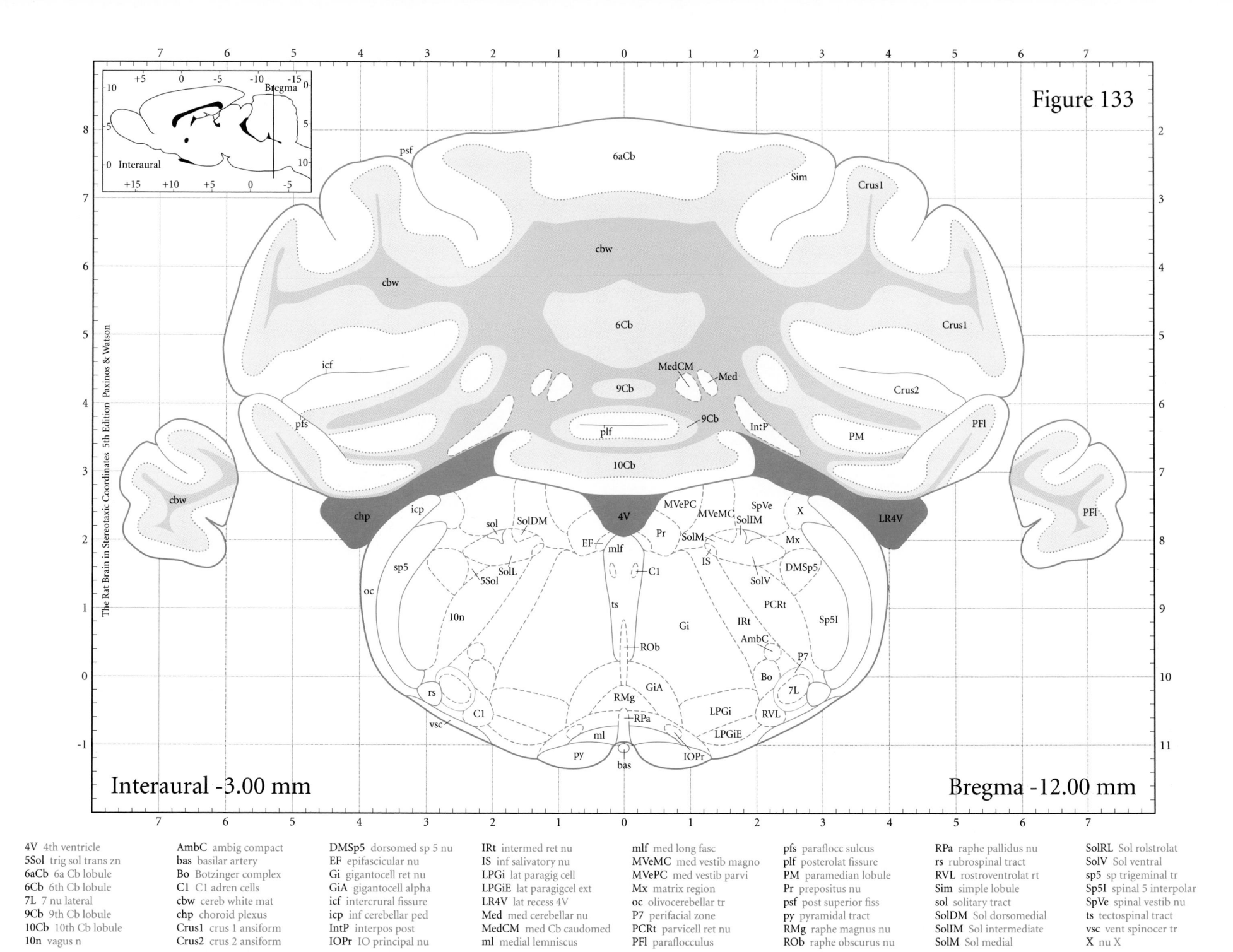

Figure 133

Interaural -3.00 mm

Bregma -12.00 mm

4V 4th ventricle	AmbC ambig compact	DMSp5 dorsomed sp 5 nu	IRt intermed ret nu	mlf med long fasc	pfs paraflocc sulcus	RPa raphe pallidus nu	SolRL Sol rolstrolat
5Sol trig sol trans zn	bas basilar artery	EF epifascicular nu	IS inf salivatory nu	MVeMC med vestib magno	plf posterolat fissure	rs rubrospinal tract	SolV Sol ventral
6aCb 6a Cb lobule	Bo Botzinger complex	Gi gigantocell ret nu	LPGi lat paragig cell	MVePC med vestib parvi	PM paramedian lobule	RVL rostroventrolat rt	sp5 sp trigeminal tr
6Cb 6th Cb lobule	C1 C1 adren cells	GiA gigantocell alpha	LPGiE lat paragigcel ext	Mx matrix region	Pr prepositus nu	Sim simple lobule	Sp5I spinal 5 interpolar
7L 7 nu lateral	cbw cereb white mat	icf intercrural fissure	LR4V lat recess 4V	oc olivocerebellar tr	psf post superior fiss	sol solitary tract	SpVe spinal vestib nu
9Cb 9th Cb lobule	chp choroid plexus	icp inf cerebellar ped	Med med cerebellar nu	P7 perifacial zone	py pyramidal tract	SolDM Sol dorsomedial	ts tectospinal tract
10Cb 10th Cb lobule	Crus1 crus 1 ansiform	IntP interpos post	MedCM med Cb caudomed	PCRt parvicell ret nu	RMg raphe magnus nu	SolIM Sol intermediate	vsc vent spinocer tr
10n vagus n	Crus2 crus 2 ansiform	IOPr IO principal nu	ml medial lemniscus	PFl paraflocculus	ROb raphe obscurus nu	SolM Sol medial	X nu X

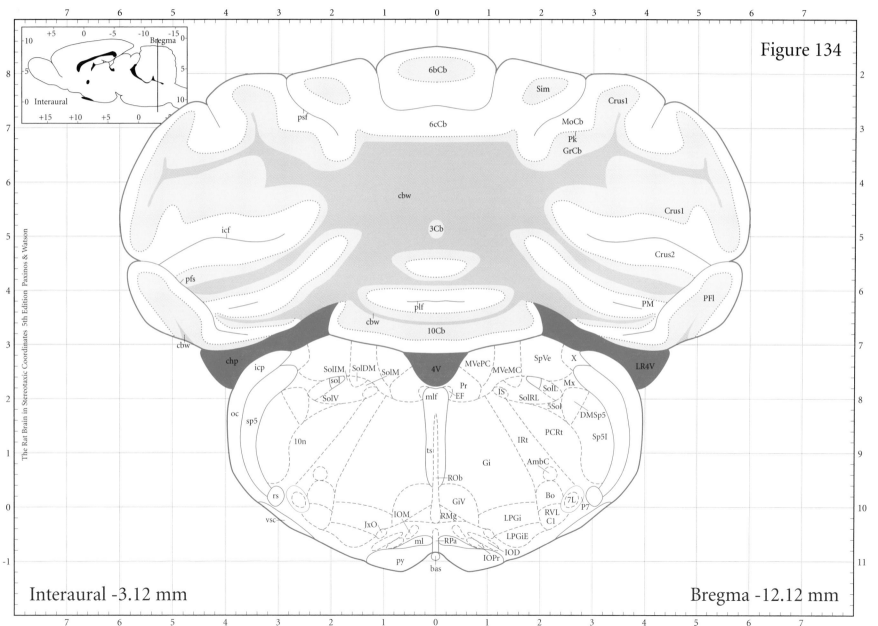

3Cb 3rd Cb lobule
4V 4th ventricle
5Sol trig sol trans zn
6bCb 6b Cb lobule
6cCb 6c Cb lobule
7L 7 nu lateral
10Cb 10th Cb lobule
AmbC ambig compact
bas basilar artery

Bo Botzinger complex
C1 C1 adren cells
cbw cereb white mat
chp choroid plexus
Crus1 crus 1 ansiform
Crus2 crus 2 ansiform
DMSp5 dorsomed sp 5 nu
EF epifascicular nu
Gi gigantocell ret nu

GiV gigantocell vent
icf intercrural fissure
icp inf cerebellar ped
IOD IO dorsal nu
IOM IO medial nu
IOPr IO principal nu
IRt intermed ret nu
IS inf salivatory nu
JxO juxtaolivary nu

LPGi lat paragig cell
LPGiE lat paragigcel ext
LR4V lat recess 4V
ml medial lemniscus
mlf med long fasc
MoCb molecul layer Cb
MVeMC med vestib magno
MVePC med vestib parvi
Mx matrix region

oc olivocerebellar tr
P7 perifacial zone
PCRt parvicell ret nu
PFl paraflocculus
pfs paraflocc sulcus
plf posterolat fissure
PM paramedian lobule
Pr prepositus nu

psf post superior fiss
py pyramidal tract
RMg raphe magnus nu
ROb raphe obscurus nu
RPa raphe pallidus nu
rs rubrospinal tract
RVL rostroventrolat rt
Sim simple lobule
sol solitary tract

SolDM Sol dorsomedial
SolIM Sol intermediate
SolL Sol lateral
SolM Sol medial
SolRL Sol rostrolat
SolV Sol ventral
sp5 sp trigeminal tr
Sp5I spinal 5 interpolar
SpVe spinal vestib nu

ts tectospinal tract
vsc vent spinocer tr
X nu X

Figure 134

Interaural -3.12 mm

Bregma -12.12 mm

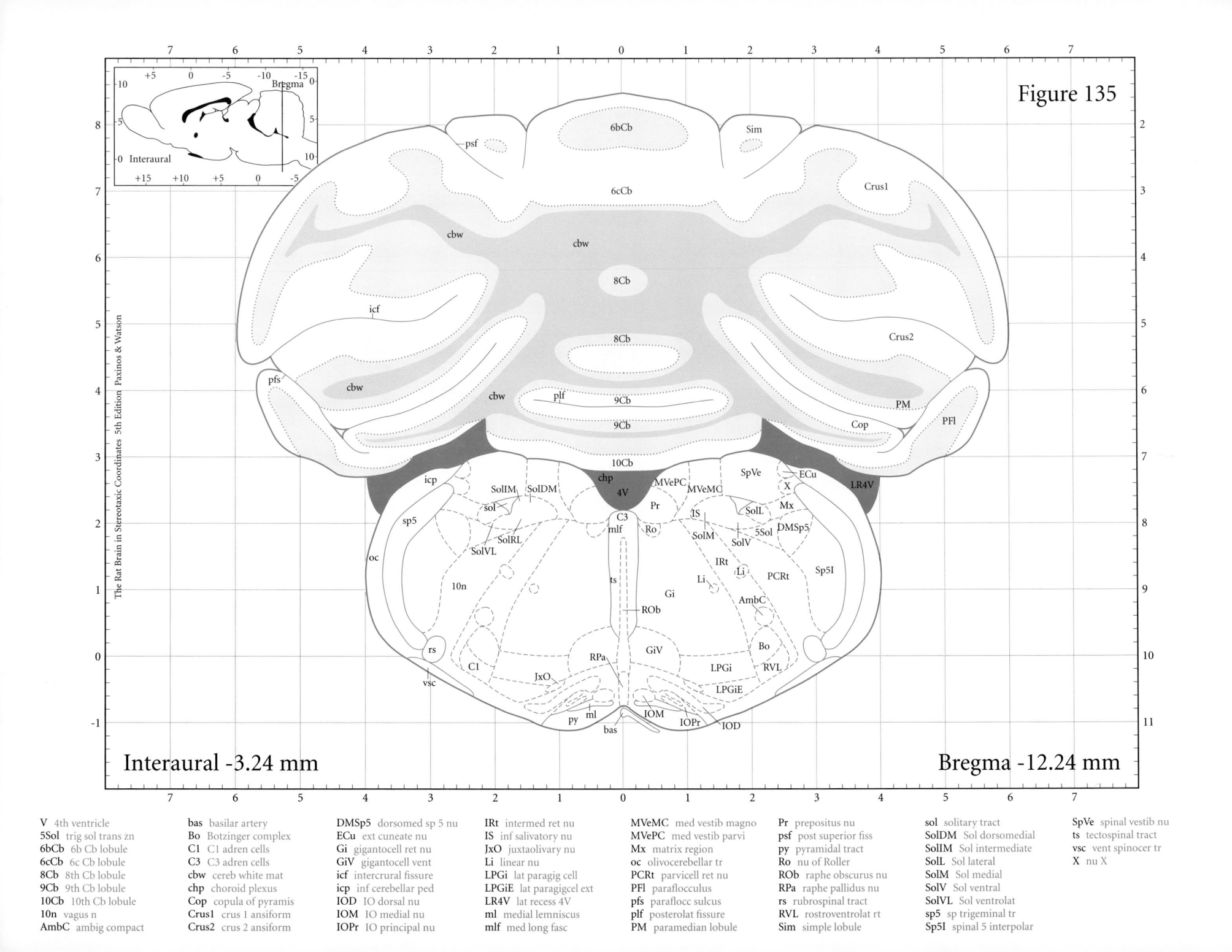

Figure 135

Interaural -3.24 mm

Bregma -12.24 mm

The Rat Brain in Stereotaxic Coordinates 5th Edition Paxinos & Watson

V 4th ventricle	bas basilar artery	DMSp5 dorsomed sp 5 nu	IRt intermed ret nu	MVeMC med vestib magno	Pr prepositus nu	sol solitary tract	SpVe spinal vestib nu
5Sol trig sol trans zn	Bo Botzinger complex	ECu ext cuneate nu	IS inf salivatory nu	MVePC med vestib parvi	psf post superior fiss	SolDM Sol dorsomedial	ts tectospinal tract
6bCb 6b Cb lobule	C1 C1 adren cells	Gi gigantocell ret nu	JxO juxtaolivary nu	Mx matrix region	py pyramidal tract	SolIM Sol intermediate	vsc vent spinocer tr
6cCb 6c Cb lobule	C3 C3 adren cells	GiV gigantocell vent	Li linear nu	oc olivocerebellar tr	Ro nu of Roller	SolL Sol lateral	X nu X
8Cb 8th Cb lobule	cbw cereb white mat	icf intercrural fissure	LPGi lat paragig cell	PCRt parvicell ret nu	ROb raphe obscurus nu	SolM Sol medial	
9Cb 9th Cb lobule	chp choroid plexus	icp inf cerebellar ped	LPGiE lat paragigcel ext	PFl paraflocculus	RPa raphe pallidus nu	SolV Sol ventral	
10Cb 10th Cb lobule	Cop copula of pyramis	IOD IO dorsal nu	LR4V lat recess 4V	pfs paraflocc sulcus	rs rubrospinal tract	SolVL Sol ventrolat	
10n vagus n	Crus1 crus 1 ansiform	IOM IO medial nu	ml medial lemniscus	plf posterolat fissure	RVL rostroventrolat rt	sp5 sp trigeminal tr	
AmbC ambig compact	Crus2 crus 2 ansiform	IOPr IO principal nu	mlf med long fasc	PM paramedian lobule	Sim simple lobule	Sp5I spinal 5 interpolar	

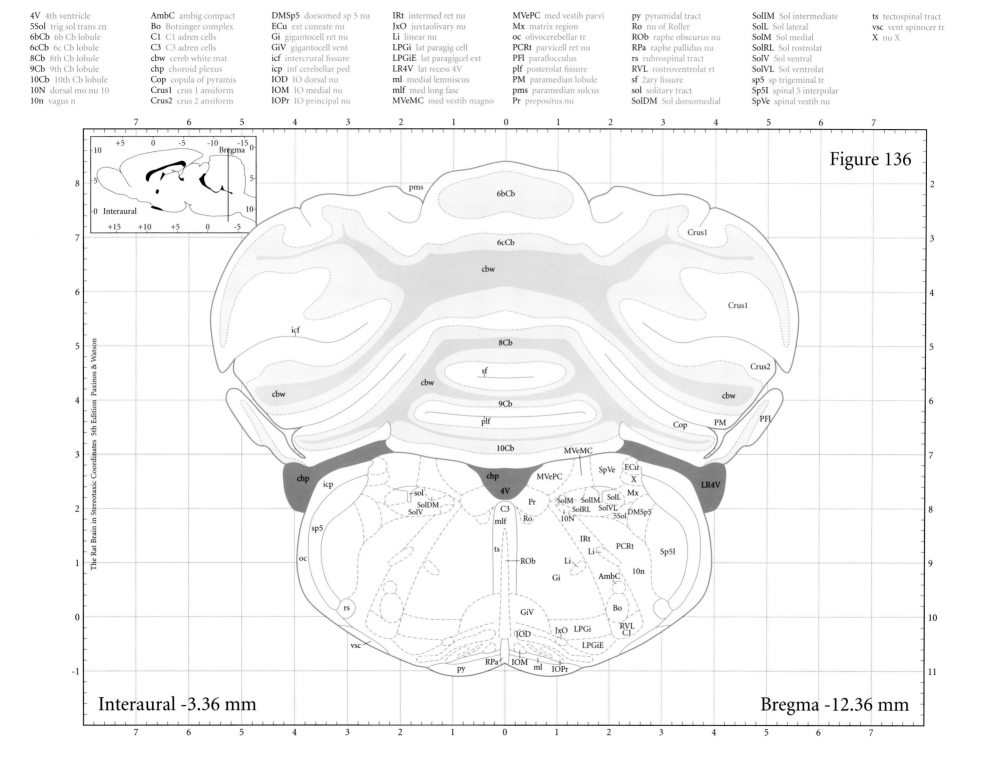

4V	4th ventricle
5Sol	trig sol trans zn
6bCb	6b Cb lobule
6cCb	6c Cb lobule
8Cb	8th Cb lobule
9Cb	9th Cb lobule
10Cb	10th Cb lobule
10N	dorsal mo nu 10
10n	vagus n
AmbC	ambig compact
Bo	Botzinger complex
C1	C1 adren cells
C3	C3 adren cells
cbw	cereb white mat
chp	choroid plexus
Cop	copula of pyramis
Crus1	crus 1 ansiform
Crus2	crus 2 ansiform
DMSp5	dorsomed sp 5 nu
ECu	ext cuneate nu
Gi	gigantocell ret nu
GiV	gigantocell vent
icf	intercrural fissure
icp	inf cerebellar ped
IOD	IO dorsal nu
IOM	IO medial nu
IOPr	IO principal nu
IRt	intermed ret nu
JxO	juxtaolivary nu
Li	linear nu
LPGi	lat paragig cell
LPGiE	lat paragigcel ext
LR4V	lat recess 4V
ml	medial lemniscus
mlf	med long fasc
MVeMC	med vestib magno
MVePC	med vestib parvi
Mx	matrix region
oc	olivocerebellar tr
PCRt	parvicell ret nu
PFl	paraflocculus
plf	posterolat fissure
PM	paramedian lobule
pms	paramedian sulcus
Pr	prepositus nu
py	pyramidal tract
Ro	nu of Roller
ROb	raphe obscurus nu
RPa	raphe pallidus nu
rs	rubrospinal tract
RVL	rostroventrolat rt
sf	2ary fissure
sol	solitary tract
SolDM	Sol dorsomedial
SolIM	Sol intermediate
SolL	Sol lateral
SolM	Sol medial
SolRL	Sol rostrolat
SolV	Sol ventral
SolVL	Sol ventrolat
sp5	sp trigeminal tr
Sp5I	spinal 5 interpolar
SpVe	spinal vestib nu
ts	tectospinal tract
vsc	vent spinocer tr
X	nu X

Figure 136

Interaural -3.36 mm

Bregma -12.36 mm

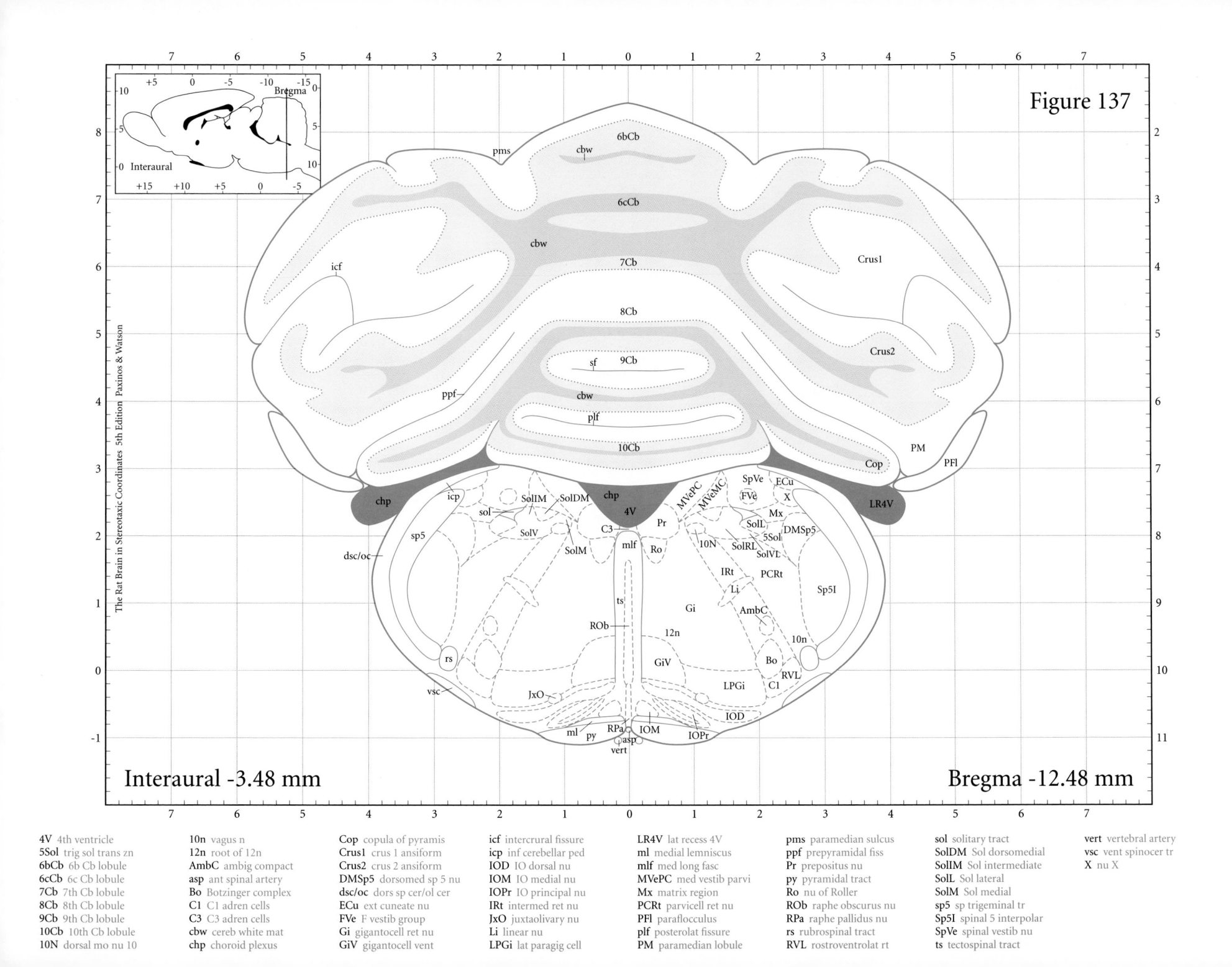

Figure 137

Interaural -3.48 mm

Bregma -12.48 mm

4V 4th ventricle
5Sol trig sol trans zn
6bCb 6b Cb lobule
6cCb 6c Cb lobule
7Cb 7th Cb lobule
8Cb 8th Cb lobule
9Cb 9th Cb lobule
10Cb 10th Cb lobule
10N dorsal mo nu 10

10n vagus n
12n root of 12n
AmbC ambig compact
asp ant spinal artery
Bo Botzinger complex
C1 C1 adren cells
C3 C3 adren cells
cbw cereb white mat
chp choroid plexus

Cop copula of pyramis
Crus1 crus 1 ansiform
Crus2 crus 2 ansiform
DMSp5 dorsomed sp 5 nu
ECu ext cuneate nu
FVe F vestib group
Gi gigantocell ret nu
GiV gigantocell vent

icf intercrural fissure
icp inf cerebellar ped
IOD IO dorsal nu
IOM IO medial nu
IOPr IO principal nu
IRt intermed ret nu
JxO juxtaolivary nu
Li linear nu
LPGi lat paragig cell

LR4V lat recess 4V
ml medial lemniscus
mlf med long fasc
MVePC med vestib parvi
Mx matrix region
PCRt parvicell ret nu
PFl paraflocculus
plf posterolat fissure
PM paramedian lobule

pms paramedian sulcus
ppf prepyramidal fiss
Pr prepositus nu
py pyramidal tract
Ro nu of Roller
ROb raphe obscurus nu
RPa raphe pallidus nu
rs rubrospinal tract
RVL rostroventrolat rt

sol solitary tract
SolDM Sol dorsomedial
SolIM Sol intermediate
SolL Sol lateral
SolM Sol medial
sp5 sp trigeminal tr
Sp5I spinal 5 interpolar
SpVe spinal vestib nu
ts tectospinal tract

vert vertebral artery
vsc vent spinocer tr
X nu X

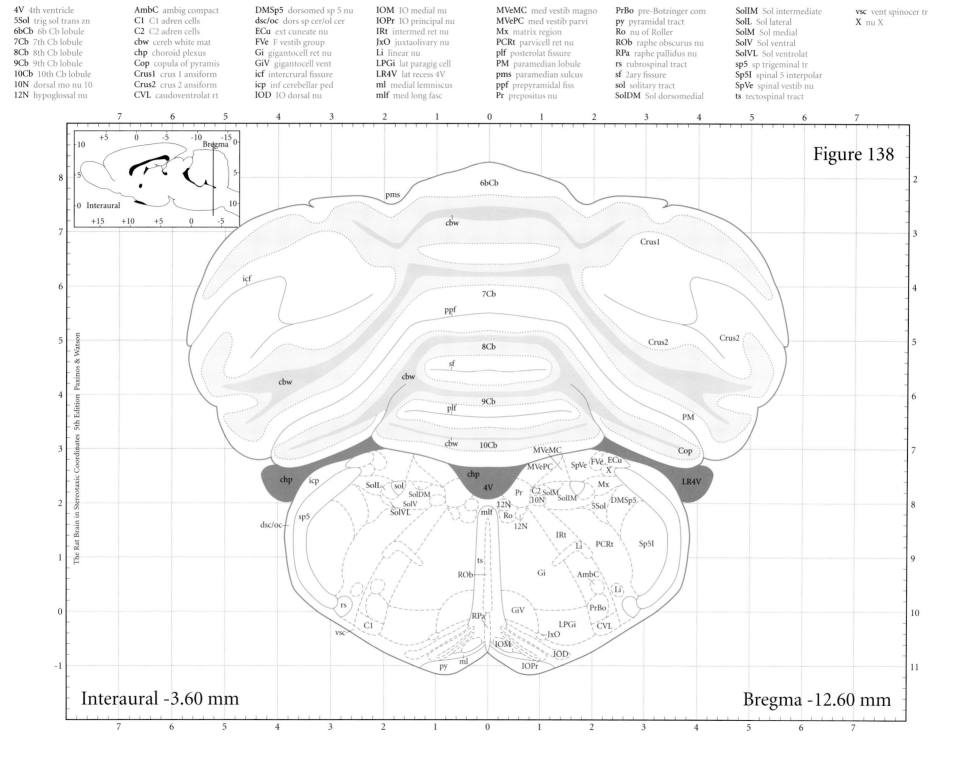

4V 4th ventricle	**AmbC** ambig compact	**DMSp5** dorsomed sp 5 nu
5Sol trig sol trans zn	**C1** C1 adren cells	**dsc/oc** dors sp cer/ol cer
6bCb 6b Cb lobule	**C2** C2 adren cells	**ECu** ext cuneate nu
7Cb 7th Cb lobule	**cbw** cereb white mat	**FVe** F vestib group
8Cb 8th Cb lobule	**chp** choroid plexus	**Gi** gigantocell ret nu
9Cb 9th Cb lobule	**Cop** copula of pyramis	**GiV** gigantocell vent
10Cb 10th Cb lobule	**Crus1** crus 1 ansiform	**icf** intercrural fissure
10N dorsal mo nu 10	**Crus2** crus 2 ansiform	**icp** inf cerebellar ped
12N hypoglossal nu	**CVL** caudoventrolat rt	**IOD** IO dorsal nu

IOM IO medial nu	**MVeMC** med vestib magno	**PrBo** pre-Botzinger com
IOPr IO principal nu	**MVePC** med vestib parvi	**py** pyramidal tract
IRt intermed ret nu	**Mx** matrix region	**Ro** nu of Roller
JxO juxtaolivary nu	**PCRt** parvicell ret nu	**ROb** raphe obscurus nu
Li linear nu	**plf** posterolat fissure	**RPa** raphe pallidus nu
LPGi lat paragig cell	**PM** paramedian lobule	**rs** rubrospinal tract
LR4V lat recess 4V	**pms** paramedian sulcus	**sf** 2ary fissure
ml medial lemniscus	**ppf** prepyramidal fiss	**sol** solitary tract
mlf med long fasc	**Pr** prepositus nu	**SolDM** Sol dorsomedial

SolIM Sol intermediate	**vsc** vent spinocer tr
SolL Sol lateral	**X** nu X
SolM Sol medial	
SolV Sol ventral	
SolVL Sol ventrolat	
sp5 sp trigeminal tr	
Sp5I spinal 5 interpolar	
SpVe spinal vestib nu	
ts tectospinal tract	

Figure 138

Interaural -3.60 mm

Bregma -12.60 mm

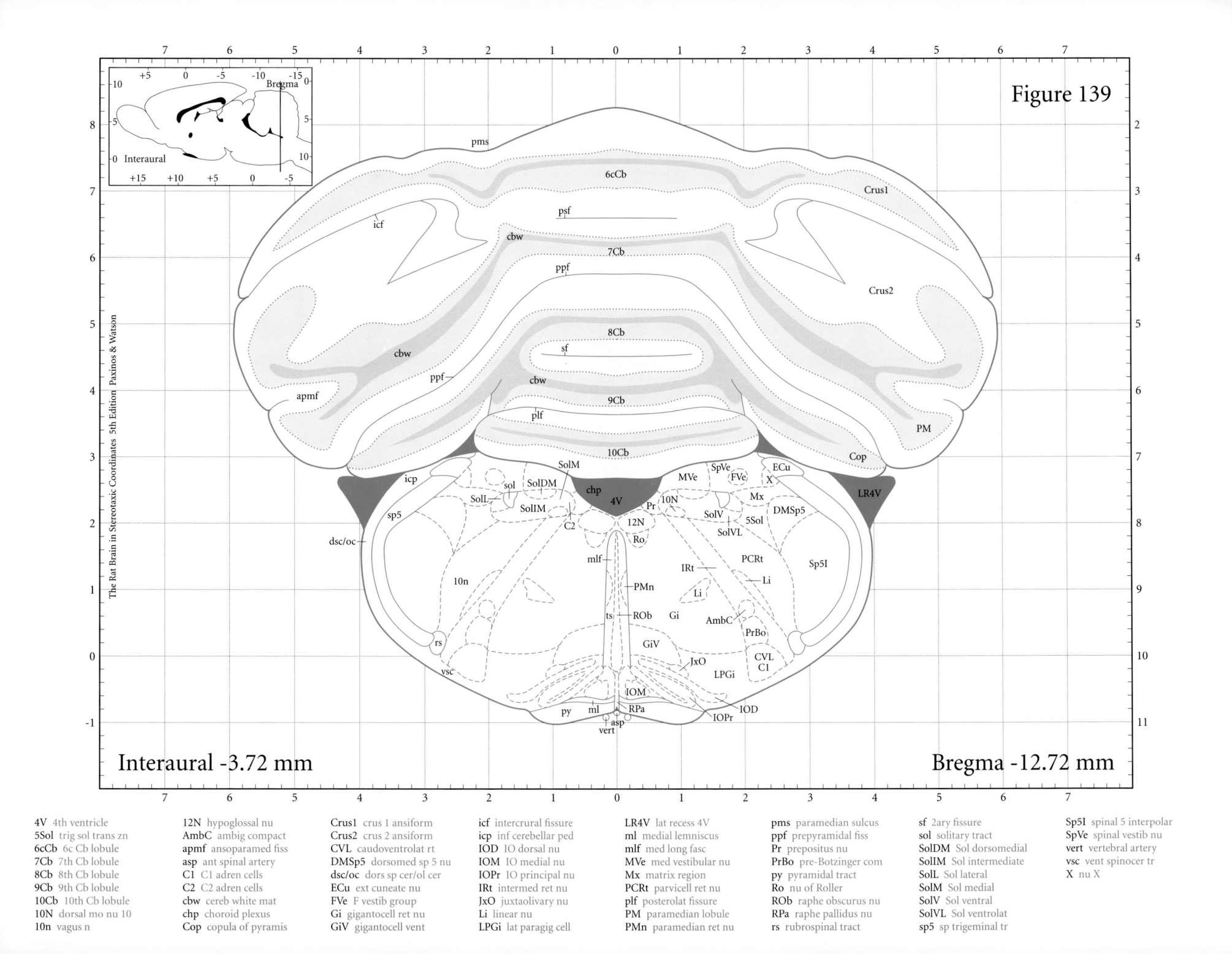

Figure 139

Interaural -3.72 mm

Bregma -12.72 mm

The Rat Brain in Stereotaxic Coordinates 5th Edition Paxinos & Watson

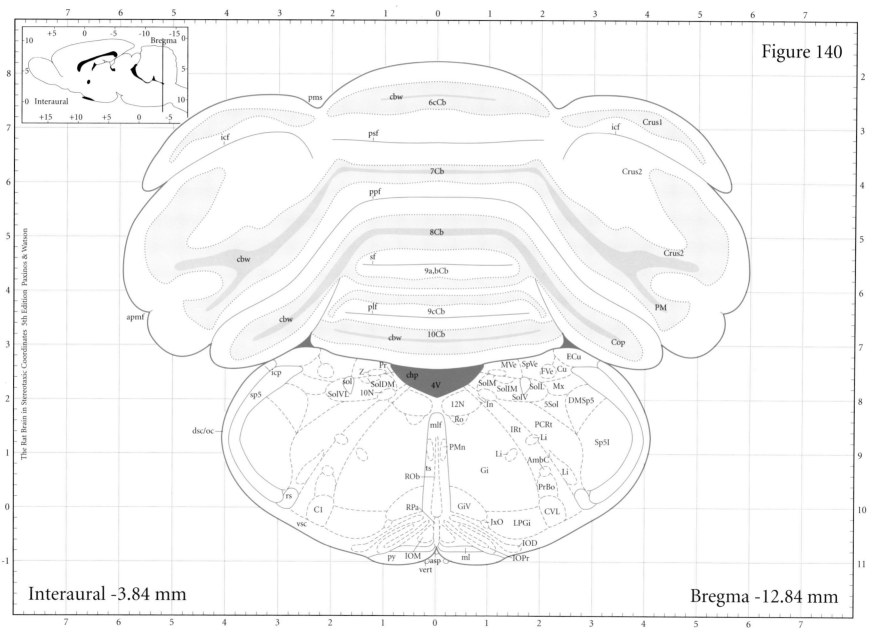

Figure 140

The Rat Brain in Stereotaxic Coordinates 5th Edition Paxinos & Watson

Interaural -3.84 mm

Bregma -12.84 mm

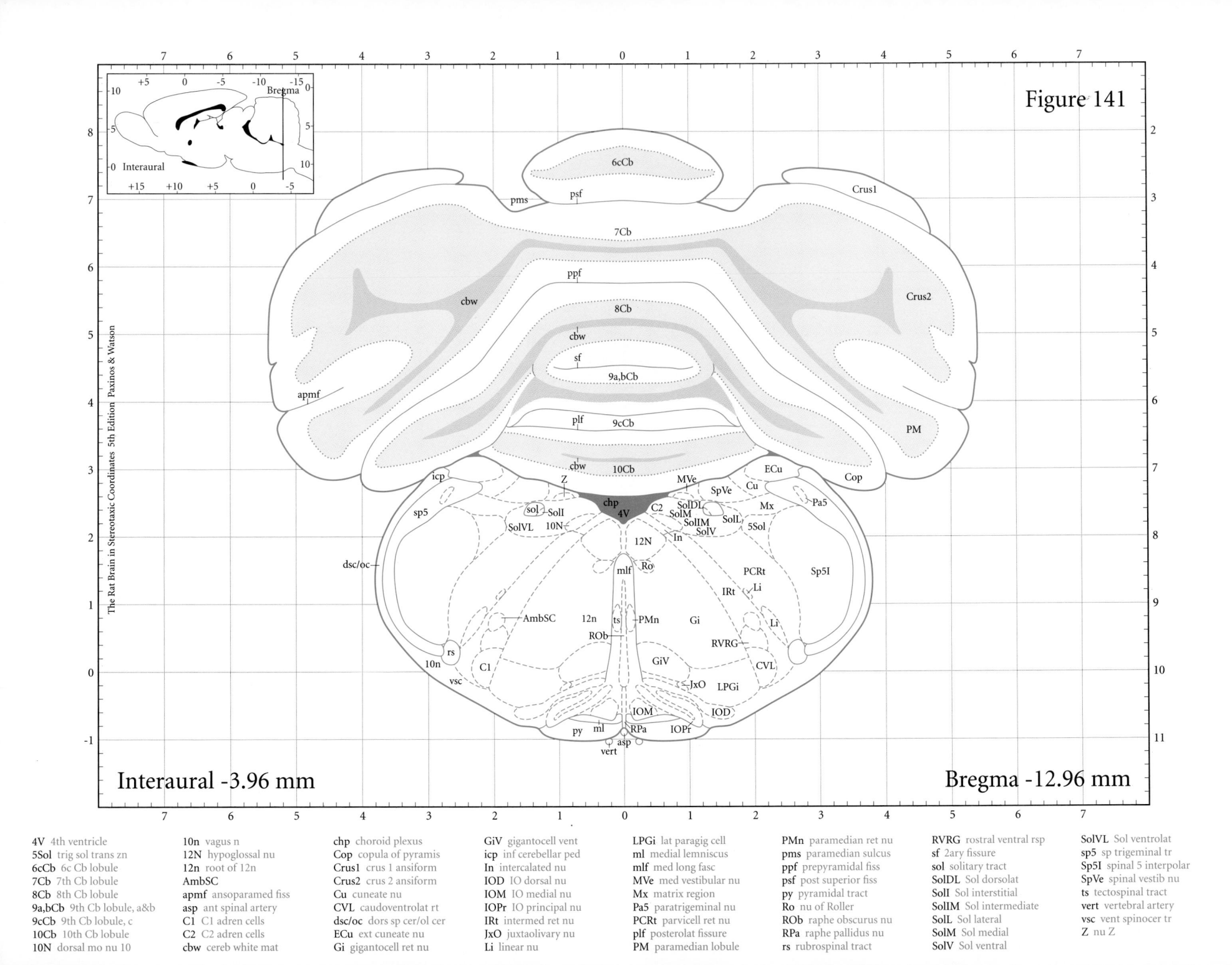

Figure 141

Interaural -3.96 mm

Bregma -12.96 mm

The Rat Brain in Stereotaxic Coordinates 5th Edition Paxinos & Watson

4V 4th ventricle	10n vagus n	chp choroid plexus	GiV gigantocell vent	LPGi lat paragig cell	PMn paramedian ret nu
5Sol trig sol trans zn	12N hypoglossal nu	Cop copula of pyramis	icp inf cerebellar ped	ml medial lemniscus	pms paramedian sulcus
6cCb 6c Cb lobule	12n root of 12n	Crus1 crus 1 ansiform	In intercalated nu	mlf med long fasc	ppf prepyramidal fiss
7Cb 7th Cb lobule	AmbSC	Crus2 crus 2 ansiform	IOD IO dorsal nu	MVe med vestibular nu	psf post superior fiss
8Cb 8th Cb lobule	apmf ansoparamed fiss	Cu cuneate nu	IOM IO medial nu	Mx matrix region	py pyramidal tract
9a,bCb 9th Cb lobule, a&b	asp ant spinal artery	CVL caudoventrolat rt	IOPr IO principal nu	Pa5 paratrigeminal nu	Ro nu of Roller
9cCb 9th Cb lobule, c	C1 C1 adren cells	dsc/oc dors sp cer/ol cer	IRt intermed ret nu	PCRt parvicell ret nu	ROb raphe obscurus nu
10Cb 10th Cb lobule	C2 C2 adren cells	ECu ext cuneate nu	JxO juxtaolivary nu	plf posterolat fissure	RPa raphe pallidus nu
10N dorsal mo nu 10	cbw cereb white mat	Gi gigantocell ret nu	Li linear nu	PM paramedian lobule	rs rubrospinal tract

RVRG rostral ventral rsp	SolVL Sol ventrolat
sf 2ary fissure	sp5 sp trigeminal tr
sol solitary tract	Sp5I spinal 5 interpolar
SolDL Sol dorsolat	SpVe spinal vestib nu
SolI Sol interstitial	ts tectospinal tract
SolIM Sol intermediate	vert vertebral artery
SolL Sol lateral	vsc vent spinocer tr
SolM Sol medial	Z nu Z
SolV Sol ventral	

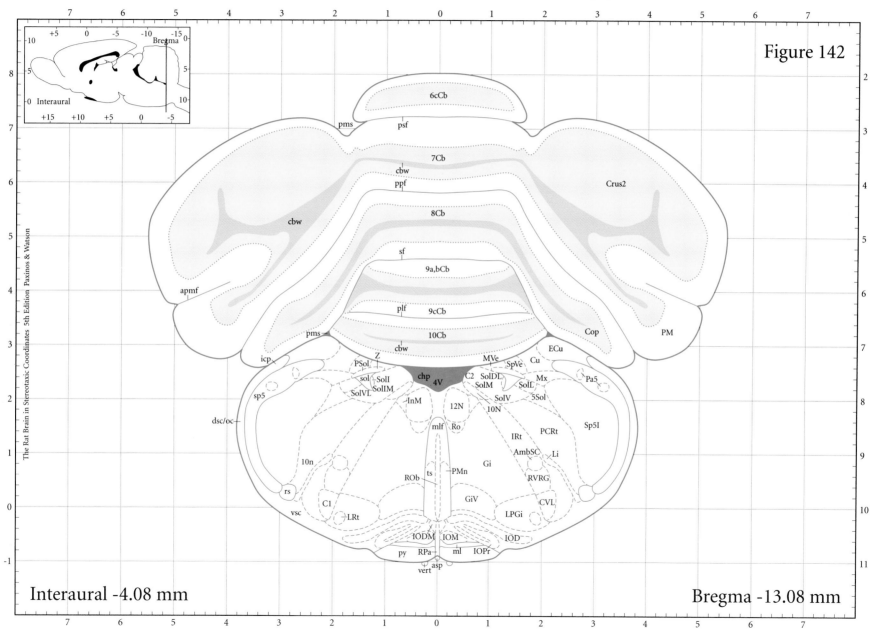

Figure 142

Interaural -4.08 mm

Bregma -13.08 mm

The Rat Brain in Stereotaxic Coordinates 5th Edition Paxinos & Watson

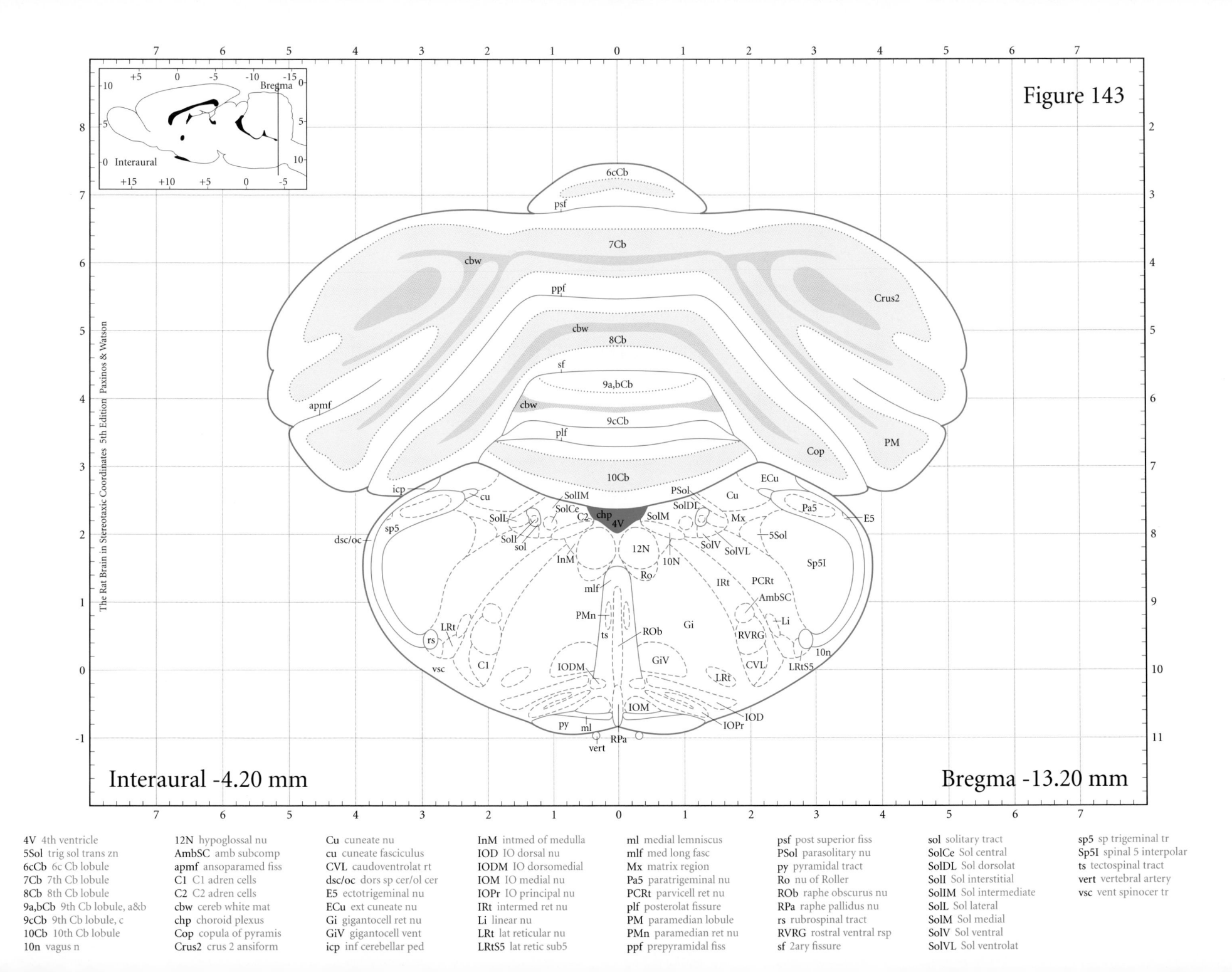

Figure 143

The Rat Brain in Stereotaxic Coordinates 5th Edition Paxinos & Watson

Interaural -4.20 mm

Bregma -13.20 mm

4V 4th ventricle	12N hypoglossal nu	Cu cuneate nu	InM intmed of medulla	ml medial lemniscus	psf post superior fiss	sol solitary tract	sp5 sp trigeminal tr
5Sol trig sol trans zn	AmbSC amb subcomp	cu cuneate fasciculus	IOD IO dorsal nu	mlf med long fasc	PSol parasolitary nu	SolCe Sol central	Sp5I spinal 5 interpolar
6cCb 6c Cb lobule	apmf ansoparamed fiss	CVL caudoventrolat rt	IODM IO dorsomedial	Mx matrix region	py pyramidal tract	SolDL Sol dorsolat	ts tectospinal tract
7Cb 7th Cb lobule	C1 C1 adren cells	dsc/oc dors sp cer/ol cer	IOM IO medial nu	Pa5 paratrigeminal nu	Ro nu of Roller	SolI Sol interstitial	vert vertebral artery
8Cb 8th Cb lobule	C2 C2 adren cells	E5 ectotrigeminal nu	IOPr IO principal nu	PCRt parvicell ret nu	ROb raphe obscurus nu	SolIM Sol intermediate	vsc vent spinocer tr
9a,bCb 9th Cb lobule, a&b	cbw cereb white mat	ECu ext cuneate nu	IRt intermed ret nu	plf posterolat fissure	RPa raphe pallidus nu	SolL Sol lateral	
9cCb 9th Cb lobule, c	chp choroid plexus	Gi gigantocell ret nu	Li linear nu	PM paramedian lobule	rs rubrospinal tract	SolM Sol medial	
10Cb 10th Cb lobule	Cop copula of pyramis	GiV gigantocell vent	LRt lat reticular nu	PMn paramedian ret nu	RVRG rostral ventral rsp	SolV Sol ventral	
10n vagus n	Crus2 crus 2 ansiform	icp inf cerebellar ped	LRtS5 lat retic sub5	ppf prepyramidal fiss	sf 2ary fissure	SolVL Sol ventrolat	

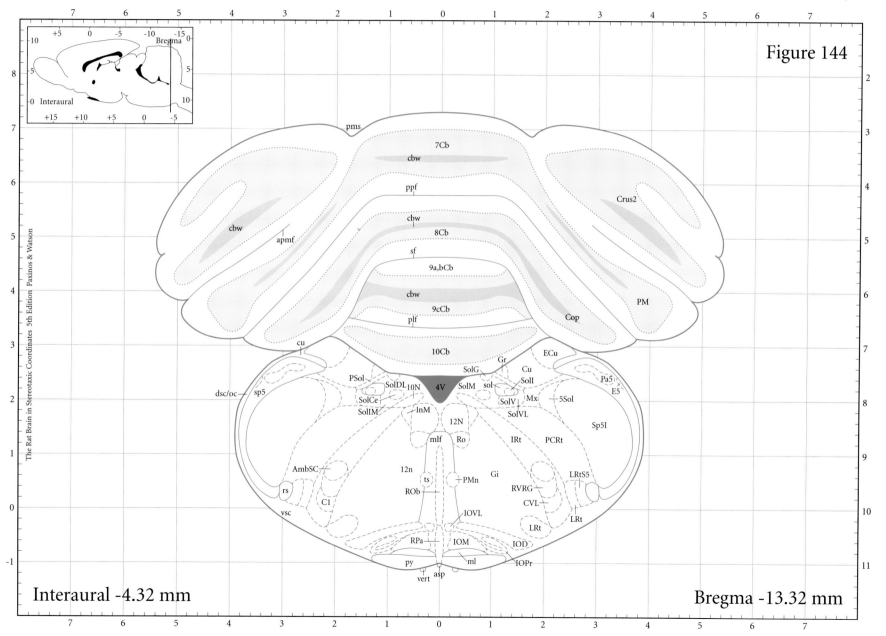

Figure 144

Interaural -4.32 mm

Bregma -13.32 mm

4V 4th ventricle
5Sol trig sol trans zn
7Cb 7th Cb lobule
8Cb 8th Cb lobule
9a,bCb 9th Cb lobule, a&b
9cCb 9th Cb lobule, c
10Cb 10th Cb lobule
10N dorsal mo nu 10

12N hypoglossal nu
12n root of 12n
AmbSC amb subcomp
apmf ansoparamed fiss
asp ant spinal artery
C1 C1 adren cells
cbw cereb white mat
Cop copula of pyramis

Crus2 crus 2 ansiform
Cu cuneate nu
cu cuneate fasciculus
CVL caudoventrolat rt
dsc/oc dors sp cer/ol cer
E5 ectotrigeminal nu
ECu ext cuneate nu
Gi gigantocell ret nu

Gr gracile nu
InM intmed of medulla
IOD IO dorsal nu
IOM IO medial nu
IOPr IO principal nu
IOVL IO ventrolat
IRt intermed ret nu
LRt lat reticular nu

LRtS5 lat retic sub5
ml medial lemniscus
mlf med long fasc
Mx matrix region
Pa5 paratrigeminal nu
PCRt parvicell ret nu
plf posterolat fissure
PM paramedian lobule

PMn paramedian ret nu
pms paramedian sulcus
ppf prepyramidal fiss
PSol parasolitary nu
py pyramidal tract
Ro nu of Roller
ROb raphe obscurus nu
RPa raphe pallidus nu

rs rubrospinal tract
RVRG rostral ventral rsp
sf 2ary fissure
sol solitary tract
SolCe Sol central
SolDL Sol dorsolat
SolG Sol gelatinous
SolIM Sol intermediate

SolM Sol medial
SolV Sol ventral
SolVL Sol ventrolat
sp5 sp trigeminal tr
Sp5I spinal 5 interpolar
ts tectospinal tract
vert vertebral artery
vsc vent spinocer tr

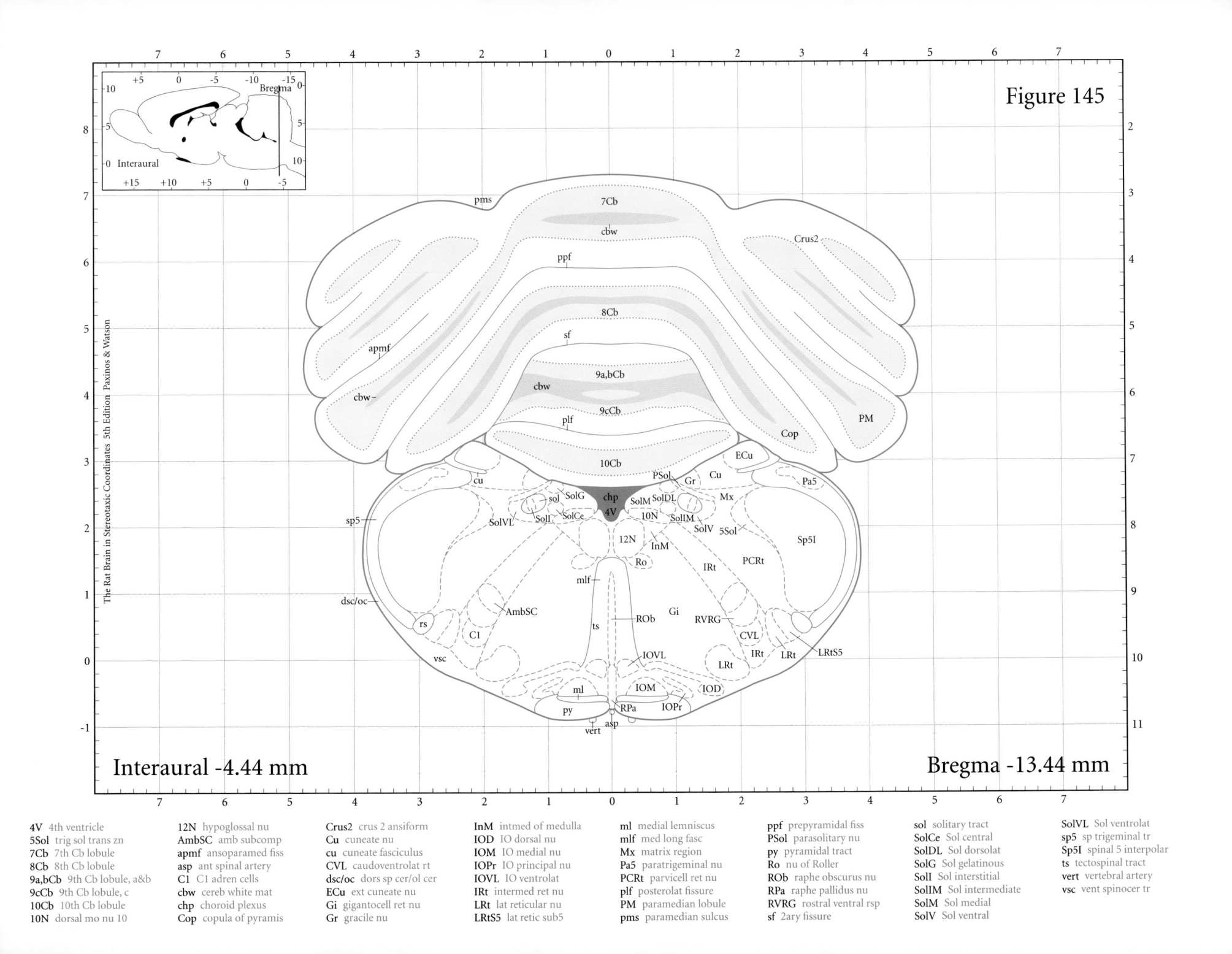

Figure 145

The Rat Brain in Stereotaxic Coordinates 5th Edition Paxinos & Watson

Interaural -4.44 mm

Bregma -13.44 mm

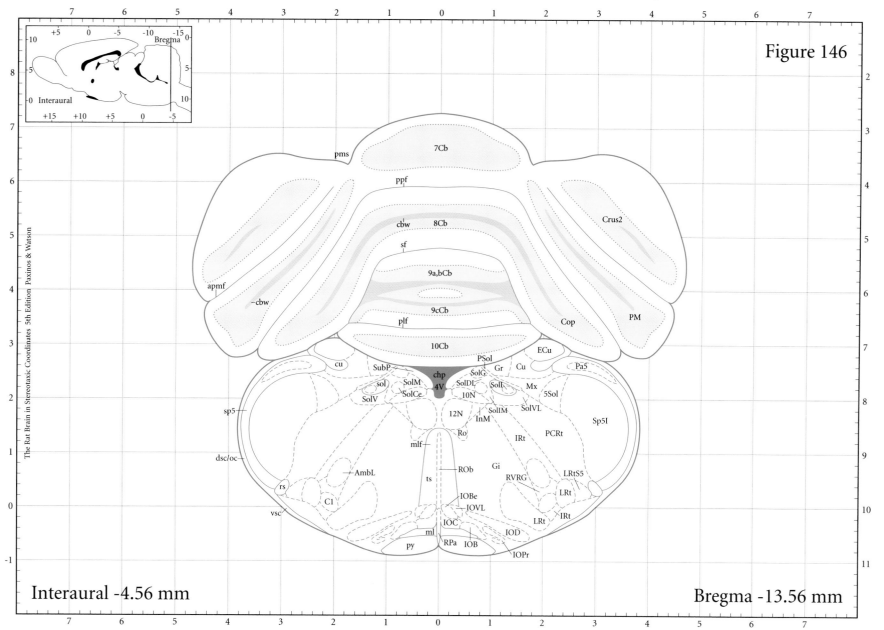

Figure 146

The Rat Brain in Stereotaxic Coordinates 5th Edition Paxinos & Watson

Interaural -4.56 mm

Bregma -13.56 mm

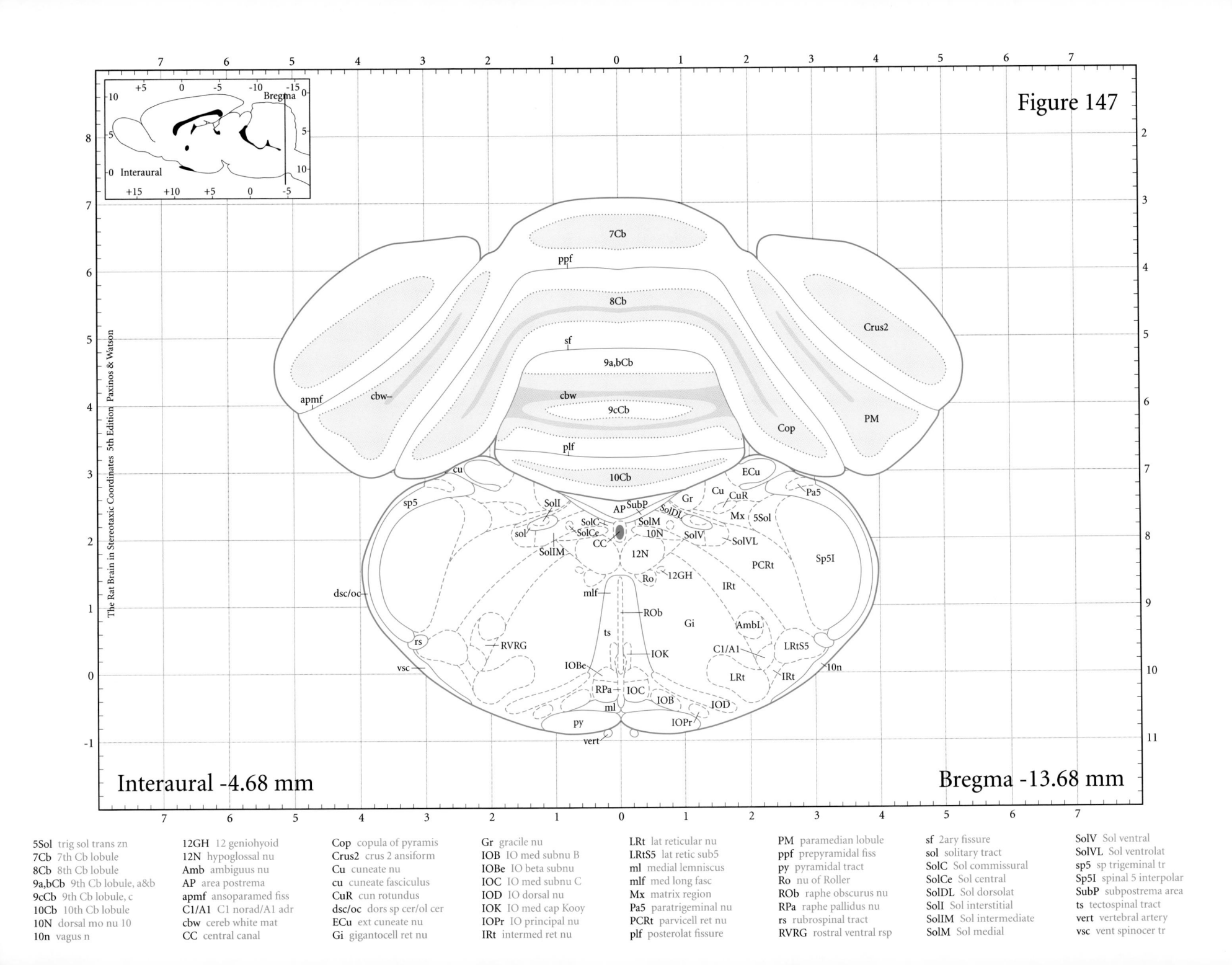

Figure 147

Interaural -4.68 mm

Bregma -13.68 mm

5Sol	trig sol trans zn	Cop	copula of pyramis
7Cb	7th Cb lobule	Crus2	crus 2 ansiform
8Cb	8th Cb lobule	Cu	cuneate nu
9a,bCb	9th Cb lobule, a&b	cu	cuneate fasciculus
9cCb	9th Cb lobule, c	CuR	cun rotundus
10Cb	10th Cb lobule	dsc/oc	dors sp cer/ol cer
10N	dorsal mo nu 10	ECu	ext cuneate nu
10n	vagus n	Gi	gigantocell ret nu

12GH	12 geniohyoid
12N	hypoglossal nu
Amb	ambiguus nu
AP	area postrema
apmf	ansoparamed fiss
C1/A1	C1 norad/A1 adr
cbw	cereb white mat
CC	central canal

Gr	gracile nu
IOB	IO med subnu B
IOBe	IO beta subnu
IOC	IO med subnu C
IOD	IO dorsal nu
IOK	IO med cap Kooy
IOPr	IO principal nu
IRt	intermed ret nu

LRt	lat reticular nu
LRtS5	lat retic sub5
ml	medial lemniscus
mlf	med long fasc
Mx	matrix region
Pa5	paratrigeminal nu
PCRt	parvicell ret nu
plf	posterolat fissure

PM	paramedian lobule
ppf	prepyramidal fiss
py	pyramidal tract
Ro	nu of Roller
ROb	raphe obscurus nu
RPa	raphe pallidus nu
rs	rubrospinal tract
RVRG	rostral ventral rsp

sf	2ary fissure
sol	solitary tract
SolC	Sol commissural
SolCe	Sol central
SolDL	Sol dorsolat
SolI	Sol interstitial
SolIM	Sol intermediate
SolM	Sol medial

SolV	Sol ventral
SolVL	Sol ventrolat
sp5	sp trigeminal tr
Sp5I	spinal 5 interpolar
SubP	subpostrema area
ts	tectospinal tract
vert	vertebral artery
vsc	vent spinocer tr

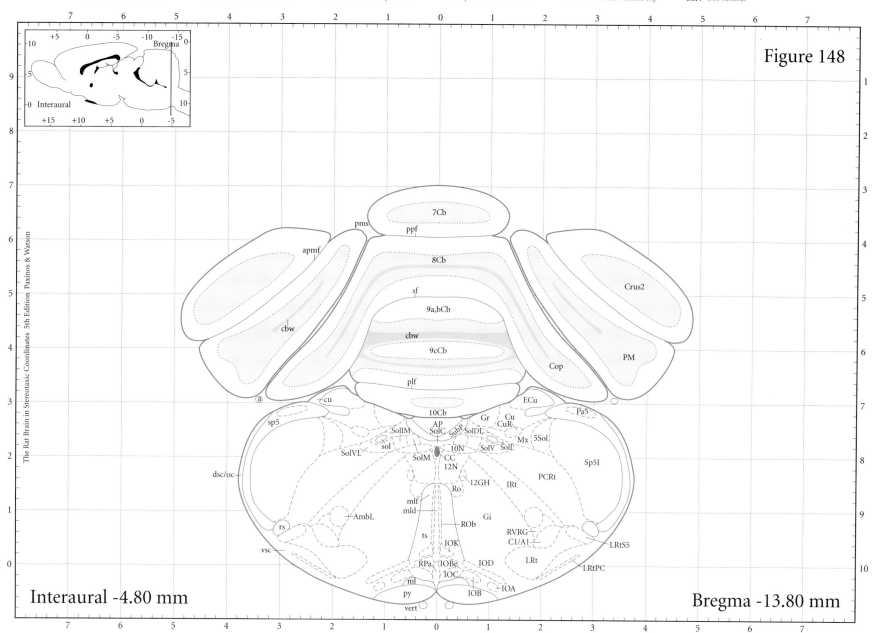

Figure 148

Interaural -4.80 mm

Bregma -13.80 mm

The Rat Brain in Stereotaxic Coordinates 5th Edition Paxinos & Watson

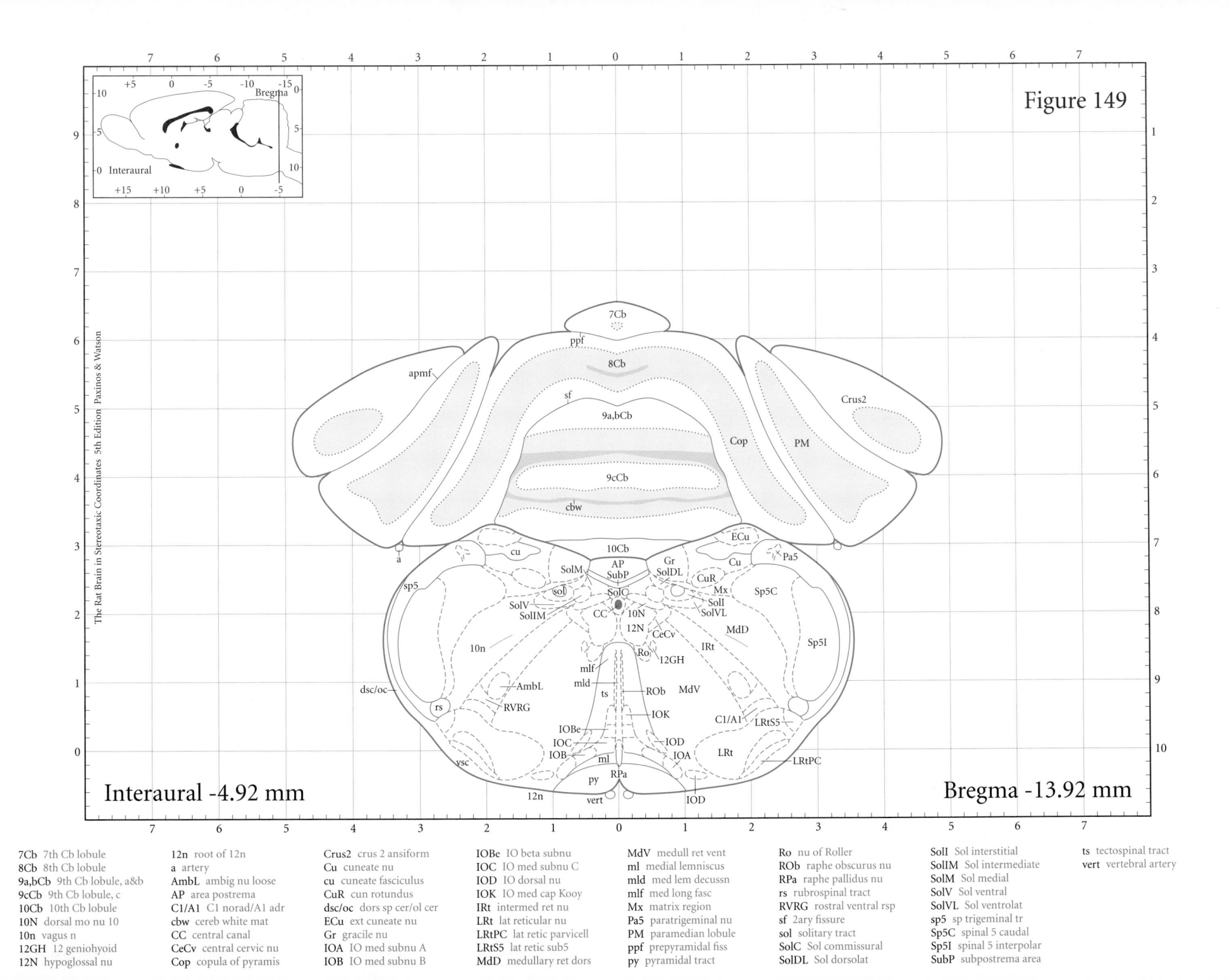

Figure 149

Interaural -4.92 mm

Bregma -13.92 mm

The Rat Brain in Stereotaxic Coordinates 5th Edition Paxinos & Watson

7Cb 7th Cb lobule
8Cb 8th Cb lobule
9a,bCb 9th Cb lobule, a&b
9cCb 9th Cb lobule, c
10Cb 10th Cb lobule
10N dorsal mo nu 10
10n vagus n
12GH 12 geniohyoid
12N hypoglossal nu

12n root of 12n
a artery
AmbL ambig nu loose
AP area postrema
C1/A1 C1 norad/A1 adr
cbw cereb white mat
CC central canal
CeCv central cervic nu
Cop copula of pyramis

Crus2 crus 2 ansiform
Cu cuneate nu
cu cuneate fasciculus
CuR cun rotundus
dsc/oc dors sp cer/ol cer
ECu ext cuneate nu
Gr gracile nu
IOA IO med subnu A
IOB IO med subnu B

IOBe IO beta subnu
IOC IO med subnu C
IOD IO dorsal nu
IOK IO med cap Kooy
IRt intermed ret nu
LRt lat reticular nu
LRtPC lat retic parvicell
LRtS5 lat retic sub5
MdD medullary ret dors

MdV medull ret vent
ml medial lemniscus
mld med lem decussn
mlf med long fasc
Mx matrix region
Pa5 paratrigeminal nu
PM paramedian lobule
ppf prepyramidal fiss
py pyramidal tract

Ro nu of Roller
ROb raphe obscurus nu
RPa raphe pallidus nu
rs rubrospinal tract
RVRG rostral ventral rsp
sf 2ary fissure
sol solitary tract
SolC Sol commissural
SolDL Sol dorsolat

SolI Sol interstitial
SolIM Sol intermediate
SolM Sol medial
SolV Sol ventral
SolVL Sol ventrolat
sp5 sp trigeminal tr
Sp5C spinal 5 caudal
Sp5I spinal 5 interpolar
SubP subpostrema area

ts tectospinal tract
vert vertebral artery

7Cb 7th Cb lobule
8Cb 8th Cb lobule
9a,bCb 9th Cb lobule, a&b
9cCb 9th Cb lobule, c
10N dorsal mo nu 10
10n vagus n
12GH 12 geniohyoid
12N hypoglossal nu

12n root of 12n
Amb ambiguus nu
AP area postrema
apmf ansoparamed fiss
C1/A1 C1 norad/A1 adr
cbw cereb white mat
CC central canal
CeCv central cervic nu

Cop copula of pyramis
Crus2 crus 2 ansiform
Cu cuneate nu
cu cuneate fasciculus
CuR cun rotundus
CVL caudoventrolat rt
dsc dorsal sp cereb tr
ECu ext cuneate nu

Gr gracile nu
IOA IO med subnu A
IOB IO med subnu B
IOBe IO beta subnu
IOC IO med subnu C
IOK IO med cap Kooy
IRt intermed ret nu
LRt lat reticular nu

LRtPC lat retic parvicell
LRtS5 lat retic sub5
MdD medullary ret dors
MdV medull ret vent
ml medial lemniscus
mlf med long fasc
Mx matrix region
Pa5 paratrigeminal nu

PM paramedian lobule
ppf prepyramidal fiss
py pyramidal tract
Ro nu of Roller
ROb raphe obscurus nu
RPa raphe pallidus nu
rs rubrospinal tract
RVRG rostral ventral rsp

sf 2ary fissure
sol solitary tract
SolC Sol commissural
SolDL Sol dorsolat
SolI Sol interstitial
SolIM Sol intermediate
SolM Sol medial
SolV Sol ventral

SolVL Sol ventrolat
sp5 sp trigeminal tr
Sp5C spinal 5 caudal
Sp5I spinal 5 interpolar
SubP subpostrema area
ts tectospinal tract
vert vertebral artery
vsc vent spinocer tr

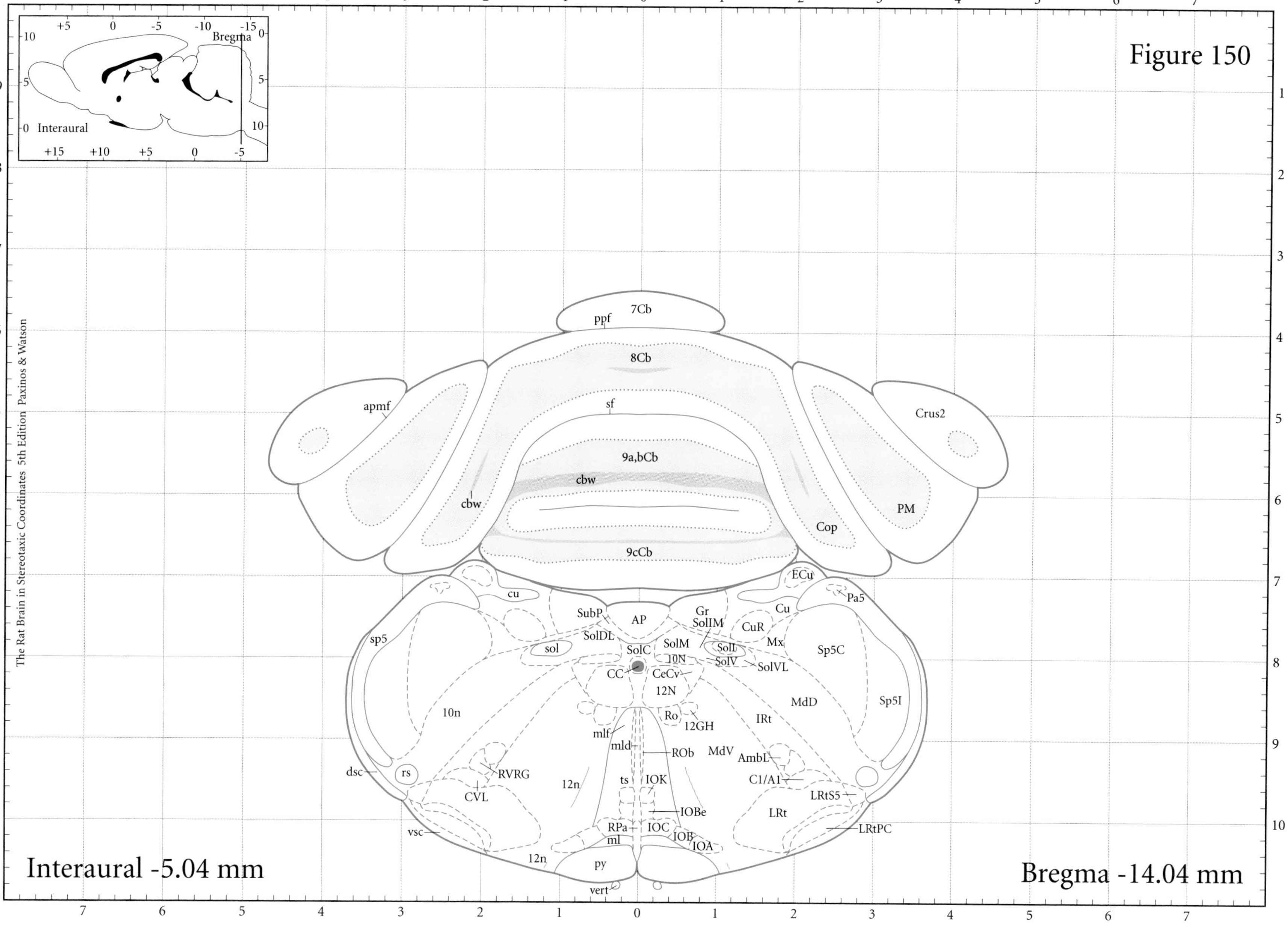

Figure 150

The Rat Brain in Stereotaxic Coordinates 5th Edition Paxinos & Watson

Interaural -5.04 mm

Bregma -14.04 mm

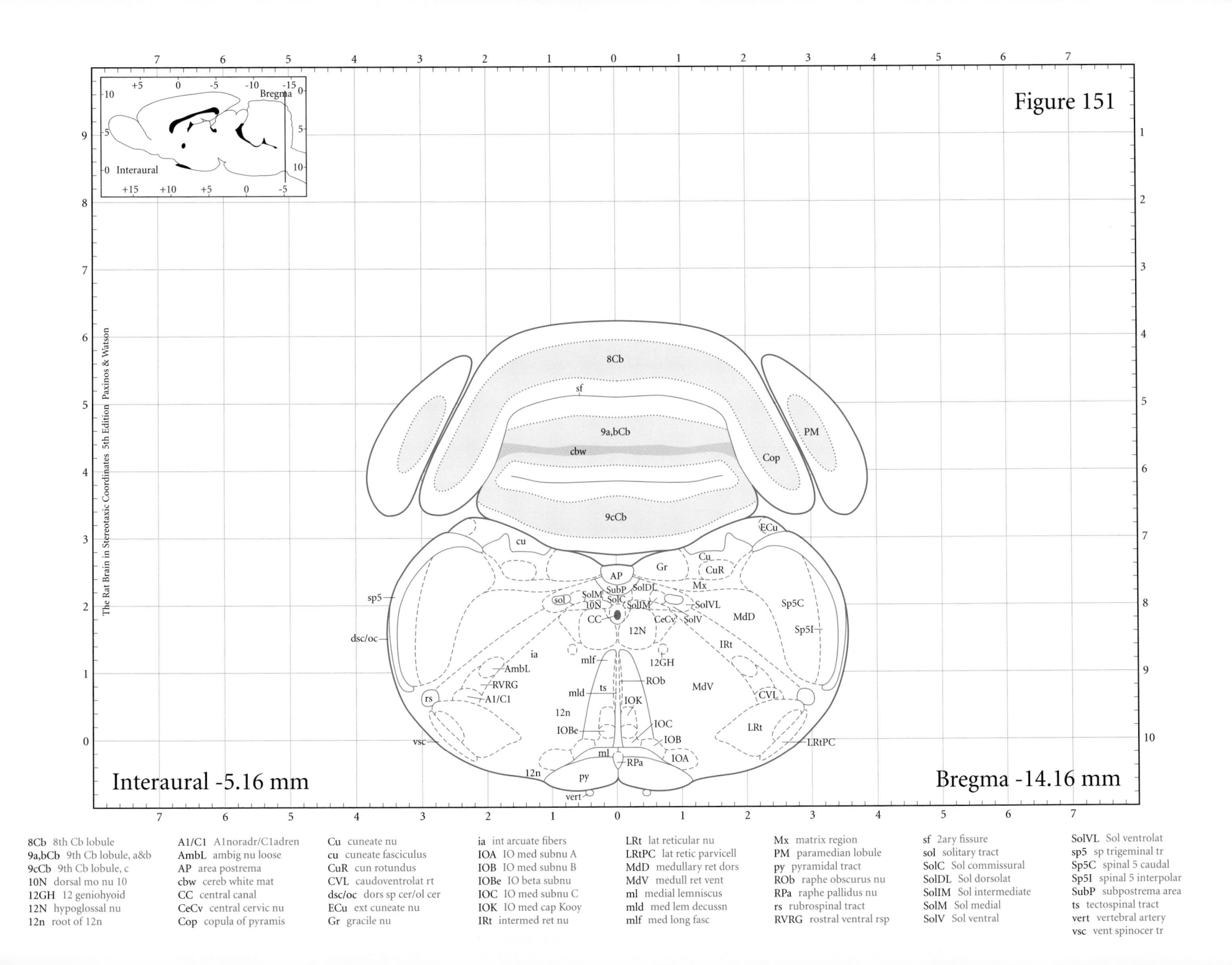

Figure 151

Interaural -5.16 mm

Bregma -14.16 mm

The Rat Brain in Stereotaxic Coordinates 5th Edition Paxinos & Watson

8Cb   8th Cb lobule
9a,bCb 9th Cb lobule, a&b
9cCb   9th Cb lobule, c
10N   dorsal mo nu 10
12GH   12 geniohyoid
12N   hypoglossal nu
12n   root of 12n

A1/C1   A1noradr/C1adren
AmbL   ambig nu loose
AP   area postrema
cbw   cereb white mat
CC   central canal
CeCv   central cervic nu
Cop   copula of pyramis

Cu   cuneate nu
cu   cuneate fasciculus
CuR   cun rotundus
CVL   caudoventrolat rt
dsc/oc dors sp cer/ol cer
ECu   ext cuneate nu
Gr   gracile nu

ia   int arcuate fibers
IOA   IO med subnu A
IOB   IO med subnu B
IOBe   IO beta subnu
IOC   IO med subnu C
IOK   IO med cap Kooy
IRt   intermed ret nu

LRt   lat reticular nu
LRtPC lat retic parvicell
MdD   medullary ret dors
MdV   medull ret vent
ml   medial lemniscus
mld   med lem decussn
mlf   med long fasc

Mx   matrix region
PM   paramedian lobule
py   pyramidal tract
ROb   raphe obscurus nu
RPa   raphe pallidus nu
rs   rubrospinal tract
RVRG   rostral ventral rsp

sf   2ary fissure
sol   solitary tract
SolC   Sol commissural
SolDL Sol dorsolat
SolIM Sol intermediate
SolM   Sol medial
SolV   Sol ventral

SolVL Sol ventrolat
sp5   sp trigeminal tr
Sp5C   spinal 5 caudal
Sp5I   spinal 5 interpolar
SubP   subpostrema area
ts   tectospinal tract
vert   vertebral artery
vsc   vent spinocer tr

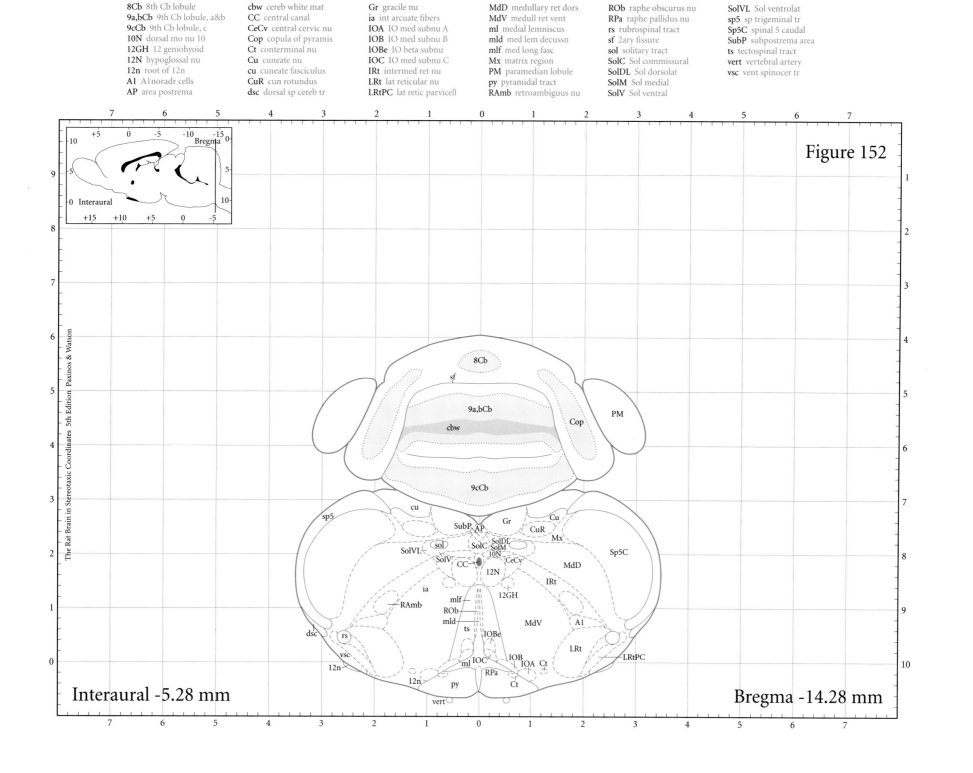

8Cb 8th Cb lobule
9a,bCb 9th Cb lobule, a&b
9cCb 9th Cb lobule, c
10N dorsal mo nu 10
12GH 12 geniohyoid
12N hypoglossal nu
12n root of 12n
A1 A1noradr cells
AP area postrema

cbw cereb white mat
CC central canal
CeCv central cervic nu
Cop copula of pyramis
Ct conterminal nu
Cu cuneate nu
cu cuneate fasciculus
CuR cun rotundus
dsc dorsal sp cereb tr

Gr gracile nu
ia int arcuate fibers
IOA IO med subnu A
IOB IO med subnu B
IOBe IO beta subnu
IOC IO med subnu C
IRt intermed ret nu
LRt lat reticular nu
LRtPC lat retic parvicell

MdD medullary ret dors
MdV medull ret vent
ml medial lemniscus
mld med lem decussn
mlf med long fasc
Mx matrix region
PM paramedian lobule
py pyramidal tract
RAmb retroambiguus nu

ROb raphe obscurus nu
RPa raphe pallidus nu
rs rubrospinal tract
sf 2ary fissure
sol solitary tract
SolC Sol commissural
SolDL Sol dorsolat
SolM Sol medial
SolV Sol ventral

SolVL Sol ventrolat
sp5 sp trigeminal tr
Sp5C spinal 5 caudal
SubP subpostrema area
ts tectospinal tract
vert vertebral artery
vsc vent spinocer tr

Figure 152

The Rat Brain in Stereotaxic Coordinates 5th Edition Paxinos & Watson

Interaural -5.28 mm

Bregma -14.28 mm

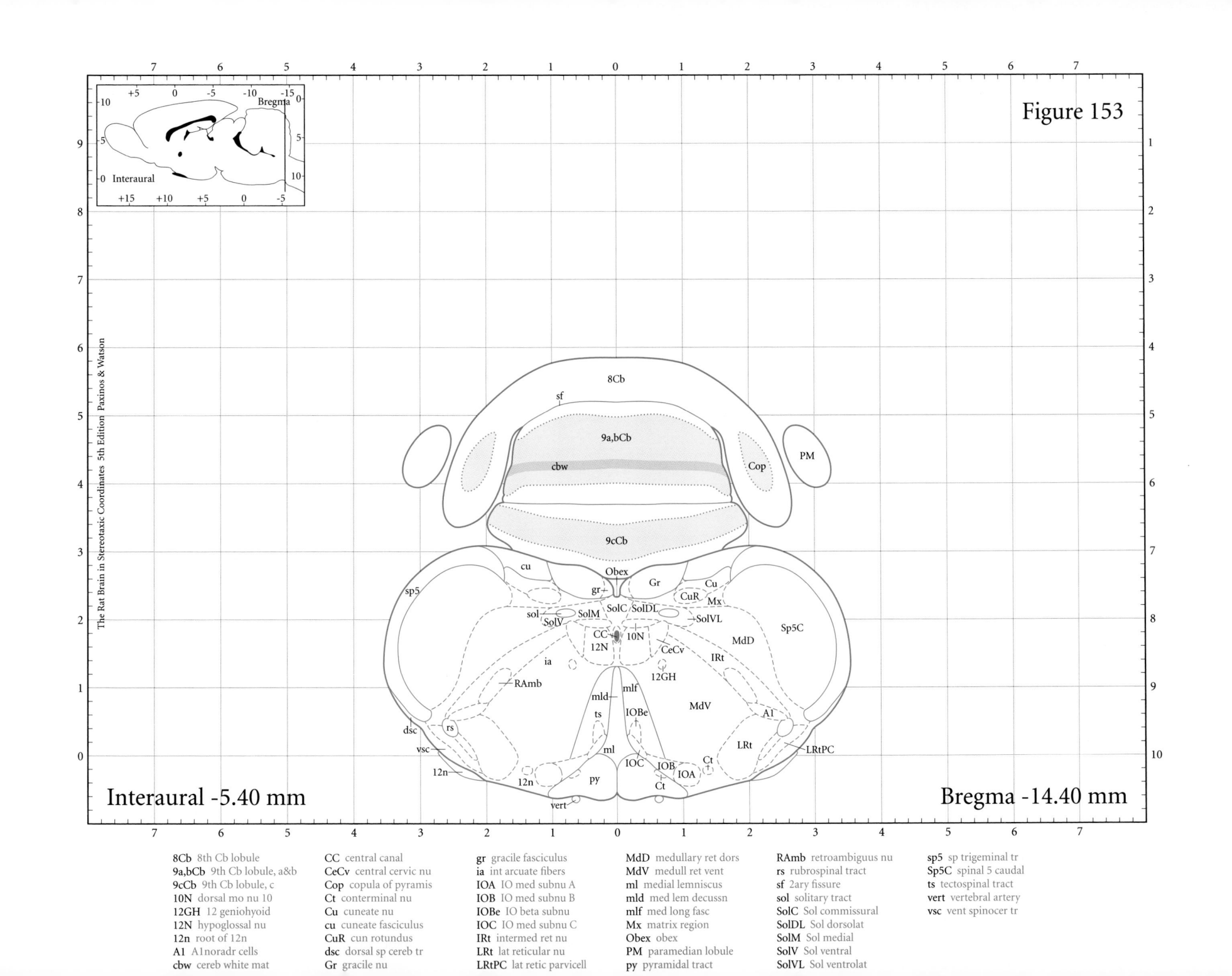

Figure 153

Interaural -5.40 mm

Bregma -14.40 mm

The Rat Brain in Stereotaxic Coordinates 5th Edition Paxinos & Watson

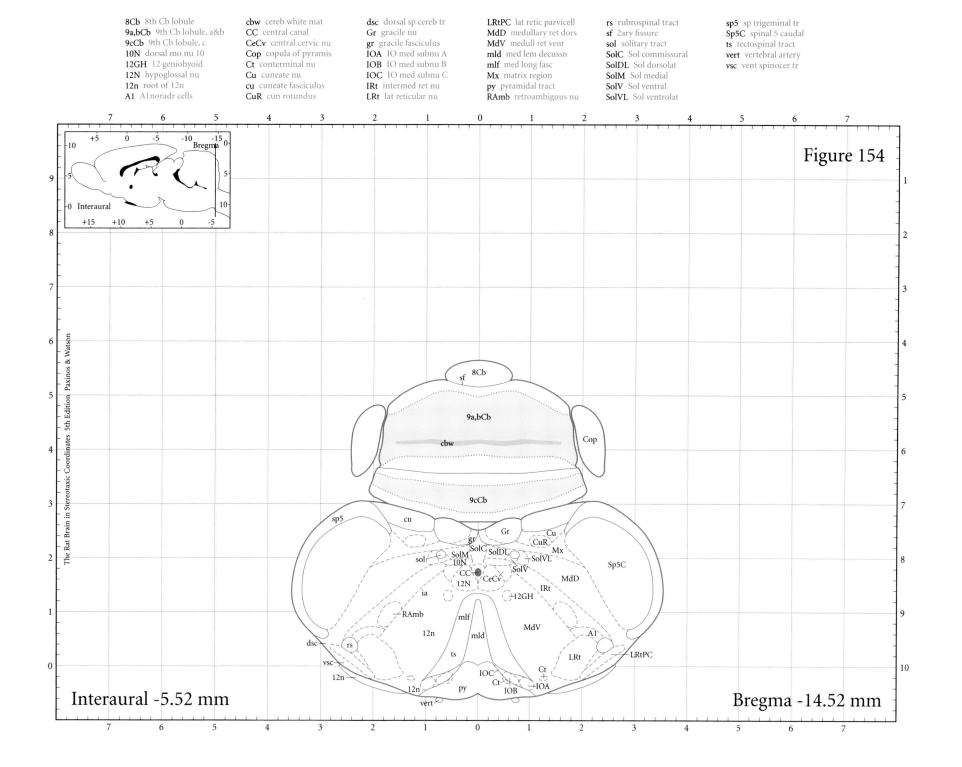

Figure 154

The Rat Brain in Stereotaxic Coordinates 5th Edition Paxinos & Watson

Interaural -5.52 mm

Bregma -14.52 mm

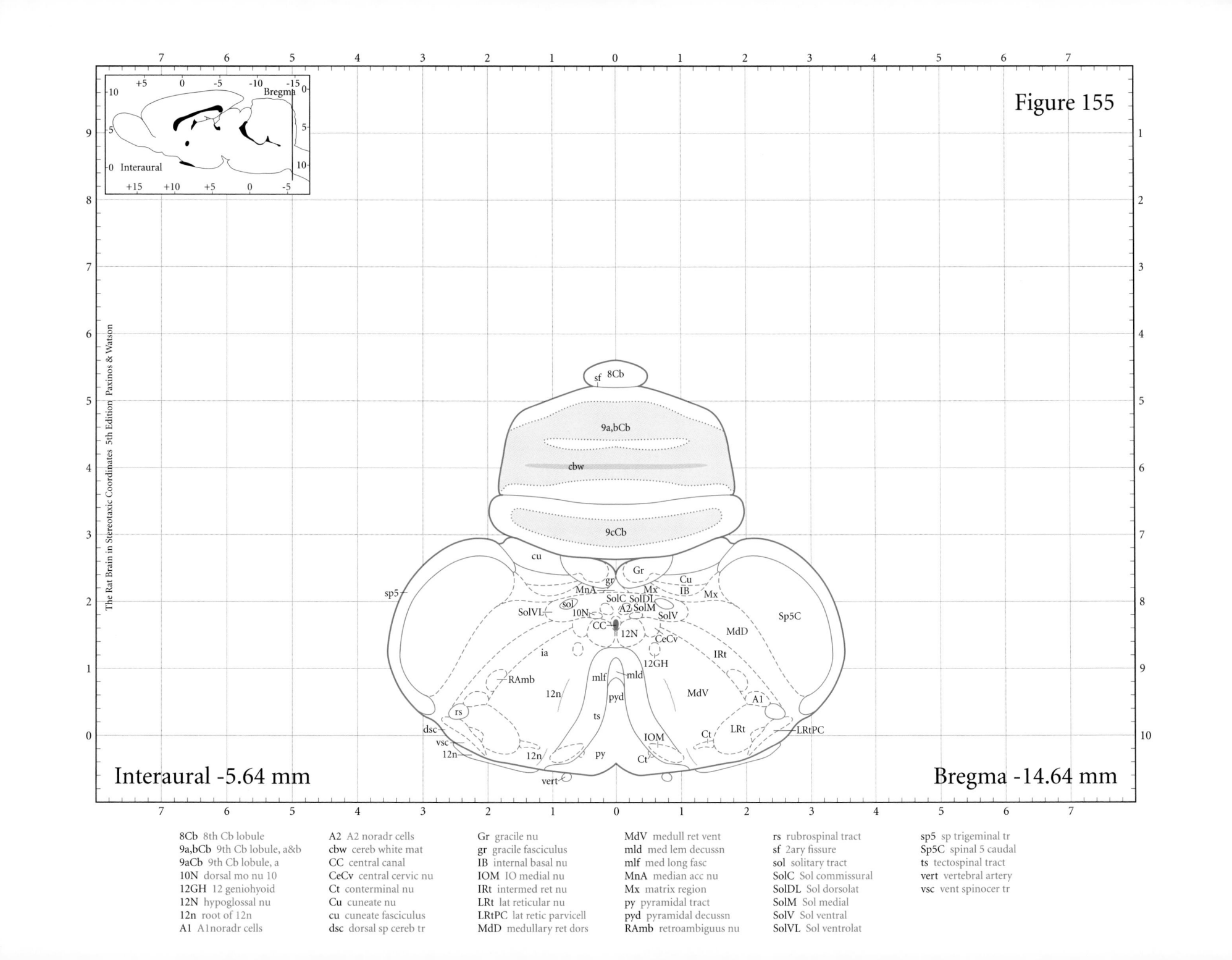

Figure 155

Interaural -5.64 mm

Bregma -14.64 mm

The Rat Brain in Stereotaxic Coordinates 5th Edition Paxinos & Watson

8Cb 8th Cb lobule	A2 A2 noradr cells	Gr gracile nu	MdV medull ret vent	rs rubrospinal tract	sp5 sp trigeminal tr
9a,bCb 9th Cb lobule, a&b	cbw cereb white mat	gr gracile fasciculus	mld med lem decussn	sf 2ary fissure	Sp5C spinal 5 caudal
9aCb 9th Cb lobule, a	CC central canal	IB internal basal nu	mlf med long fasc	sol solitary tract	ts tectospinal tract
10N dorsal mo nu 10	CeCv central cervic nu	IOM IO medial nu	MnA median acc nu	SolC Sol commissural	vert vertebral artery
12GH 12 geniohyoid	Ct conterminal nu	IRt intermed ret nu	Mx matrix region	SolDL Sol dorsolat	vsc vent spinocer tr
12N hypoglossal nu	Cu cuneate nu	LRt lat reticular nu	py pyramidal tract	SolM Sol medial	
12n root of 12n	cu cuneate fasciculus	LRtPC lat retic parvicell	pyd pyramidal decussn	SolV Sol ventral	
A1 A1noradr cells	dsc dorsal sp cereb tr	MdD medullary ret dors	RAmb retroambiguus nu	SolVL Sol ventrolat	

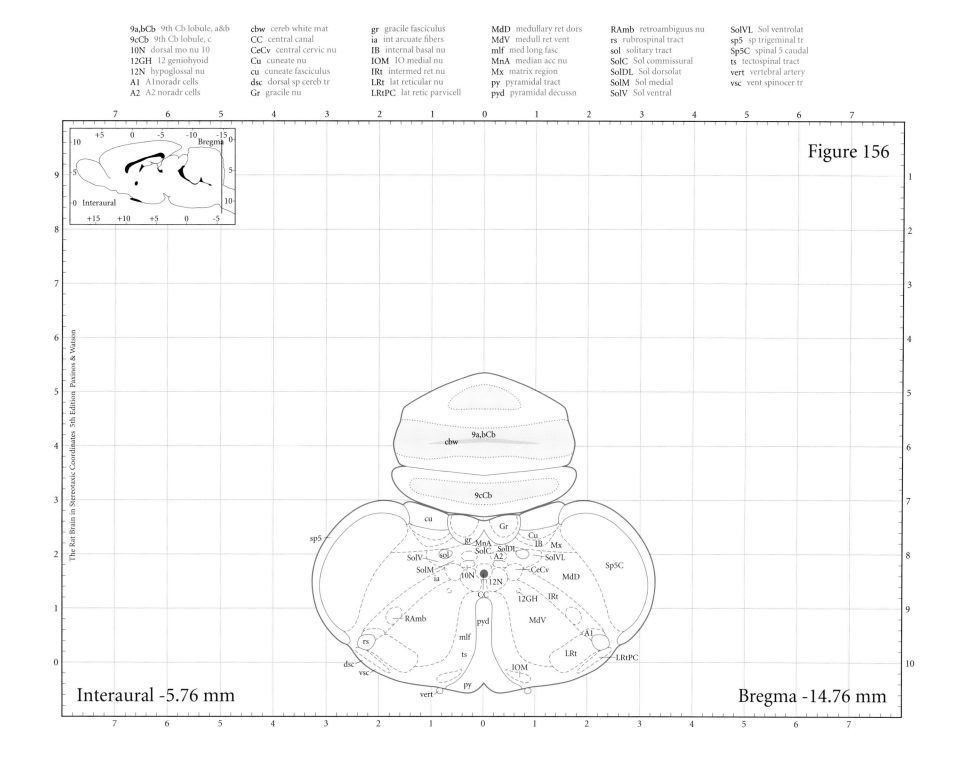

Figure 156

Interaural -5.76 mm

Bregma -14.76 mm

The Rat Brain in Stereotaxic Coordinates 5th Edition Paxinos & Watson

9a,bCb 9th Cb lobule, a&b	cbw cereb white mat	gr gracile fasciculus	MdD medullary ret dors	RAmb retroambiguus nu
9cCb 9th Cb lobule, c	CC central canal	ia int arcuate fibers	MdV medull ret vent	rs rubrospinal tract
10N dorsal mo nu 10	CeCv central cervic nu	IB internal basal nu	mlf med long fasc	sol solitary tract
12GH 12 geniohyoid	Cu cuneate nu	IOM IO medial nu	MnA median acc nu	SolC Sol commissural
12N hypoglossal nu	cu cuneate fasciculus	IRt intermed ret nu	Mx matrix region	SolDL Sol dorsolat
A1 A1noradr cells	dsc dorsal sp cereb tr	LRt lat reticular nu	py pyramidal tract	SolM Sol medial
A2 A2 noradr cells	Gr gracile nu	LRtPC lat retic parvicell	pyd pyramidal decussn	SolV Sol ventral

SolVL Sol ventrolat
sp5 sp trigeminal tr
Sp5C spinal 5 caudal
ts tectospinal tract
vert vertebral artery
vsc vent spinocer tr

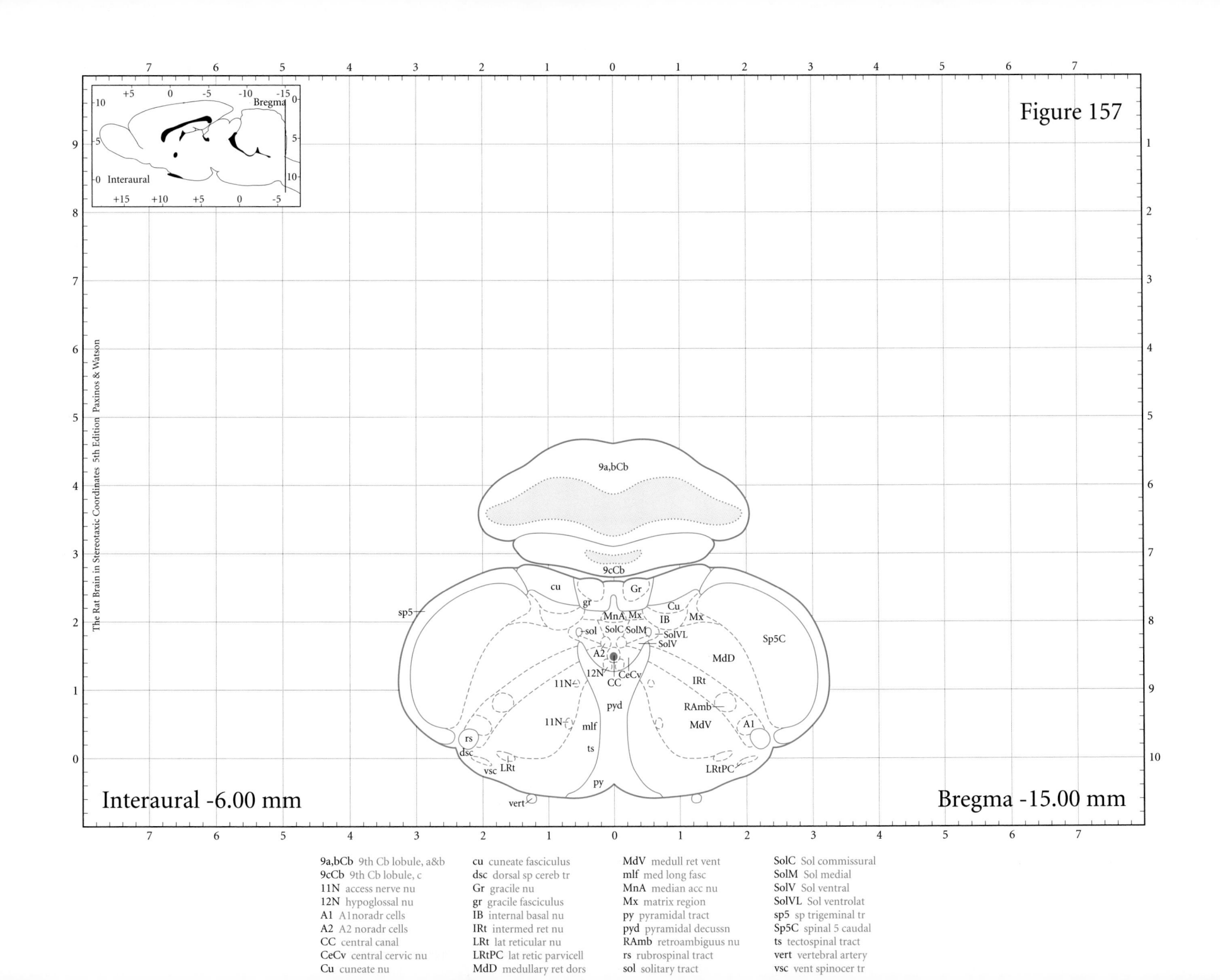

Figure 157

Interaural -6.00 mm

Bregma -15.00 mm

The Rat Brain in Stereotaxic Coordinates 5th Edition Paxinos & Watson

9a,bCb 9th Cb lobule, a&b
9cCb 9th Cb lobule, c
11N access nerve nu
12N hypoglossal nu
A1 A1 noradr cells
A2 A2 noradr cells
CC central canal
CeCv central cervic nu
Cu cuneate nu

cu cuneate fasciculus
dsc dorsal sp cereb tr
Gr gracile nu
gr gracile fasciculus
IB internal basal nu
IRt intermed ret nu
LRt lat reticular nu
LRtPC lat retic parvicell
MdD medullary ret dors

MdV medull ret vent
mlf med long fasc
MnA median acc nu
Mx matrix region
py pyramidal tract
pyd pyramidal decussn
RAmb retroambiguus nu
rs rubrospinal tract
sol solitary tract

SolC Sol commissural
SolM Sol medial
SolV Sol ventral
SolVL Sol ventrolat
sp5 sp trigeminal tr
Sp5C spinal 5 caudal
ts tectospinal tract
vert vertebral artery
vsc vent spinocer tr

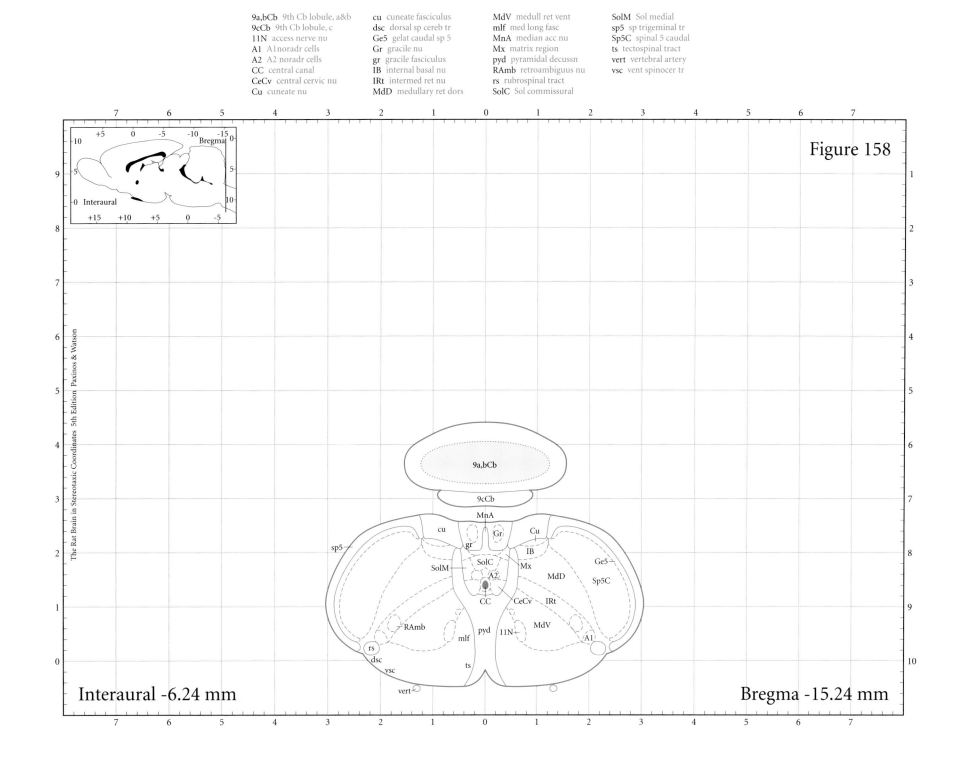

Figure 158

9a,bCb	9th Cb lobule, a&b
9cCb	9th Cb lobule, c
11N	access nerve nu
A1	A1noradr cells
A2	A2 noradr cells
CC	central canal
CeCv	central cervic nu
Cu	cuneate nu
cu	cuneate fasciculus
dsc	dorsal sp cereb tr
Ge5	gelat caudal sp 5
Gr	gracile nu
gr	gracile fasciculus
IB	internal basal nu
IRt	intermed ret nu
MdD	medullary ret dors
MdV	medull ret vent
mlf	med long fasc
MnA	median acc nu
Mx	matrix region
pyd	pyramidal decussn
RAmb	retroambiguus nu
rs	rubrospinal tract
SolC	Sol commissural
SolM	Sol medial
sp5	sp trigeminal tr
Sp5C	spinal 5 caudal
ts	tectospinal tract
vert	vertebral artery
vsc	vent spinocer tr

Interaural -6.24 mm

Bregma -15.24 mm

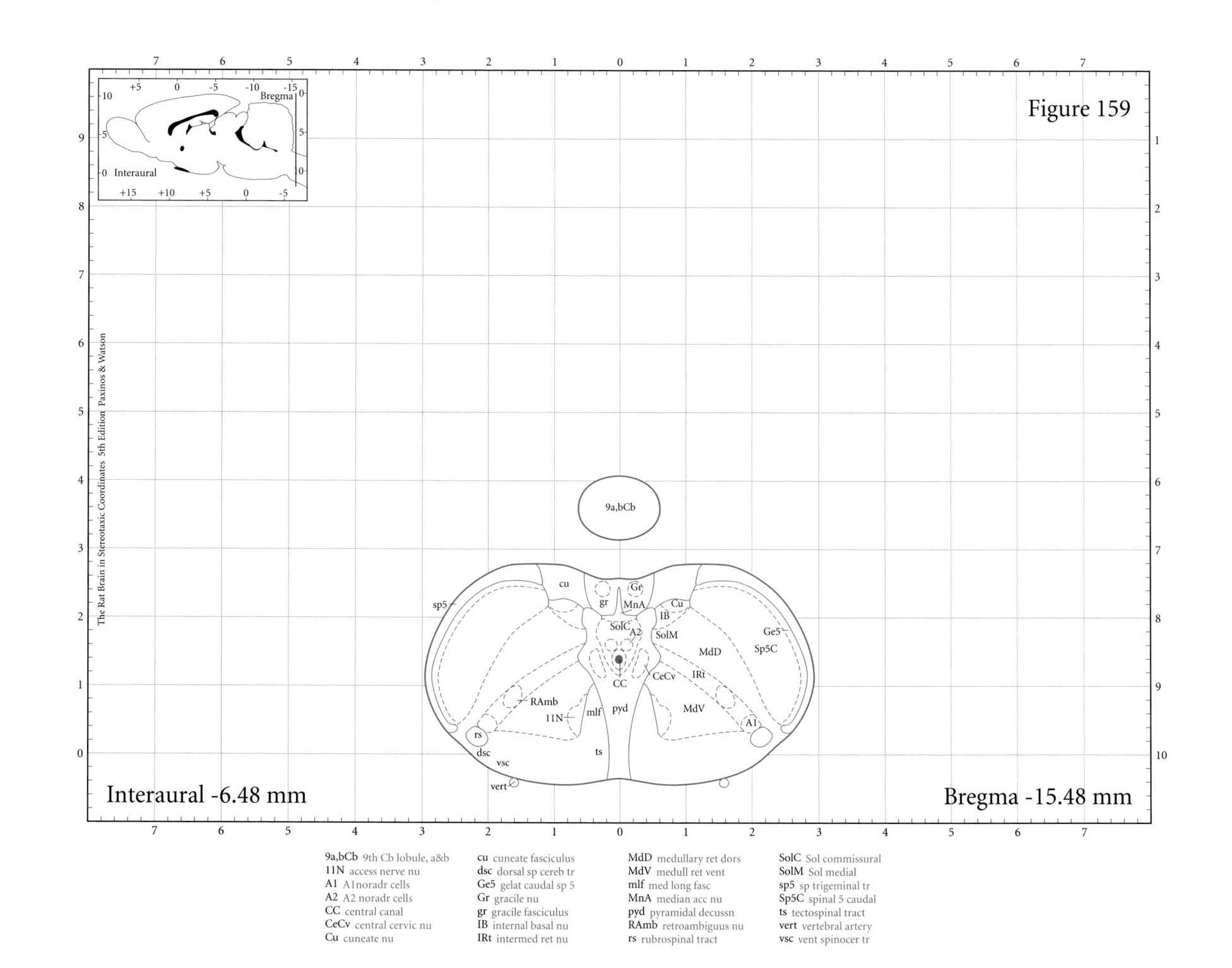

Figure 159

Interaural -6.48 mm

Bregma -15.48 mm

The Rat Brain in Stereotaxic Coordinates 5th Edition Paxinos & Watson

9a,bCb 9th Cb lobule, a&b	cu cuneate fasciculus	MdD medullary ret dors	SolC Sol commissural
11N access nerve nu	dsc dorsal sp cereb tr	MdV medull ret vent	SolM Sol medial
A1 A1noradr cells	Ge5 gelat caudal sp 5	mlf med long fasc	sp5 sp trigeminal tr
A2 A2 noradr cells	Gr gracile nu	MnA median acc nu	Sp5C spinal 5 caudal
CC central canal	gr gracile fasciculus	pyd pyramidal decussn	ts tectospinal tract
CeCv central cervic nu	IB internal basal nu	RAmb retroambiguus nu	vert vertebral artery
Cu cuneate nu	IRt intermed ret nu	rs rubrospinal tract	vsc vent spinocer tr

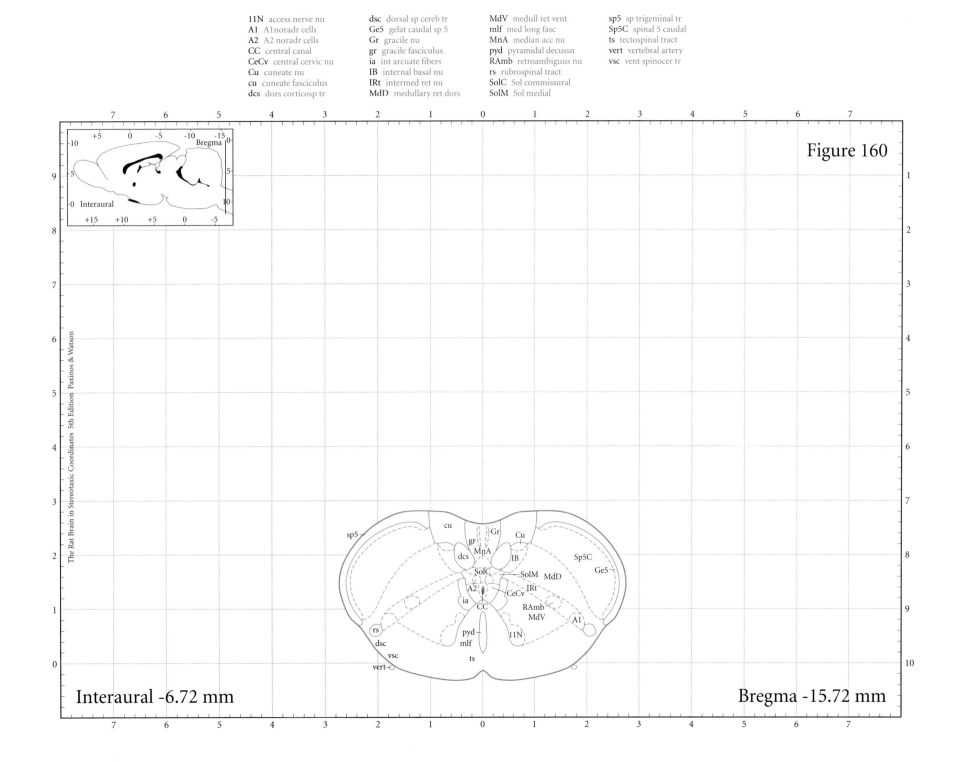

11N access nerve nu	dsc dorsal sp cereb tr
A1 A1noradr cells	Ge5 gelat caudal sp 5
A2 A2 noradr cells	Gr gracile nu
CC central canal	gr gracile fasciculus
CeCv central cervic nu	ia int arcuate fibers
Cu cuneate nu	IB internal basal nu
cu cuneate fasciculus	IRt intermed ret nu
dcs dors corticosp tr	MdD medullary ret dors

MdV medull ret vent	sp5 sp trigeminal tr
mlf med long fasc	Sp5C spinal 5 caudal
MnA median acc nu	ts tectospinal tract
pyd pyramidal decussn	vert vertebral artery
RAmb retroambiguus nu	vsc vent spinocer tr
rs rubrospinal tract	
SolC Sol commissural	
SolM Sol medial	

Figure 160

Interaural -6.72 mm

Bregma -15.72 mm

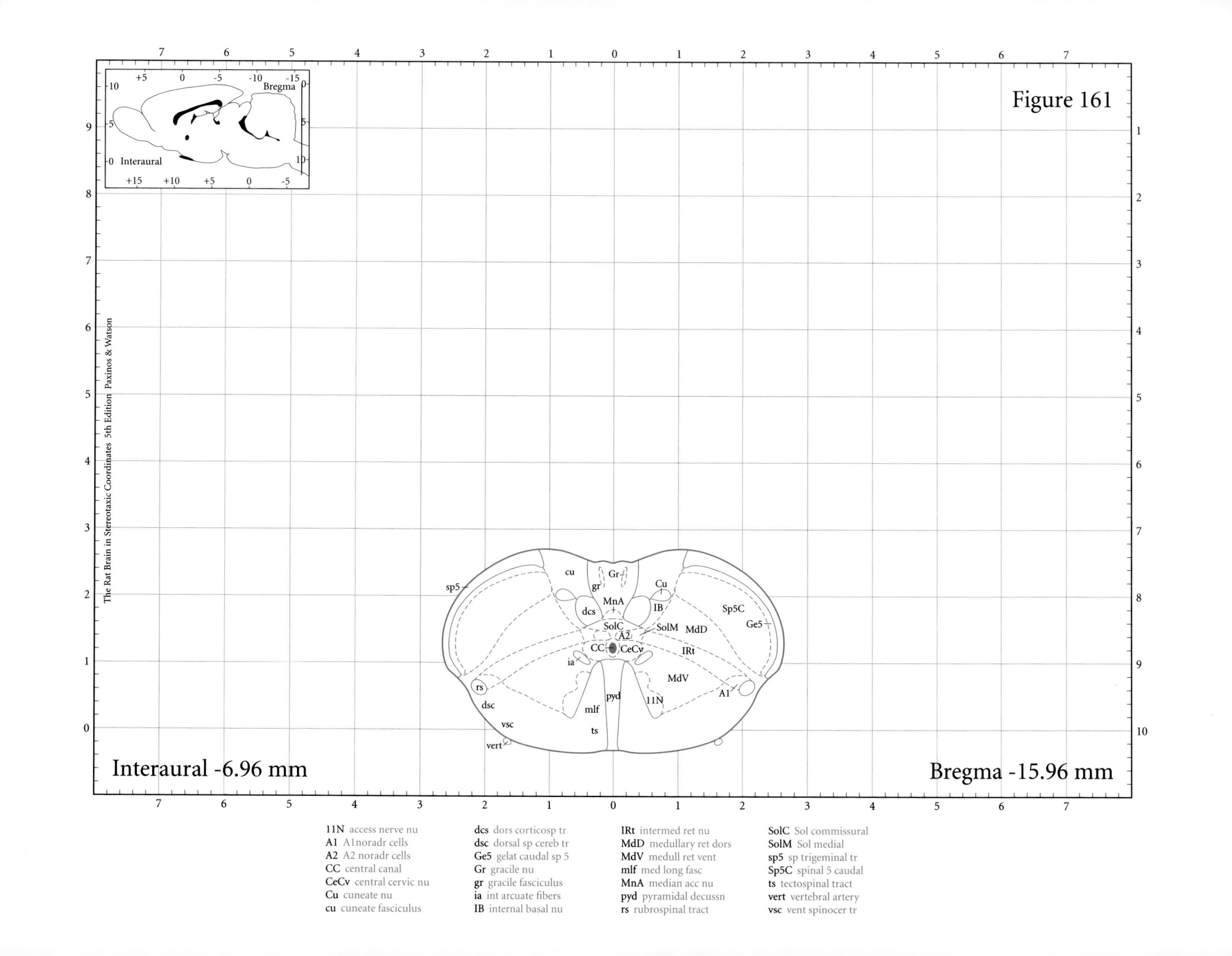

Figure 161

The Rat Brain in Stereotaxic Coordinates 5th Edition Paxinos & Watson

Interaural -6.96 mm

Bregma -15.96 mm

11N access nerve nu
A1 A1noradr cells
A2 A2 noradr cells
CC central canal
CeCv central cervic nu
Cu cuneate nu
cu cuneate fasciculus
dcs dors corticosp tr
dsc dorsal sp cereb tr
Ge5 gelat caudal sp 5
Gr gracile nu
gr gracile fasciculus
ia int arcuate fibers
IB internal basal nu
IRt intermed ret nu
MdD medullary ret dors
MdV medull ret vent
mlf med long fasc
MnA median acc nu
pyd pyramidal decussn
rs rubrospinal tract
SolC Sol commissural
SolM Sol medial
sp5 sp trigeminal tr
Sp5C spinal 5 caudal
ts tectospinal tract
vert vertebral artery
vsc vent spinocer tr

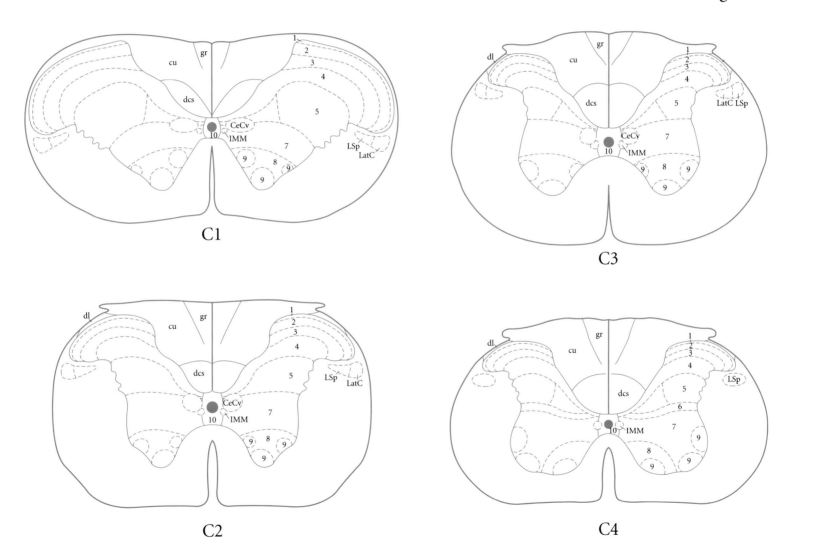

1-10 spinal cord layers
CeCv central cervical nucleus
cu cuneate fasciculus
dl dorsolateral fasciculus
gr gracile fasciculus

IML intermediolateral cell column
IMM intermediomedial cell column
LatC lateral cervical nucleus
LSp lateral spinal nucleus
dcs dorsal corticospinal tract

*Fig 162a, 162b, 162c, 162d, 162e are reproduced from Molander and Grant (1995) with permission of the authors. Users of these figures should cite Molander, C. and Grant, G., 1995, Spinal cord cytoarchitecture.

In G. Paxinos(Ed), *The Nervous System*, Second Edition, Academic Press, SanDiego.

Figure 162a*

C1

C2

C3

C4

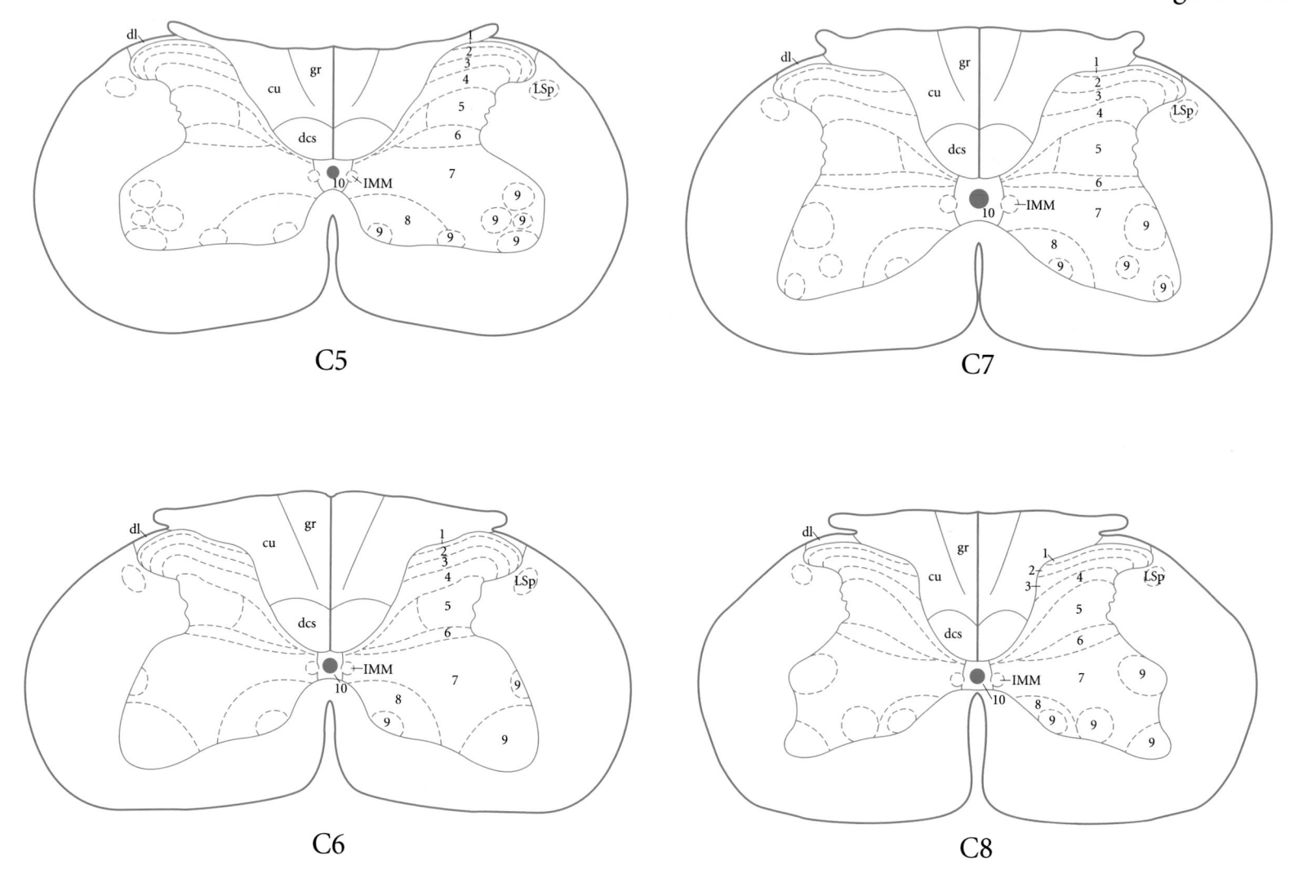

C5

C7

C6

C8

1-10 spinal cord layers
CeCv central cervical nucleus
cu cuneate fasciculus
dl dorsolateral fasciculus
gr gracile fasciculus

IML intermediolateral cell column
IMM intermediomedial cell column
LatC lateral cervical nucleus
LSp lateral spinal nucleus
dcs dorsal corticospinal tract

*Fig 162a, 162b, 162c, 162d, 162e are
reproduced from Molander and Grant
(1995) with permission of the authors.
Users of these figures should cite
Molander, C. and Grant, G., 1995,
Spinal cord cytoarchitecture.

In G. Paxinos(Ed), *The Nervous System*,
Second Edition, Academic Press,
SanDiego.

Figure 162c*

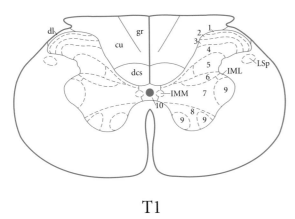

T1

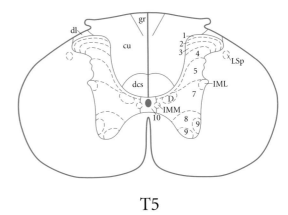

T5

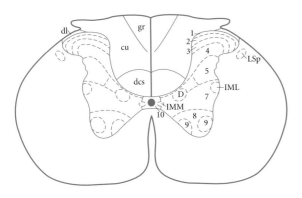

T3

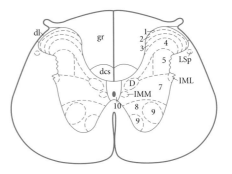

T10

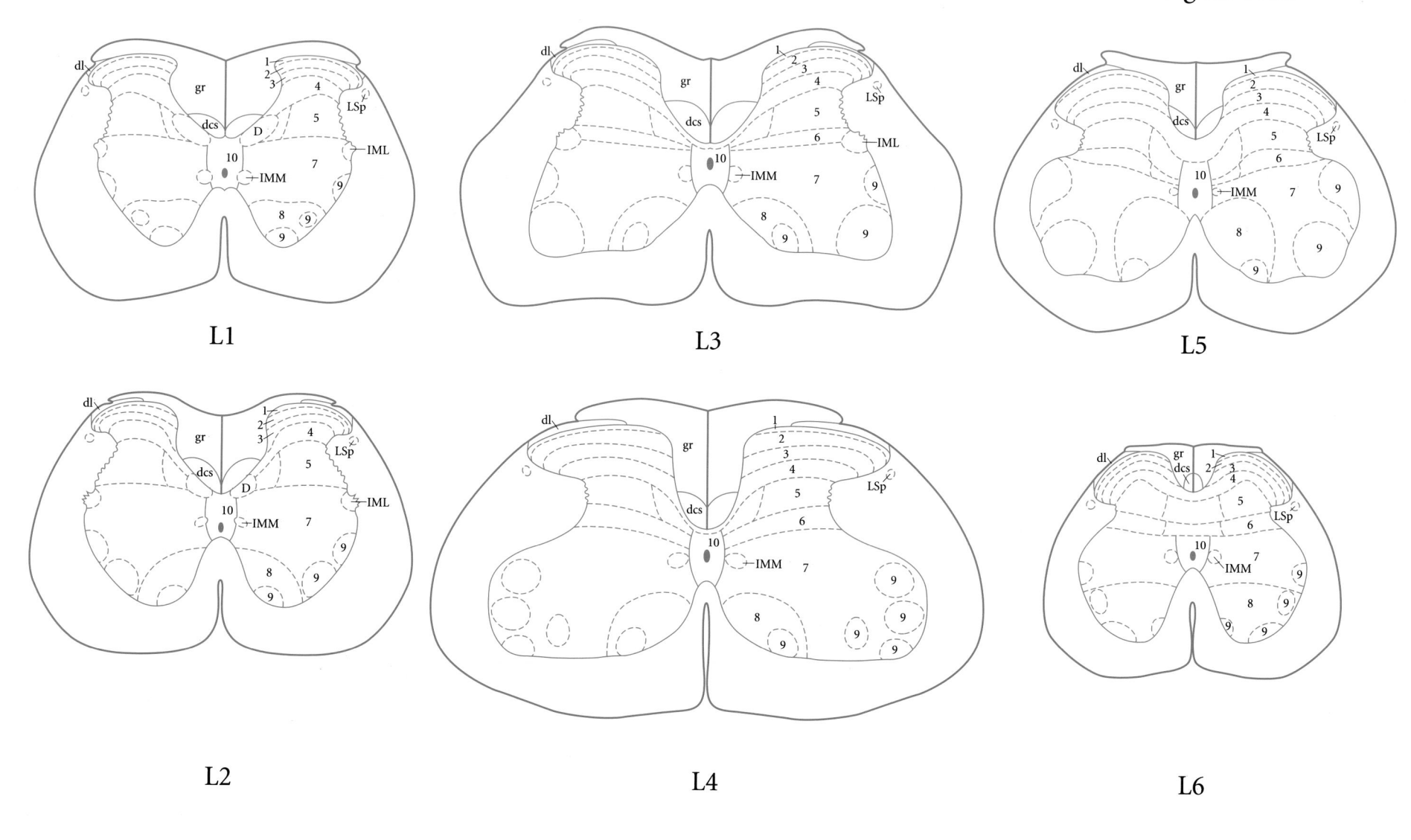

Figure 162d*

L1

L3

L5

L2

L4

L6

1-10 spinal cord layers
CeCv central cervical nucleus
D dorsal nucleus (Clarke)
dl dorsolateral fasciculus
gr gracile fasciculus

IML intermediolateral cell column
IMM intermediomedial cell column
LatC lateral cervical nucleus
LSp lateral spinal nucleus
dcs dorsal corticospinal tract

*Fig 162a, 162b, 162c, 162d, 162e are
reproduced from Molander and Grant
(1995) with permission of the authors.
Users of these figures should cite
Molander, C. and Grant, G., 1995,
Spinal cord cytoarchitecture.

In G. Paxinos(Ed), *The Nervous System*,
Second Edition, Academic Press,
SanDiego.

Figure 162e*

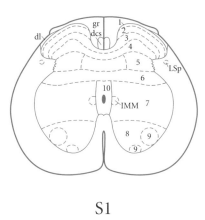

S1

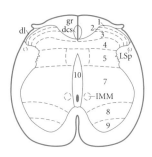

S3

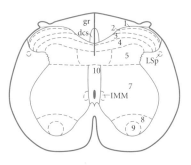

S2

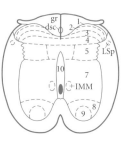

S4